Advances in Intelligent Systems and Computing

Volume 216

Series Editor

J. Kacprzyk, Warsaw, Poland

For further volumes:
http://www.springer.com/series/11156

Swamidoss Sathiakumar · Lalit Kumar Awasthi
M. Roberts Masillamani · S. S. Sridhar
Editors

Proceedings of International Conference on Internet Computing and Information Communications

ICICIC Global 2012

Springer

Editors
Swamidoss Sathiakumar
School of Electrical and Information Engineering
University of Sydney
Sydney, NSW
Australia

Lalit Kumar Awasthi
Department of Computer Science and Engineering
National Institute of Technology
Hamirpur, Himachal Pradesh
India

M. Roberts Masillamani
School of Computing Sciences
Hindustan University
Chennai, Tamil Nadu
India

S. S. Sridhar
Department of CSE
SRM University
Chennai, Tamil Nadu
India

ISSN 2194-5357 ISSN 2194-5365 (electronic)
ISBN 978-81-322-1298-0 ISBN 978-81-322-1299-7 (eBook)
DOI 10.1007/978-81-322-1299-7
Springer New Delhi Heidelberg New York Dordrecht London

Library of Congress Control Number: 2013946910

Printed on acid-free paper

Springer is part of Springer Science+Business Media (www.springer.com)

Preface

ICICIC Global is an offshoot of a collaborative research leading to technical paper writing and presentation in leading international conferences and journals, between professors from Australian universities, Indian universities, and the students. It was initiated in the year 2006, where quite a few of the students from Indian universities could learn to write quality paper and get them accepted to be published.

ICICIC Global, a nonprofit research consortium started to promote research and development in all the areas of Engineering, Technology, Architecture, Arts and Science in collaboration with Springer. Again, this consortium is the fruit of endeavor and interest taken by the professors from Australia, USA, UK, India, Malaysia, Singapore, and the Middle East countries. To commemorate this endeavor the ICICIC Global organized the First International Conference on Internet Computing and Information Communications 2012, from 12 to 14 February 2012, at Chennai, India in collaboration with Saveetha University, Chennai and supported by Mr. Ashok Verghese, Director, Hindustan University. Mr. Anand Karuppaih, MD and SCBIT, Sydney, Australia were the overall sponsor of this International Conference.

The Sakthi Mariamman Engineering College and Dr. Ramachandran, Vice President of Private Colleges Association of Tamil Nadu supported the Second International Conference ICCIC Global which was conducted from 27 to 28 July 2012.

The papers selected and presented in the above two conferences have been included for publishing in the Springer book series, *Advances in Intelligent Systems and Computing* (AISC).

The ICICIC Global mission is to provide an effective and established international forum for discussion and dissemination of recent advances and innovations in use of Technology, Internet Computing, and Information Communications in various facets of our day-to-day life and in all the Engineering fields and professions for the developing country's students and researchers.

The other mission of this consortium is to provide avenues for incubation facility for researchers who would like to apply their researches for industries. Many of the advisors and industries have voiced out their assistance in this venture as well. This forum will be the stepping stone for the students' research community.

Since, the research culture had to be inculcated into every professional student and researcher from various areas of Engineering and Non-Engineering fields, the consortium took a decision to include other technology areas and render the privilege to all the students and researchers.

This preface would be incomplete without our expression of thanks and gratitude to the general and conference chairs, members of the technical and advisory committees, and external reviewers for their excellent reviewing.

We will be failing in our duty if we don't thank the advisory committee chairman, Prof. P. S. Manisundaram, Former and Founder Vice Chancellor, Bharathidasan University, and Founder Principal of REC Trichy, India.

We also thank Springer and their professionals for the strong support in bringing this proceeding of the International Conference to reality. Last but not the least, we thank all the authors who have contributed to the success of this conference. We hope that the participants and authors who had attended the conference were benefitted academically and in their research works.

Dr. S. Sathia Kumar
Dr. M. Roberts Masillamani
Dr. Lalit Kumar Awasthi
Prof. S. S. Sridhar

Editorial Committee and Advisory Committee

Advisory Committee

Prof. K. Thulasiraman	Professor and Hitachi Chair, University of Oklahoma
Prof. San Murugesan	Director, BRITE Professional Services, Sydney, Australia
Dr. Paul Chellakumar	Managing Director, CAMPUS Abroad, Chennai, India
Prof. R. C. Joshi	Chancellor, Graphics Era University, Dehradun, India
Dr. Ir. Sim Kok Swee	Professor, Multi Media University, Malaysia
Dr. Bindu Rana	Director (Research), Educomp Solutions Ltd., India
Mr. Hank Magnuski	Ph.D., President and CEO, NCast Corporation Sunnyvale, California
Mr. Ashwin R. Desai	MD, A & T Network Systems Private Ltd., Madurai
Dr. Karanjeet Singh Kahlon	Dean, Guru Nanak Dev University, Amritsar, India
Dr. N. Parameswaran	Faculty, University of New South Wales, Sydney, Australia
Dr. Lim Way Soong	Deputy Dean, Multi Media University, Malaysia
Dr. Robin Doss	Associate Head of School, Deakin University, Melbourne, Australia
Dr. Raja Subramanian	Adjunct Professor of IIIT, Bangalore and BITS, Pilani
Dr. P. Latha	Professor, Pondicherry College of Engineering, India
Prof. (Dr.) Manoj Kumar	Group Director, CT Group of Institutions, Jalandhar
Dr. C. R. Rene Robin	Professor and Head, Jerusalem College of Engineering, Chennai
Mr. Shyam Prasad Shetty	Vice President, Impelsys (eLearning Initiatives), India
Dr. J. R. Alex Noel	Associate Professor, Saveetha Engineering College, Chennai

Contents

Secure Text Steganography

P. Akhilandeswari and Jabin G. George

Abstract Steganography is an art in which the data can be hidden in other data as cover, the text files are commonly used for hiding data. The main aspects of steganography are the capacity and security, where the capacity refers to how much data can be hidden in the cover carrier, while the security concerns with the ability of disclosing or altering the data by unauthorized party. The aim of this project is to implement an algorithm to reduce the size of objects created using steganography. In addition, the security level of each approach is made more secured. This project presents an overview of text steganography and various existing text-based steganography techniques. Highlighted are some of the problems inherent in text steganography as well as issues with existing solutions. A new approach is proposed in information hiding using interword spacing which reduces the amount of information to hide. This method offers generated stego text of maximum capacity according to the length of the secret message.

Keywords Cover · Security · Interword spacing · Stego text · Capacity

Introduction

In the field of Data Communication, security issues have got the top priority. So, of late, the degree of security provided by a security tool has become the main evolutionary criteria of it. Classical cryptography is one of the ways to secure plain text messages. Along with that, at the time of data transmission, security is also

P. Akhilandeswari (✉) · J. G. George
Department of Computer Science, SRM University, Chennai, India
e-mail: akilasharanju@gmail.com

J. G. George
e-mail: jabing28@gmail.com

S. Sathiakumar et al. (eds.), *Proceedings of International Conference on Internet Computing and Information Communications*, Advances in Intelligent Systems and Computing 216,
DOI: 10.1007/978-81-322-1299-7_1,

implemented by introducing the concept of steganography, watermarking, etc. In these types of combined approach, there exist some drawbacks.

In remote networking, at the time of transmission of hidden encrypted text message, if the eavesdroppers get the track of the hidden text, then they could easily get the encrypted text [1]. Now breaking of encrypted text message can be achieved by applying some brute force technique. So, there remains some probability of snooping of information. So, these types of techniques incur another level of security which can route the Cryptanalyzer or Steganalyzer in a different direction [2].

Steganography

Steganography or Stego as it is often referred to in the IT community, literally means, "Covered writing" which is derived from the Greek language. Steganography is defined by Markus Kahn [3] as follows: "Steganography is the art and science of communicating in a way which hides the existence of the communication. In contrast to Cryptography, where the unauthorized party is allowed to detect, intercept and modify messages without being able to violate certain security premises guaranteed by a cryptosystem, the goal of Steganography is to hide messages inside other harmless messages in a way that does not allow any unauthorized party to even detect that there is a second message present" [4].

Existing System

Peter Wayner's Mimicry Algorithm

Peter Wayner [5] proposed a mimicry algorithm that aimed at text in his book Mimic Functions, Cryptologia XVI-3. His approach is to produce mimicked text that looks similar to the real structure of the original text. Peter Wayner used a set of grammatical rules to generate stego text and the choice of each word determines how secret message bits are encoded.

Merits:

- Use of grammar makes easy to debug and reduce the errors.
- Secret message can be encoded into something innocent looking, in a form of spam, which nobody will notice that there is a secret message being concealed.

Demerits:

- Complex logic
- The stego object file is larger.

Brassil's Document Coding Method

Brassil et al. [6] gave the initial idea of document coding methods in their paper by proposing life-shift coding, word-shift coding, and feature coding (character coding) to discourage illicit dissemination of document distributed by computer network [7]. Line-shift coding is a method of altering a document by vertically shifting the locations of text lines to uniquely encode the document. Word-shift coding is a method to alter a document by horizontally shifting the locations of words within text lines to uniquely encode the document. Character coding or feature specific coding is a coding method that is applied only to the bitmap image of the document and can be examined for chosen character features, and those features are altered, or not altered, depending on the codeword.

A document is marked in an indiscernible way by a codeword identifying the registered owner to whom the document is sent. If a document copy is found that is suspected to have been illicitly disseminated, that copy can be decoded and the registered owner identified [8].

Merits:

- Easy to implement
- Large number of characters can be embedded.

Demerits:

- This can be done in hard copy documents only.
- It can be observed easily.

Proposed System

Objective

In the proposed system, I have started an overview of text steganography and various existing text-based steganography techniques. Highlighted are some of the problems inherent in text steganography as well as issues with existing solutions.

A new approach is proposed in information hiding using interword spacing, which reduces the amount of information to hide. This method offers generated stego-text of maximum capacity according to the length of the secret message. This proposed system also analyzed the significant drawbacks of each existing method and how this new approach could be recommended as a solution.

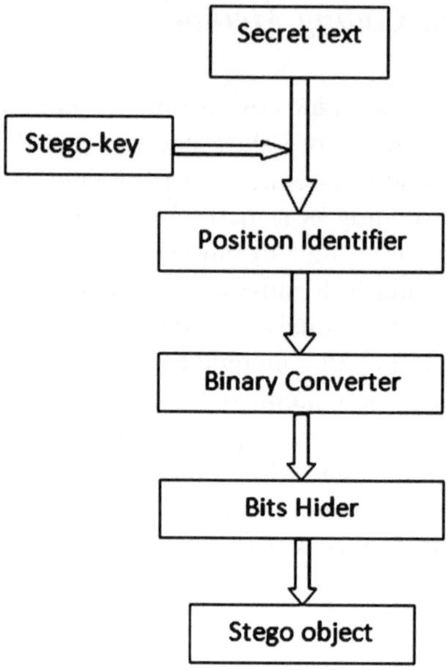

Sender Side Manipulation

In sender side, the sender has to give the information which is to be hidden and also the stego-key which is usually a password. The following figure illustrates the overall concept of this proposal. The flow diagram shows the general blocks present in the flow of hiding process. Each block is explained as follows:

Algorithm

Step 1: Get the secret message file and stego-key from sender in order to hide the information.
Step 2: Find the size of the file. According to the size, generate the cover text file dynamically.
Step 3: Give the cover text file along with the password and secret information to Position Identifier which returns the position vector which contains the occurrence byte number in the cover text file of each character present in secret file.
Step 4: Give the position vector to binary convertor which returns the binary value of each element of the position vector.
Step 5: Then Manchester encoding is performed for the sequence of bits.
Step 6: Finally, those bits are hided in the cover text file using different hiding methods by Bits Hider.

An Approach to Reduce the Size of Stego Object and a Secure Text Steganography

Now a days the text steganography is widely used along with the cryptography or stand-alone method [9]. Depending on the type and size of the information to hide, the various methods have been proposed and implemented. This has several cases as follows:

If the secret information is very small (like any important personal message) where secrecy plays vital role, the message is hidden in text file because comparing to other medium like image, audio or video, hidden information in text file is hard to detect.

- If the secret information is medium, the capable size image is taken as the cover medium.
- If the secret information is large, the audio or video is taken as the cover medium depending upon the relative size of information.

In this project work, a very secure text steganography has been implemented that is explained in two phases.

Phase I Sender side process like encryption in cryptography
Phase II Receiver side process like decryption in cryptography

Process of Position Identifier

The main work of Position Identifier is to read each character from the secret file and to find the occurrence of that character in the cover text file. It returns the number of bytes traversed from the start of the cover text file as the first occurrence. It forms all byte number in the form of vector. The following flow diagram explains the concept clearly and the algorithm also given for ease of understanding [10].

Process of Manchester Encoding

The Manchester encoding is the traditional concept but it includes extra security layer in the flow of hiding process. The process Manchester encoding is that if it encounters high to low transition it encodes it as low transition. That is if the sequence of bits is '10' then '0' is encoded. If it encounters low to high transition it encodes it as high transition. That is if it is '01' then '1' is encoded. Other then these two sequences (00 and 11) simply discarded. The following flow diagram illustrates the concept clearly and algorithm also given.

Process of Bits Hider

The Bits Hider can use different hiding methods to hide the two bits namely 0 and 1. The very popular methods are as follows:

Receiver Side Manipulation

The receiver side process is opposite to sender side process. That is first Bits Extractor extracts the bits from stego object file. Then Decimal Converter converts the binary number to decimal number which is position vector values. It assembles the values and returns the position vector. Next Character Identifier identifies the character at the position value from position vector. It returns sequence of character. Then the Character Assembler assembles the character as secret message. The following flow diagram illustrates the concept and algorithm also given.

Algorithm

Step 1: Read the stego object file character.
Step 2: Give these characters to Bits Extractor which identifies the bits and return the Bit Sequence. The process of Bits Extractor can be understood from the opposite process of Bits Hider in the sender side manipulation.
Step 3: Give this Bit Sequence to Decimal Converter which returns the Position vector consisting of all decimal numbers. The process of Decimal Converter can be understood from the opposite process of Binary Converter in the sender side manipulation.
Step 4: Give this position vector to Character Identifier which identifies the character at the positions in the position vector. The process of Character Identifier can be understood from the opposite process of Position Identifier.
Step 5: Finally, Character Assembler gets the character from Character Identifier, and assembles them. It returns the Secret Text.

Process of Bits Extractor

The Bits Extractor can use different extracting methods to extract the single space, double space, and triple space. The very popular methods are as follows:

Process of Manchester Decoding

The Manchester decoding is the traditional concept but it includes extra security layer in the flow of extracting process. The purpose of Manchester decoding is that if it encounters high to low transition it encodes it as low transition.

Expected Output

The experiment on various sizes of input file our method shows reduced usage of space to hide. This is explained as follows:

For example, consider the following line of text we are going to hide: "I will come on Monday". This sentence has 22 characters, that is, size of the file is 22 bytes. In order to hide these characters by this proposed method, the cover medium is taken as pangram sentences, which have all 26 letters in English. So the position vector contains 22 entries where each one entry for one character. The value ranges from 0 to 32 because the first sentence itself contains all 26 letters and we are looking for first occurrence. So to hide the numbers 0 to 32 requires 5 bits. So for 22 numbers totally 110 bits needed to hide. So it consumes around 14 bytes in memory. By traditional schemes they are directly hiding the character in binary form so it requires ($22 \times 8 = 176$) 176 bits to hide. It consumes 22 bytes. So the amount of space saved is 8 bytes.

References

1. Por, L.Y., Delina, B.: Information hiding: a new Approach in text steganography. Paper presented at the 7th WSEAS international conference on applied computer & applied computational science'08, Hangzhou, China, 6–8 April 2008
2. Chapman, M., Davida, G., Rennhard, M.: A practical and effective approach to large-scale automated linguistic steganography. In: Proceedings of the Information Security Conference, October 2001, pp. 156–165. Malaga, Spain, October 1-3
3. Johnson, N.F.: Steganography. http://www.jjtc.com/stegdoc/index2.html. Accessed November 1995
4. Bennett, K.: Linguistic steganography: survey, analysis, and robustness concerns for hiding information in text. Purdue University, CERIAS Technical Report, 2004
5. Wayner, P.: Disappearing Cryptography: Being and Nothingness on the Net. Academic Press Inc., New York (1996)
6. Brassil, J., Low, S., Maxemchuk, N.F., O'Garman, L.: Electronic marking and identification techniques to discourage document copying. IEEE J. Sel. Areas Comm. **13**, 1495–1504 (1995)
7. Spammimic. http://www.spammimic.com (2000). Accessed 1 July 2009
8. Bender, W., Gruhl, D., Morimoto, N., Lu, A.: Techniques for data hiding. IBM Syst. J. **35**, 313–336 (1996)
9. Provos, N., Honeyman, P.: Hide and seek: an introduction to steganography. Security & Privacy, IEEE **1**(2), 32–44 (2003)
10. Mohammed Al-Mualla and Hussain Al-Ahmad.: Information hiding: steganography and watermarking. http://www.emirates.org/ieee/information_hiding.pdf (2009). Accessed 1 July 2009

Policy-Based Energy Management in Smart Homes

T. K. Anandalakshmi, S. Sathiakumar and N. Parameswaran

Abstract This paper proposes the use of policies in smart homes to manage energy efficiently and reduce peak energy demand. In peak hours, demand increases and supply providers bring additional power plants online to supply more power, which results in higher operating costs and carbon emission. In order to meet peak demand, utility companies have to build additional power plants, which may be operated only for short period of time. Therefore, reducing peak load will reduce the need for building additional power plants and decrease carbon emission. Our policy-based framework achieves peak shaving so that power consumption adapts to available power while ensuring the comfort level of the inhabitants and taking device characteristics into account at the same time. Our simulation results on Matlab indicate that the proposed policy driven homes can effectively contribute to demand side power management.

Keywords Agents · Policy · House agent · Energy management · Policy-based modeling

Introduction

Global demand for energy is increasing fast as the population increases and the number of household appliances also increases as the technology improves. The international energy agency says the world's energy needs could be 50% higher

T. K. Anandalakshmi · S. Sathiakumar (✉)
School of Electrical and Information Engineering, The University of Sydney, Sydney, Australia
e-mail: s.sathiakumar@sydney.edu.au

N. Parameswaran
School of Computer Science and Engineering, University of New South Wales, Sydney, Australia
e-mail: paramesh@cse.unsw.edu.au

S. Sathiakumar et al. (eds.), *Proceedings of International Conference on Internet Computing and Information Communications*, Advances in Intelligent Systems and Computing 216, DOI: 10.1007/978-81-322-1299-7_2, © Springer India 2014

in 2030. Thus, there is a need to use less energy or use energy more efficiently and smartly. To opt out of more power generation, smart energy management is required.

Instead of managing energy at the supply side only which is expensive, it may also help to have control on energy consumption at the distribution side. For a given amount of power generation with insignificant energy storage options, whatever energy is generated must be transmitted and consumed by the consumers at once. If power generation is not equal to power consumption, then the stability problem arises. In order to overcome peak demand and power system stability problems, we propose in this paper an approach where the power consumption behavior of the household appliances in a smart home are controlled and managed by a set of policy rules such that the average power consumption of the house is always less than the available power for that house over a period of time and that the average comfort level of the residents is not excessively compromised.

Traditional energy management techniques and studies always ignore social behavior or treat it as a single parameter describing the society as a whole. Consumers' choice and interests on energy use is one of the main factors which affects energy demand and should be considered in power management. Policy-based simulation technique views electric appliances as agents whose behaviors are controlled by policy rules. The agents react to the changes in their social and physical environments.

Policy-based management is acknowledged as a promising approach for dealing with automated management of large-scale distributed systems and networks. Policies which are sets of rules can alter the behavior of objects within the system [1]. Policy-based approach attempts to model specific behaviors of a given individual whereas the traditional macro simulation techniques, which are based on mathematical models, attempt to capture the overall characteristics of a given population. Thus, policy-based modeling is appropriate for domains characterized by a high degree of localization and distribution whereas equation-based modeling is suitable for systems that can be modeled centrally in which the dynamics are dominated by physical laws rather than using information processing [2].

This paper is organized as follows: In Section "Related Work", we review the existing energy management techniques. We discuss the user behavior with regard to energy consumption in Section "User's Behavior and Energy Management". In Section "Power Consumption Management and Policies", we explain power consumption management and policies. In Section "Power System Model", we discuss how our policy-based model works in different situations using simulation results. Concluding remarks are given in Section "Conclusion".

Related Work

Smart home refers to a home where energy is managed smartly and efficiently while ensuring the comfort level of the inhabitants. Several research efforts have focused on energy saving through intelligent management of house appliances. In

[3], the authors present a Multi Agent Home Automation System (MAHS) which allows the agents to cooperate and coordinate their actions so that power is managed efficiently. Tabu search algorithm is used to reduce the complexity of the problem by dividing into independent sub problems. In [4], the authors propose an agent-based approach to reduce energy consumption and carbon emission. They mainly focus on how to reduce energy wastage and to develop "good" habits and life style toward energy saving and CO2 reducing. Simulation results demonstrate how energy consumption and CO2 emission can be calculated. In [5], the authors propose a hybrid social model for accurate power demand estimation which extends traditional models by adding a social simulation layer to capture social responsiveness on power conservation policies. In this, they develop a simulator called Residential Power Demand Simulator for evaluating power-pricing policies, in which a consumer's behavior and social interactions are considered. They run a variety of scenarios and observe the impact of how policies may affect total demand. In [6], the authors propose a demand side management system for a household by using Keiko University Network oriented Intelligent and Versatile Energy System (KNIVES) and a smart circuit breaker box. It controls electric power by using a distributed and cooperative power control algorithm. In this paper, we focus our research on how to reduce peak energy demand by managing the household appliances using policy while taking care of device characteristics. We also evaluate the user satisfaction factor while executing policies for the appliances and make sure that their comfort level is maintained.

User's Behavior and Energy Management

Household appliances account for 30% of the total energy consumption. White goods such as refrigerators, washing machines, and dish washers are the largest contributors to household energy use consuming 34% of all energy consumed by household appliances as shown in Fig. 1.

An average household in Australia supposedly uses around 15–20 kWh electricity per day[7]. For most utilities, electricity demand peaks between 3 p.m. and 8 p.m. when people come home from work and start cooking dinner, washing clothes, running dishwasher, and turning on their big screen TVs. Thus, targetting household appliances will be a useful strategy in residential energy management.

The total energy savings from all appliances (which could amount to several millions typically in a city) could be significant and can effectively contribute to substantial energy savings if consumers learn to manage their energy consumption behavior (which may also help reduce the peaks in the power consumption curve).

Consumer behavior changes may be achieved by incentives such as vouchers, tax credits, and utility rebates. Consumers may have preferences in their usage of electricity and their appliances. They also have the authority and responsibility for device level management within their homes. Consumers have the right to employ any instruments and tools to manage energy.

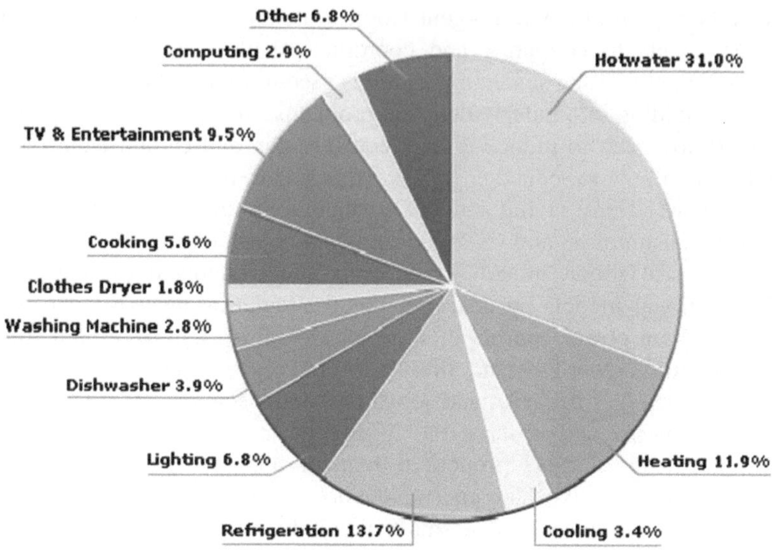

Fig. 1 Energy use within an average NSW household [7]

Using policies to control electric appliances is one of the ways of managing the overall power consumption behavior of a consumer. Policies are set of rules that turn on or off an appliance depending on a set of conditions. In general, rules in a policy can examine the current state of an appliance and decide upon the next state to which the appliance can be switched to depending on the current situation. Policies can be modified dynamically thus offering flexibility. Policies also allow communications with other appliances using standard protocols over a network (Wi-Fi, Ethernet, and Home area networks such as Zigbee, Z-Wave, and Homeplug) and are leveraged to support grid applications such as meter reading, DR, and energy management [8].

Power Consumption Management and Policies

Our power consumption management framework consists of a house agent that monitors and controls the electric appliances in the house using a policy implemented using a set of rules. (See Fig. 2.) Communication between the house agent and the appliance is implemented using existing power lines. The house agent accesses the states of the appliances using a standard Zigbee protocol, and decides upon the new state to which the appliance must be taken to depending on the time and total power consumption at that time.

Let us consider an example. Suppose the air conditioners (A/Cs) in a house are permitted to be switched on between 10:00 a.m. and 4:00 p.m. in summer

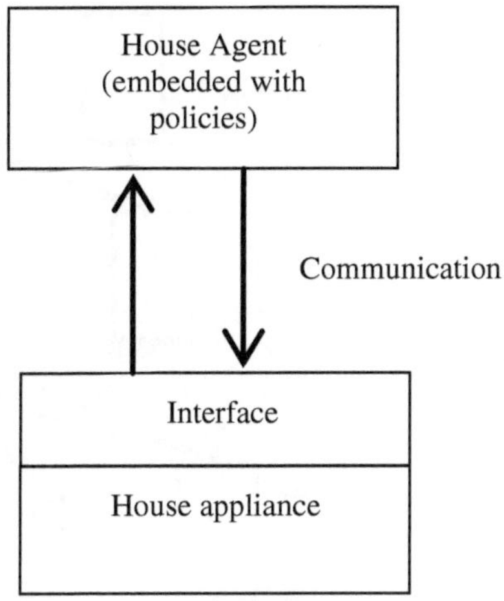

Fig. 2 Communication between house agent and appliance

according to a policy. During this time, the A/Cs will be on if the user wants them on and if the total power consumption does not exceed the available power; otherwise they will be off.

Let t be the sampling time of a day and RonAC is the request time of the user to turn on the A/C. Then, the following rules denote a policy specifying when the A/Cs can be on or off:

if (t = RonAC & 10 $\leq t \leq$ 16 & total power $\leq$ available power) **then** turn on AC;
if (t = RonAC & 10 $\leq t \leq$ 16 & total power $\geq$ available power) **then** turn off A/C;

When a home agent receives a request to turn on the A/Cs from the consumer, the home agent invokes the policy rules. Policy rules which reside in the home agent propose an action for the appliance depending upon the time, the state of the appliance and actual power consumption at that time. Action is sent to the interface using an agreed upon protocol. The Interface upon receiving the action name and the parameters, passes the action to the appliance and the appliance executes the action after which the state changes.

Power System Model

Electricity distribution is a complex process where demand fluctuates quickly and loads that are responsive in real time can be highly valuable to retailers and networks. This domain is ideal for the adoption of policy-based power consumption management. Fig. 3 shows the power system from generating station to household appliances.

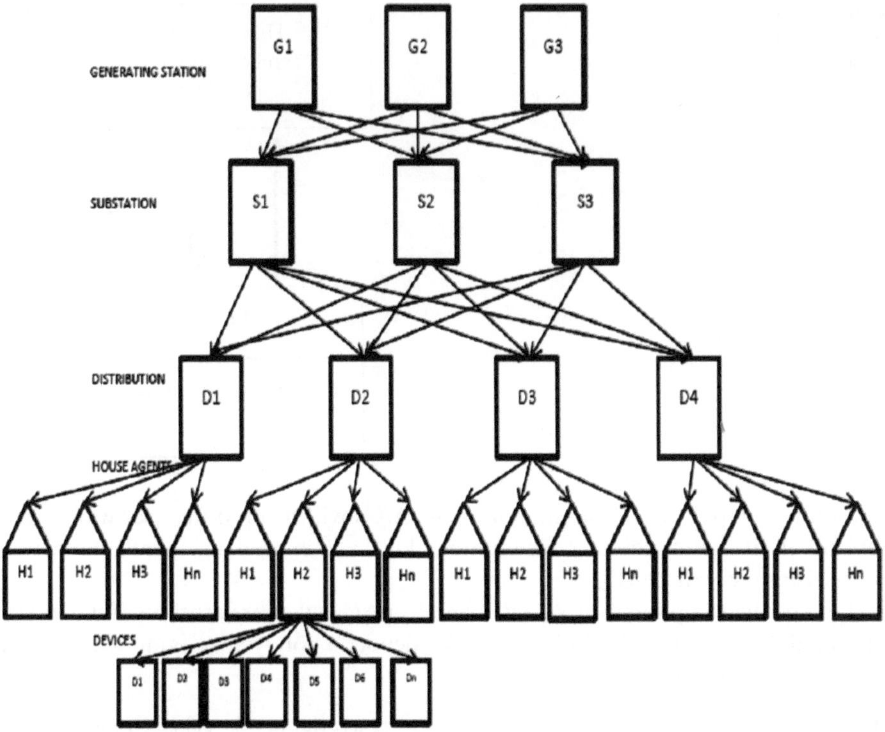

Fig. 3 Electric power system model from generation to appliances

Hierarchy in Power System

- Generating station

Electricity is produced close to supplies of energy such as coal and water, which are used to drive equipment used to generate power.

- Substation

The electric power carried over the extra high voltage transmission lines is delivered to regional and neighborhood substations where the electricity is stepped down from high voltage to a voltage that can be used in homes and offices.

- Distribution

This is the last stage in the delivery of electric power to the users. Distribution circuits start from a transformer located in the electrical distribution substation.

- Home agents

Each home has an "electrical service" connection and a meter for billing. The home agent monitors and manages energy effectively throughout the day by controlling the appliances and ensures that the total power does not exceed the available power while maintaining the comfort level of the user.

- Appliances

Appliances are the real consumers of electricity in the houses. Power produced at the generating plant finally goes to the appliances so that the inhabitants benefit from the services provided by the appliances. In this paper, we assume that the appliances do not interact amongst themselves and they can only share information with the home agent.

Policy-Based Implementation

We assume that the house agent wants to maintain total energy consumption below an available power of 5,000 W.

Power Consumption Behavior without Policies

Figure 4 shows the overall power consumption of appliances in a typical home where the consumption is not monitored or managed. Note that it has peaks exceeding over 5,000W line. The appliances are turned on any time any number of times and any longer. Since appliances like refrigerators, although appear to be on all day, are actually running between 12 and 15 h a day, we assume that on an average a refrigerator is on for 10 h a day. The peaks in Fig. 4 are due to many appliances being turned on during the same interval.

Fig. 4 Actual power consumption curve

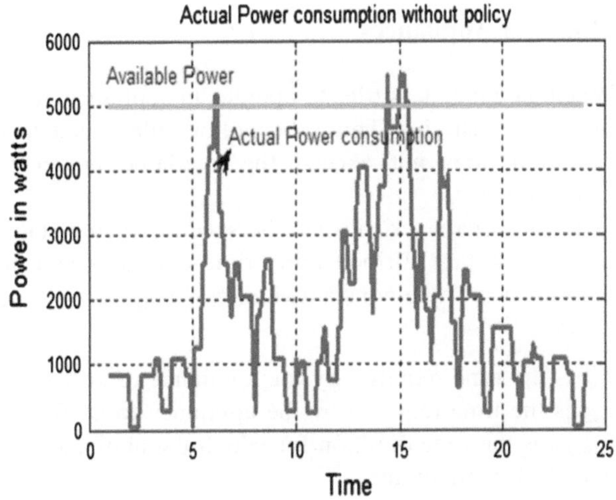

Table 1 Classification of appliances

Grade 1 (essential appliances)	Grade 2 (entertainment appliances)	Grade 3 (optional appliances)
Microwave (1,000 W)	Air conditioner (1,000 W)	Washing machine (800 W)
Computer (250 W)	Nintendo (500 W)	Clothes dryer (1,000 W)
Water heater (1,000 W)	TV (500 W)	Dish washer (1,200 W)
Lights (200 W) 4 rooms	Music (400 W)	Vacuum cleaner (1,000 W)
Iron box (1,000 W)		
Coffee maker (800 W)		
Fridge (800 W) (permanent service)		

Power Consumption with Monitoring Policies

The appliances are categorized according to their importance to the inhabitants. A typical classification of appliances is given in Table 1.

(a) *Grade 1 appliances*

These appliances are essential to sustain the day-to-day activities of the inhabitant. Agents or policies should not turn them off while the inhabitant is using them.

(b) *Grade 2 appliances*

These appliances are turned on most of the time, and yet not considered essential.

(c) *Grade 3 appliances*

The operation of these appliances is required only rarely.

Modeling Appliances

When the inhabitant tries to operate an appliance, the maximum available power is the first constraint. Thus, the policy rules need to ensure that the power consumption does not exceed the available power while following the device characteristics.

Device characteristics can be specified using states along with their attributes. Some important attributes are: time constants and power consumption. Some of the important time constants are: τ_{on}, $\tau_{min\text{-}on}$, τ_{off}, and $\tau_{min\text{-}off}$,
where

τ_{on} is the time required by the appliance to go from OFF state to ON state;
τ_{off} is the time required by the appliance to go from ON state to OFF state;
τ_{minoff} is the time duration for which the appliance stays in its OFF state before it goes to ON state; and

τ_{minon} is the time duration for which the appliance stays in its ON state before it goes to OFF state.

These time constants are important since they can determine how quickly the system can move from one state to another. State diagram is used to describe the dynamic behavior of a device or an appliance. When a rule operates on a particular appliance, it consults the state diagram of that appliance before the rule fires.

Figure 5 describes the simple state diagram of a light. We model this device with two states: light on (s0) and light off (s1). The arrows between the states are called transitions and will happen when the switch is flipped.

A more general state diagram of an appliance with multiple states is shown in Fig. 6. The states are s0 (On), s1 (turning off), s2 (Off), and s3 (turning on).

Once the appliance is switched on, it goes to warming up state, and after τ_{on} units of time, it goes to the ON state. The appliance then stays in the working state at least for τ_{minon} period. During the ON state, if a policy rule says that the appliance should be turned on, the appliance is turned on to which event the

Fig. 5 State diagram of a light

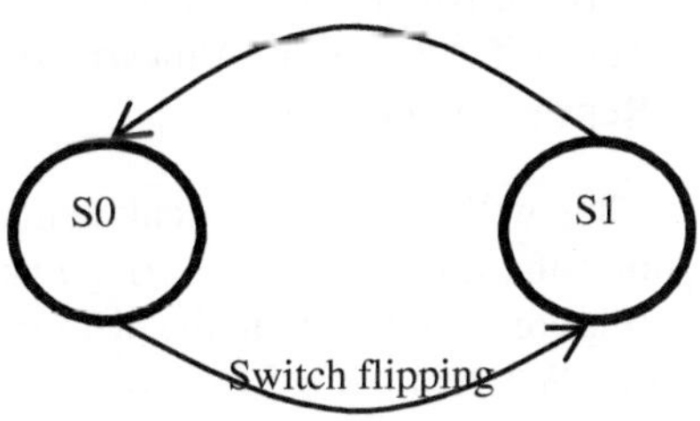

Fig. 6 State diagram of an appliance with multiple states

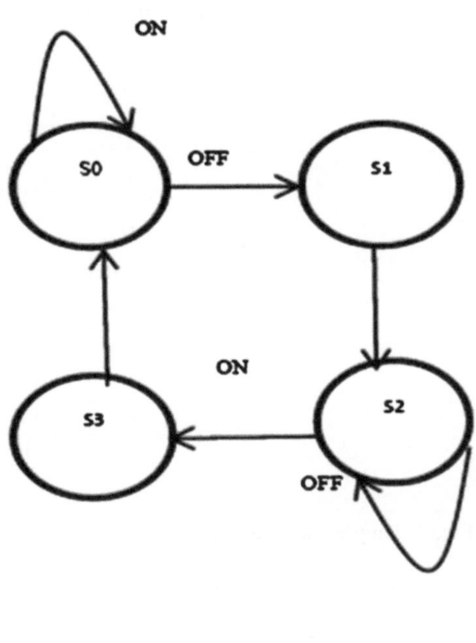

appliance responds by continuing to stay in the ON state. When it is switched off, it starts the shutting down operation and after τ_{off} it goes to the OFF state. It stays in the OFF state for at least τ_{minoff} period. During the OFF state, if the rule requires it to turn off, the appliance continues to stay in the OFF state.

Rules for the House Agent

Before defining a set of rules for the house agent to monitor and control the appliances, we define a term called *on-time* which denotes the interval over which we need the appliance to remain ON once it is turned on. For example, the customer may need for an A/C appliance, a minimum on-time of 4 h, say between 10 a.m. and 4 p.m. A rule that uses the on-time notion may look like:

if RontimeAC ≥ 10 & $t \leq 16$
then

Turn off fridge and turn on A/C for the first five sampling periods;
Turn off A/C and turn on fridge for next five sampling periods;
Repeat this until A/C is on;

This will help reduce peak demand during, for example, hot summer. Figure 7 shows the duty cycle of a fridge before policy is implemented.

Figure 8 shows that the demand for energy suddenly increases from 800 to 1,800 W when A/C (1,000 W) is turned on and when there is no policy executed.

Figure 9 shows that the power consumption by the A/C and the fridge when operated together with the policy above is only 1,000 W. If the duty cycle is not adjusted, then the power during the A/C time would have reached 1,800 W.

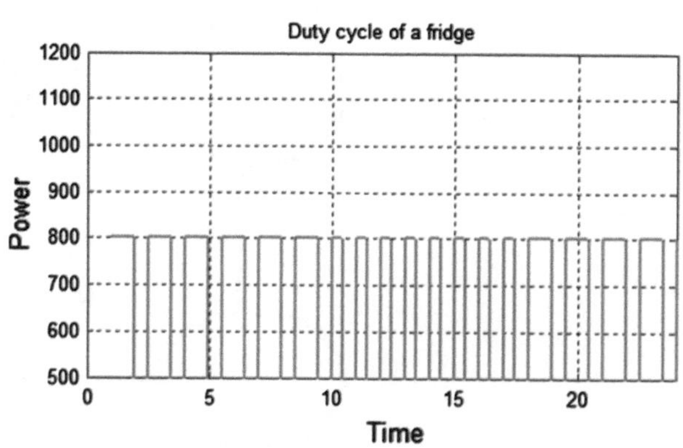

Fig. 7 Duty cycle of a fridge

Fig. 8 A/C and fridge working together without policy

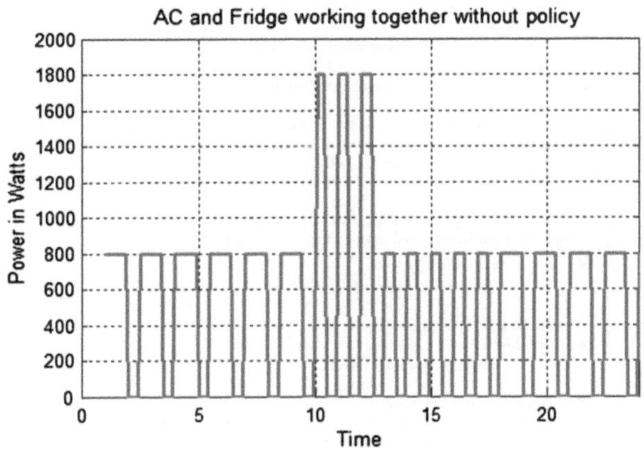

Fig. 9 Duty cycle of Fridge and A/C during the operation of A/C

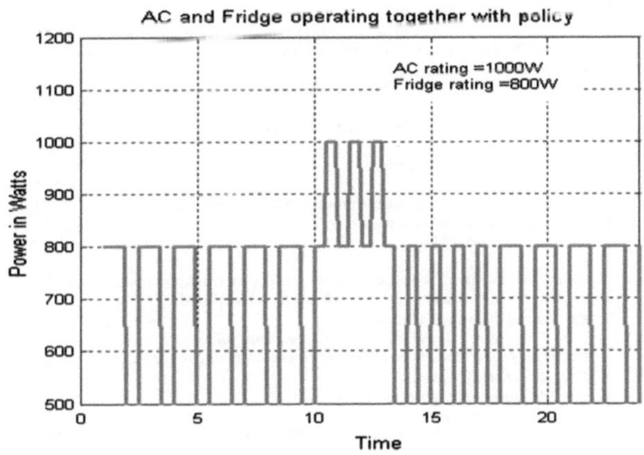

Rules for House Agent
t =sampling time starts at 0 and incremented by 0.1;
ontime(wash) = requestontime for washing machine from user;
durationwash = operation time for washingmachine;

// only one grade3 appliance will be allowed to operate at any time.
R1: *ontime(wash)= ontime(dish)=ontime(vacuum) →*
ontime(dish)=ontime(wash)+durationwash+0.1,
ntime(vacuum)=ontime(wash)+durationwash+0.3;

//once a grade 3 appliance is on, then no other grade 3
//appliances can be operated until that appliance is off.
R2: *t ≥ ontimegrade3appliance →*
turn off other grade3 appliances

// postpone grade 3 appliances
R3: *t = ontimegrade1 appliance →*
turn on grade1 appliance;

R4: *power > available power*
& t=ontimegrade3appliance →
ontimegrade3appliance =
ontimegrade1appliance+durationgrade1appliance;

t=ontimewaterheater & t = ontimewash or ontimedish
or ontimevacuum →
postpone grade 3 appliances;

//to prevent power consumption to go high when A/C is
// ON, the iron box is turned off
R5: *t = ontimeironbox & t = AC time →*
turn off Iron box;

//TV and Nintendo are not operated together.
R6: *ontimeTV = ontimeNintendo →*
turn off TV and turn on Nintendo;

//Nintendo is not allowed before school hours.
R7: *ontimeNintendo <15 →*
turn off Nintendo;

//if total power exceeds available power music is turned off.
R8: *total power >available power →*
turn off Music;

//to ensure quality of the process once grade 3 appliance is turned on it will run until the process(washing, drying, dish wash) is completed.
R9: *t >ontimegrade 3 appliances →*
grade 3 appliances cannot be turned off;
//TV is preferred than music when there is a power constraint.
R10: *t=ontimeTV=ontimemusic*
& power>available power →
turn off music;
//if lights are on during the day and if actual power exceeds //available power then lights will be turned off.
R11: *t = ontimelights & 10≤t≤16*
& power >available power →
turn off lights.
R12: *if t<10 &t>16 turn on fridge for 10 sampling periods and*
turn off for 5 sampling periods.
//Outside AC time fridge will work normally.
R13: *If t <10 or t>16 & if power>available power →*
turn off AC.
//Outside AC time if A/C is on and if power
//exceeds the available power then A/C will be switched off.
R14: *total power consumption < available power*
& t<6 & t<22 & available power-actual power >500 →
turn on music
available power-power >1200 →
turn on AC;
Entertainment appliances like music and AC are turned on depending upon the remaining power and the actual time to enhance the comfort level of the inhabitant.
(For simplicity reasons, we did not write rules to check the time constants of appliances each time.)

Performance with Policy

Figure 10 demonstrates the fact that the rules in the house agent make sure that the total power does not exceed the available power and at the same time they try to satisfy the inhabitant by postponing the operation of grade 3 appliances later in the day. By repeating the simulation several times, we realized that our policy-based framework works well in other scenarios as well.

Figure 11 shows the status of appliances before and after the rules were applied. Since the power goes above the available power level, the on-time of the

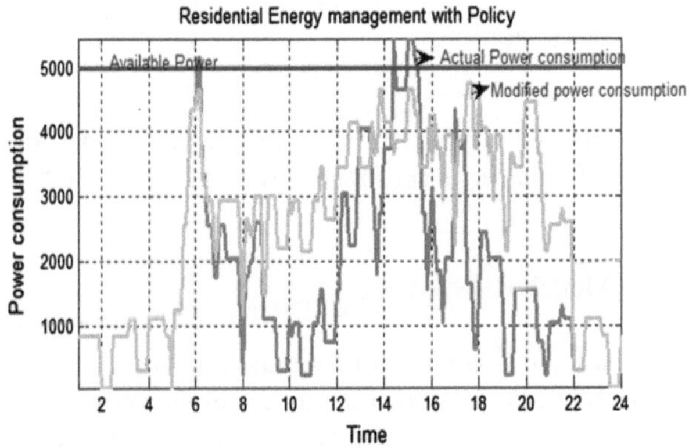

Fig. 10 Residential energy management with policy

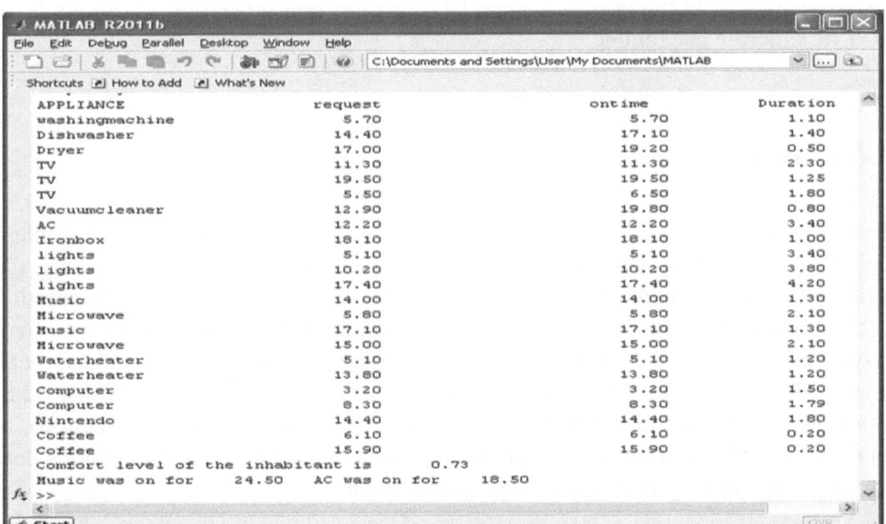

APPLIANCE	request	ontime	Duration
washingmachine	5.70	5.70	1.10
Dishwasher	14.40	17.10	1.40
Dryer	17.00	19.20	0.50
TV	11.30	11.30	2.30
TV	19.50	19.50	1.25
TV	5.50	6.50	1.80
Vacuumcleaner	12.90	19.80	0.80
AC	12.20	12.20	3.40
Ironbox	18.10	18.10	1.00
lights	5.10	5.10	3.40
lights	10.20	10.20	3.80
lights	17.40	17.40	4.20
Music	14.00	14.00	1.30
Microwave	5.80	5.80	2.10
Music	17.10	17.10	1.30
Microwave	15.00	15.00	2.10
Waterheater	5.10	5.10	1.20
Waterheater	13.80	13.80	1.20
Computer	3.20	3.20	1.50
Computer	8.30	8.30	1.79
Nintendo	14.40	14.40	1.80
Coffee	6.10	6.10	0.20
Coffee	15.90	15.90	0.20

Comfort level of the inhabitant is 0.73
Music was on for 24.50 AC was on for 18.50

Fig. 11 Status of appliances (on-time request and actual on-time)

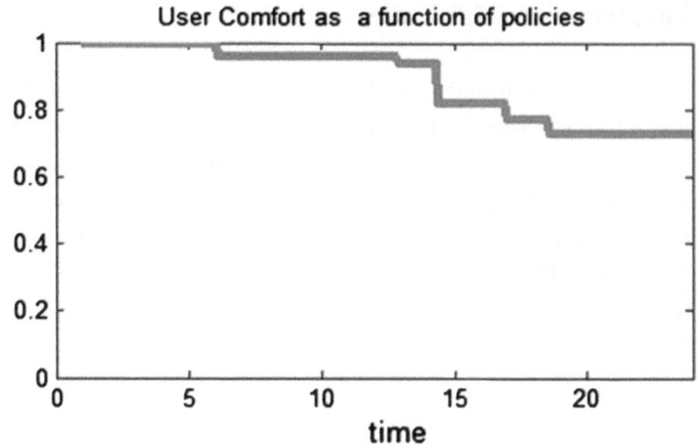

Fig. 12 User comfort model

dishwasher is postponed to 17:10 h. Further, TV was postponed from 5:50 to 6:50 h and the on-time of the dryer is postponed from 17:00 to 19:20 h.

User's Satisfaction Model

To estimate the customer's satisfaction, we assume that the actual power consumption behavior "b" of the customer as, for example, shown in Fig. 4 above is a reasonable model of his satisfaction behavior over time. Any deviation from this behavior is interpreted as a measure of dissatisfaction. When a policy is used to manage the power consumption behavior, deviations are measured and plotted as shown in Fig. 12. When the deviation exceeds a threshold, say 0.5, the user may require a change in the policy. Figure 12 illustrates that inhabitant's comfort level is 0.72 which is acceptable.

Conclusion

In this paper, we have proposed a policy-based smart home energy management approach to monitor and control the state of appliances to consume only the available power and manage energy efficiently throughout the day. Simulation results demonstrate that policy-based smart home can perform peak shave effectively taking electrical characteristics of appliances into account. Since peak energy demand is minimized in this approach, there is no need to operate additional power plants and therefore carbon emission can be significantly reduced. In our future work, we plan to extend our approach to a community of agents where agents cooperatively perform energy management.

References

1. Sloman, M.: Policy driven management for distributed systems. J. Netw. Syst. Manag. **2**, 333–360 (1994)
2. Davidsson, P.: Multi agent based simulation: beyond social simulation. In: Moss, S., Davidsson, P. (eds.) Multi agent based simulation, vol. 1979. Springer Verlag LNCS series, Heidelberg (2000)
3. Abras, S., Pesty, S., Ploix, S., Jacomino, M.: An anticipation mechanism for power management in a smart home using multi agent systems, ICTTA 2008. Paper presented at the 3rd international conference, pp. 1–6
4. Huang, R., Itou, M., Tamura, T., Ma, J.: Agent based approach for smart eco –home environments. In: Proceedings of the International Joint Conference in Neural Networks, IJCNN 2010, pp. 1–8
5. Lin, F., Zhang, D., Shi, Z., Xu, M., Zhou, Y.: A novel simulation approach for estimating residential power demand based on multi agent society, Cognitive informatics. Paper presented at the 6th IEEE international conference, pp. 450–455, 2007.
6. Suhara, Y., Nakabe, T., Mine, G. Distributed demand side management system for home energy management. Paper presented at the 36th annual conference on IEEE industrial electronics society, pp. 2430–2435, 2010.
7. IPART Research Paper 29; November 2007: Residential energy and water use in Sydney, the Blue Mountains and Illawarra; Results from the 2006 household survey; Electricity, Gas and Water. http://www.savepower.nsw.gov.au/get-the-facts/power use-in-sw.aspx. http://www.savepower.nsw.gov.au/get-the-facts/power-use-in-sw.aspx
8. Perry, R., Wacks, K.: Creating a robust market for residential energy management through an open energy management architecture. Cable Labs pp. 1–13 (2010)

References

[illegible]

Saturation Throughput and Delay Analysis of IEEE 802.11 Broadcast Transmissions Scheme

Gayathri Narayanan

Abstract In this paper, the performance of IEEE 802.11-based single hop broadcast network assuming saturated nodes is studied by determining the threshold parameters for both saturation throughput as well as delay. An upper limit for the achievable throughput is derived by maximizing the broadcast saturation throughput. An analytical expression for optimal contention window size required to meet the desired objective is derived. Expression for the maximum saturation throughput is also obtained. The results reveal that the maximum saturation throughput is independent of the number of contending stations. This work also addresses the computation of the minimum delay in a wireless ad-hoc network with saturated nodes. The optimal contention window that minimizes the saturation delay under broadcast is also determined. The minimum delay parameter is significant in such a network since the network applications are generally not delay tolerant.

Keywords IEEE 802.11 · Broadcast · Medium access control · Performance analysis · Saturation throughput and delay

Introduction

Broadcasting is an integral communication technique widely used in both ad-hoc and infrastructure networks. In particular, ad-hoc networks depend heavily on the MAC layer's broadcast for routing, neighbor discovery, and data dissemination. Broadcast services are indispensable in the exchange of safety-related alarm and beacon messages in a vehicular communication scenario. The most popular MAC

G. Narayanan (✉)
Department of Electronics and Communication Engineering, Amrita School of Engineering, Amritapuri Campus, Amrita Vishwa Vidyapeetham, Clappana P.O, Kollam, Kerala, India
e-mail: gayathrin@am.amrita.edu

S. Sathiakumar et al. (eds.), *Proceedings of International Conference on Internet Computing and Information Communications*, Advances in Intelligent Systems and Computing 216, DOI: 10.1007/978-81-322-1299-7_3, © Springer India 2014

protocols for ad-hoc wireless networks is the distributed coordination function (DCF) which is a random access scheme based on carrier sense multiple access with collision avoidance (CSMA/CA) and binary exponential backoff as deployed in 802.11 standard [1].

The 802.11 MAC protocol uses the request-to-send/clear-to-send (RTS/CTS) mechanism in order to communicate unicast transmissions effectively. This four-way handshaking mechanism accounts for the reliability of unicast transmission since the use of RTS/CTS frames reserves the channel for transmission and reduces the possibility of collisions. In contrast, broadcast data are transmitted without any control frames. According to the IEEE 802.11 protocol, no acknowledgement (ACK) will be transmitted by the receiving node, no RTS/CTS exchange will be used and no MAC layer recovery or re-transmission will be performed. As a result, the broadcast transmission scheme is less reliable compared to unicast transmissions [1]. In addition, repeated broadcasts can flood the network due to a lot of redundant packets in the network and results in the Broadcast Storm Effect. Work in [2] addresses this issue and provides some suppression techniques.

Related Work

The study of broadcast schemes was not of primary focus in the past as broadcast transmissions were considered a trivial component of a wireless local area network. As a result, most of the early analytical studies focused on the performance analysis of unicast transmissions. The two-dimensional Markov chain model developed by Bianchi [3] is the pioneering work which computes the saturation throughput of the IEEE 802.11 unicast traffic. Subsequently, this accurate model has been extended to address further issues like freezing of the backoff counter. Chen et al. [4] extended the results of [3] for finding the saturation throughput for a lossy channel. The saturation performance of a broadcast network was analyzed in [5–7] in terms of throughput and delay by modeling the backoff counter as a one-dimensional Markov chain for the broadcast case and as a two-dimensional Markov chain for the unicast case. The performance analysis assuming nonsaturated nodes and with Poisson arrival process at each node was presented in [8]. The same model has been used to evaluate the performance metrics for broadcast service in vehicle-to-vehicle networks [9]. Even though the works in [5–7] evaluate the performance metrics for a saturated broadcast network, they do not address the additional considerations on the maximum throughput and minimum delay theoretically achievable.

This paper considers an 802.11 single hop network where nodes have the capability to communicate with each other using broadcast messages. The one-dimensional Markov chain model of [7] which characterizes the operation of the backoff counter procedure in an 802.11 network is the model adopted for analysis. First of all, the value of the optimal transmission probability and contention window size which maximizes the saturation throughput is derived analytically. The expression for the maximum broadcast saturation throughput is also obtained. Second, the broadcast delay is also minimized and the optimal transmission probability and contention window size required to meet this objective is derived. The rest of the paper is organized as follows. The analysis of saturation throughput and delay is presented in Section "Performance Analysis". The numerical and simulation results are presented in Section "Numerical and Simulation Results". Concluding remarks are given in Section "Conclusion".

Performance Analysis

The DCF basic access method can be described as follows. An active station, with a packet to transmit, senses the channel activities for an idle period of time equal to a distributed inter frame space (DIFS). After the DIFS medium idle time, the station generates a random initial backoff time as an additional deferral time before transmitting. This accounts for collision avoidance. The selection of a random backoff counter value is not required if the backoff counter already contains a nonzero value. The initial back-off time is generated as

$$T_{\text{ib}} = \text{uniform}(0, w - 1) * \sigma \tag{1}$$

According to the IEEE standards, a station transmits when its backoff counter value decrements and reaches zero. If the medium is detected idle by a station which senses the channel, the backoff counter will be decremented by one slot time. If the medium is detected busy at any time during a backoff slot, the backoff counter will be frozen and it will be reactivated only on sensing the medium idle again. In broadcast transmissions, since there are no acknowledgements and retransmissions, the current backoff window size is constant and equal to initial minimum backoff window size.

Consider a WLAN with n number of stations. The network is assumed to be saturated which implies that each station always has a packet available for transmission immediately after the completion of the current transmission. It is also assumed that the number of contending stations is large but finite and that there are no hidden terminal problems. Further, the channel is assumed to be error-free. The backoff counter process is characterized using a one-dimensional Markov chain model where the backoff counter does not choose zero as the initial backoff time [7]. Let W_0 denote the initial backoff window size for each station and let τ be the probability that a station transmits during a slot time. These quantities are related as follows [7]:

$$\tau = \frac{2}{W_0 + 1} \tag{2}$$

Let T_s and T_c be the average time for which the channel is sensed busy due to successful transmission and collision respectively and let σ be the slot duration. The average length of a packet is denoted as $E[P]$ and the packet header includes the physical and MAC layer headers. In broadcast scheme, since a station cannot differentiate between a successful transmission and a collision, we have $T_s = T_c$, and is given by

$$T = T_s = T_c = \frac{L_H + E[P]}{R_d} + \text{DIFS} + \delta \tag{3}$$

where δ is the propagation delay and $L_H = PHY_{hdr} + MAC_{hdr}$. The probability that the channel is busy and is given by

$$p_b = 1 - (1 - \tau)^n \tag{4}$$

The probability that a successful transmission occurs in a slot is given by

$$p_s = n\tau(1 - \tau)^{n-1} \tag{5}$$

The probability that a collision occurs is

$$p_c = p_b - p_s = 1 - (1 - \tau)^n - n\tau(1 - \tau)^{n-1} \tag{6}$$

The normalized saturation throughput of the system is [7]

$$S = \frac{p_s E[P]}{(1 - p_b)\sigma + p_s T_s + p_c T_c} \tag{7}$$

The average delay for a successfully transmitted packet is [7]

$$D = \frac{W_0 + 1}{2}[(1 - p_b)\sigma + p_s T_s + p_c T_c] \tag{8}$$

Maximizing the Broadcast Saturation Throughput

In this subsection, the expressions for the optimal transmission probability and optimal contention window size which achieve the maximum saturation throughput are derived. We also obtain expression for the maximum saturation throughput. On rearranging (7), the expression for saturation throughput can be written as: $S = \frac{E[P]}{T_s - T_c + \frac{\sigma + p_b(T - \sigma)}{p_s}}$. The saturation throughput is maximized by maximizing the term $\frac{p_s}{\sigma + p_b(T - \sigma)}$ On substituting the values of p_b and p_s differentiating with respect to τ and equating to zero, the expression for the optimal transmission probability is obtained as

$$\tau_{\text{opt}} = \frac{1}{n}\sqrt{\frac{2\sigma}{\tau}} \tag{9}$$

Corresponding to (2), the optimum contention window size is given by:

$$W_{0,\text{opt}} = n\sqrt{\frac{2T}{\sigma}} \tag{10}$$

Substituting (9) in (7), and assuming $n \gg 1$ and $\tau \ll 1$, the expression for maximum saturation throughput is obtained as follows:

$$S_{\max} = \frac{E[P]}{T + \sqrt{\frac{T\sigma}{2}}} \tag{11}$$

It is observed that the maximum saturation throughput is independent of the number of contending stations.

Minimizing the Broadcast Saturation Delay

This section determines the optimal contention window that minimizes the broadcast saturation delay. Differentiating (8) with respect to τ and equating to zero, the optimum transmission probability is obtained as

$$\tau^* \frac{1}{n} \tag{12}$$

Using (2), the optimum contention window size is obtained as

$$W_0^* = 2n - 1 \tag{13}$$

Accordingly, the expression for the minimum saturation delay is given by

$$D_{\min} = n\left[T - \frac{T}{e}\right] \tag{14}$$

Numerical and Simulation Results

In this section, the numerical and simulation results for a saturated single hop broadcast network are presented. The numerical results are from the analytical model presented in previous sections. The simulation results are obtained using Matlab. Table 1 summarizes the values of parameters specified by IEEE 802.11 that have been used for numerical computation. Table 2 shows the numerical evaluation of the saturation throughput. In order to establish the maximization, the

Table 1 Parameters for MAC and PHY layers in IEEE 802.11

Parameter	Value
Packet payload, P	8184 bits
MAC header	272 bits
PHY header	128 bits
Channel bit rate	1 Mbps
Propagation delay, δ	1 μs
DIFS	128 μs
Slot time, σ	50 μs
Number of stations	5 ~ 50

Table 2 Throughput Comparison

n	Optimum transmission probability	Optimum CW	S_{max}	Saturation throughput	
				$W = 32$	$W = 64$
1	0.1071	19	0.8915	0.8626	0.7955
2	0.0536	37	0.8915	0.8723	0.849
5	0.0214	93	0.8915	0.8129	0.8526
10	0.0107	187	0.8915	0.693	0.8003
15	0.0071	280	0.8915	0.582	0.7408
20	0.0054	373	0.8915	0.4853	0.6823
25	0.0043	467	0.8915	0.4009	0.6264

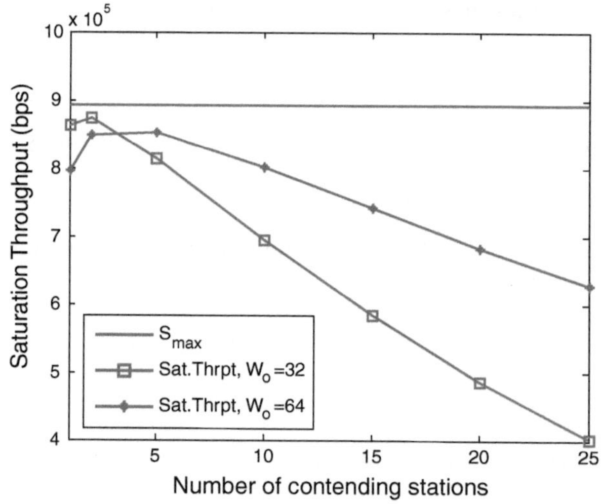

Fig. 1 Saturation throughput versus number of stations

saturation throughput is computed for two arbitrarily chosen window sizes over varying number of contending stations.

It is evident from Fig. 1, which shows the saturation throughput against the number of stations that the throughput is maximized for the optimal contention window case. These throughput values are greater than the achievable throughput obtained for each arbitrary window size. Further, it is observed that the maximum throughput is a constant and is independent of the number of contending stations. For $n = 10$, with optimal contention window, there is 28.6 % improvement in throughput as compared to that is obtained for a fixed window size $W_0 = 32$. When $W_0 = 64$, the improvement is 11.4 %. For $n = 20$, there is an improvement of 83.7 % when the window size is set to 32 and an improvement of 30.7 % when the window size is set to 64.

Figure 2 shows the saturation delay plotted against the number of stations. In general, the delay increases as the number of stations increase. Table 3 shows the numerical computation of saturation delay. In order to ascertain that the delay is minimized, the optimum delay is compared with the delay obtained for varying number of contending stations when the window sizes are fixed. Two values of

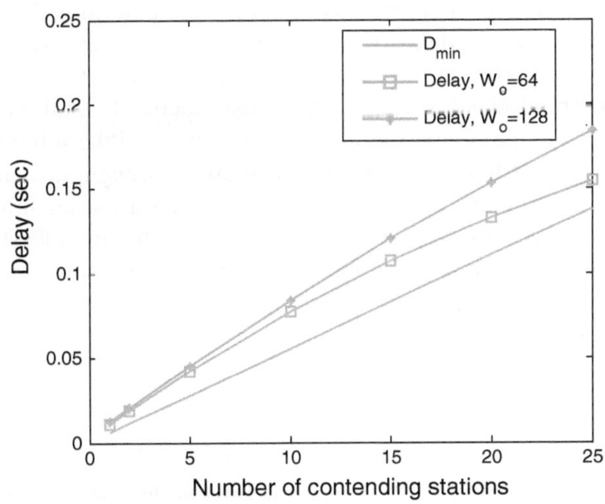

Fig. 2 Saturation delay versus number of stations

Table 3 Delay comparison

n	Optimum transmission probability	Optimum CW	$D_{\min}$	Saturation delay	
				$W = 64$	$W = 128$
1	1	1	0.0055	0.0103	0.0119
2	0.5	3	0.011	0.0187	0.0204
5	0.2	9	0.0275	0.0424	0.0452
10	0.1	19	0.055	0.0772	0.0841
15	0.067	29	0.0826	0.107	0.12
20	0.05	39	0.1102	0.1325	0.1532
25	0.04	49	0.1377	0.1543	0.1839

window sizes are chosen as $W_0 = 64$ and $W_0 = 128$. The percentage reduction in delay is observed for two randomly chosen values for n, the number of contending stations. For $n = 10$, it is seen that with optimal contention window, the delay is reduced by 28.7 % for $W_0 = 64$ and by 34.6 % for $W_0 = 128$. Considering a network size of 20 contending stations, it may be observed that the delay is reduced by 16.8 % for $W_0 = 64$ and by 28.1 % for $W_0 = 128$.

Conclusion

In this paper, the IEEE 802.11-based broadcast transmission scheme was analyzed for a single hop broadcast network with saturated nodes. The upper limit performance measures of the network were derived. To achieve the objective, the optimal transmission probability and optimal contention window size, which maximized the broadcast saturation throughput, were determined. The problem of minimizing the broadcast delay was also considered independently. Analytical expressions for the optimal transmission probability and optimal contention window size, which minimized the delay, were also obtained. Analytical expressions were derived for the maximum saturation throughput and the minimum saturation delay. Results showed that the maximum saturation throughput is independent of the number of contending stations. The maximum saturation throughput is an indication of the maximum achievable throughput for the broadcast network. The minimum broadcast delay is a significant performance measure since the network applications are generally delay sensitive.

References

1. IEEE 802.11 Working Group, Part 11: wireless LAN medium access control (MAC) and physical layer (PHY) specifications ANSI/IEEE Std. 802, 1999.
2. Tonguz, O.K., Wisitpongphan, N.: On the broadcast storm problem in ad-hoc wireless networks. Paper presented at 3rd international conference on broadband communications, networks and systems '06, BROADNETs, San Jose, 2006.
3. Bianchi, G.: Performance analysis of the IEEE 802.11 distributed coordination function. IEEE J. Sel. Areas Comm. **18**, 535–547 (2000)
4. Dong, X.J., Varaiya, P.: Saturation throughput analysis of IEEE 802.11 wireless LANs for a lossy channel. IEEE Commun. Lett. **9**, (2005)
5. Ma, X., Chen, X.: Saturation performance of IEEE 802.11 broadcast networks. IEEE Comm. Lett. **11**, 686–688 (2007)
6. Chen, X., Refai, H.H., Ma, X.: Saturation performance of IEEE 802.11 broadcast scheme in ad-hoc wireless LANs. Paper presented at IEEE vehicular technology conference, Baltimore, pp. 1897–1901, VTC Fall, 2007
7. Ma, X., Chen, X.: Performance analysis of IEEE 802.11 broadcast scheme in ad hoc wireless LANs. IEEE Trans. Veh. Technol. **57**, 3757–3768 (2008)

8. Wang, J.C.P., Abolhasan, M., Franklin, D.R., Safaei, F.: Characterizing the behavior of IEEE 802.11 broadcast transmissions in ad hoc wireless LANs. Proceedings of IEEE International Conference on Communications, Istanbul, 2009.
9. Chen, X., Refai, H.H., Ma, X.: Unsaturated performance of IEEE 802.11 broadcast service in vehicle-to-vehicle networks. Paper presented at IEEE vehicular technology conference, Baltimore, pp. 1957–1961, VTC Fall, 2007.

Spatial Query Monitoring in Wireless Broadcast Environment

Koenni Naresh, J. Thangakumar and Durgamahesh Pannem

Abstract Wireless data broadcast is a promising technique for information dissemination that leverages the computational capabilities of the mobile devices in order to enhance the scalability of the system. Under this environment, the data are continuously broadcast by the server, interleaved with some indexing information for query processing. Clients may then tune in the broadcast channel and process their queries locally without contacting the server. It performs location updates only when they would likely alter the query results through monitoring process. Previous work on spatial query processing for wireless broadcast systems has only considered snapshot queries over static data. Here, we use the simple of K-Means clustering algorithm for making clusters of sensor node data in a wireless sensor network. It is used to monitor the objects continuously.

Keywords kNN · WSN · LDIS · LAN · BGI · GPS

Introduction

The use of repetitive broadcast as a way of augmenting the memory hierarchy of clients in an asymmetric communication environment. We describe a new technique called "Broadcast Disks" for structuring the broadcast in a way that provides improved performance for non-uniformly accessed data. The Broadcast Disk superimposes multiple disks spinning at different speeds on a single broadcast

K. Naresh (✉) · J. Thangakumar · D. Pannem
Department of Computer Science and Engineering, Hindustan University, Padur, India
e-mail: nareshkoenni@gmail.com

J. Thangakumar
e-mail: thang.kumar@gmail.com

D. Pannem
e-mail: durgamagesh@gmail.com

S. Sathiakumar et al. (eds.), *Proceedings of International Conference on Internet Computing and Information Communications*, Advances in Intelligent Systems and Computing 216,
DOI: 10.1007/978-81-322-1299-7_4,

channel—in effect creating an arbitrarily fine-grained memory hierarchy. In addition to proposing and defining the mechanism, a main result of this work is that exploiting the potential of the broadcast structure requires a reevaluation of basic cache management policies. We examine several "pure" cache management policies and develop and measure implementable approximations to these policies. Previous work on location-dependent spatial query processing for wireless broadcast systems has only considered snapshot queries over static data. Suppose moving client wants to know the route from one place to another place. In old GPS application systems client can send the query to server like where am I? the request will go to server and do some process and send the reply to client. Mean while suppose client may reach to another place and he can get the wrong data from the server. It means there is no proper communication between servers and clients. The processing load at the server side increases with the number of queries. In applications involving numerous clients, the server may be overwhelmed by their queries or take prohibitively long time to answer them. Spatial monitoring techniques do not apply to the broadcast environment because they assume that the server is aware of the client locations and processes their queries centrally (Fig. 1).

Proposed System can handle multiple queries at a time. Server will continuously receives the spatial queries from the client side. Because there are more than one client can access the same server at time. In this situation server may react slowly because it has to process lot of queries and filtering. At the client Side, every time it receives the spatial data and continuously accessing server for new data. For that we may have problem with battery consumption in travelling. To avoid this they proposed Air Indexes. The main motivation behind the air indexes is to minimize the power consumption at the mobile client.

Implementation of Air Indexes includes when the query is issued by the client, it tunes to the broadcast channel and it goes to Sleep mode until the next index segment arrives. The client traverses the index and determines when the data objects satisfying its query will be broadcast, The client goes to sleep and returns

Fig. 1 Broadcasting process

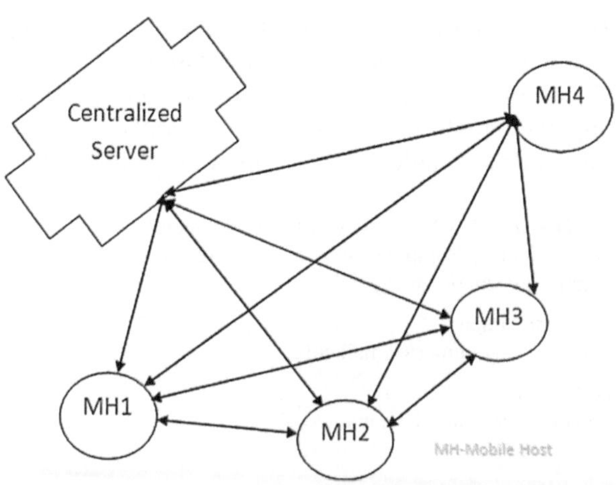

to the receive mode only to retrieve the corresponding data objects. We propose an air indexing framework that outperforms the existing techniques in terms of energy consumption while achieving low access latency and constitutes the first method supporting efficient processing of continuous spatial queries over moving objects.

Literature Survey

Area of sensor network is very wide and used in various applications [1]. There has been much work done in data mining, sensor network, data stream etc. but very less work has been done at the combination of these areas. So we introduce the combination of wireless sensor network and data mining to get some interesting and fresh results [2]. Research in WSNs area has focused on two separate aspects of such networks, namely networking issues, such as capacity, delay, and routing strategies; and application issues [3]. The beauty of sensor networking protocols is that they attracted a tremendous amount of research effort For large sensor network the management of sensor database is itself a big task. Query processing techniques have been proposed for acquiring and managing sensor data. One major research goal of this problem in the database community is to efficiently detect outliers in a large-scale database [4]. To collect the data we treat the Sensor networks as databases and SQL queries. Clustering of sensor network is related with network topology, not on the sensor data. In our study we are basically concentrating towards Spatial-Temporal data [5]. Traditional data mining works on Association Rule, looking for patterns of the form: "Customers who buy bread, also buy milk, with probability x %." There is significant literature.

Mobile Location-Dependent Information Services (LDISs) have drawn a lot of attention from wireless data industries in the past few years. This growth in mobile communications presents new aspects to the resource assignment problem as well as new applications [6]. In these services, information provided to mobile users' reflects their current geographical locations. Location dependent data is a data whose value depends on the location. The answer to a query depends on the geographical location where the query originates. Let's consider an example in which a user drives a car and wants to find the nearest gas stations. The user sends a query, such as, "what are the names and locations of the gas stations near to my current location?", using his mobile device [7]. Once the user gets the answer from the server, he will visit the gas stations in order of the nearest to his location based on price. To handle such a query, the positions of the objects and the user must be found. In this paper, we propose the broadcast-based LDIS scheme under a geometric location model. We first introduce the broadcast based location dependent data delivery scheme (BBS). In this scheme, the server periodically broadcasts reports, which contains the IDs of the data items (e.g., building names) and the values of the location coordinates to the clients. The broadcasted data objects are sorted sequentially, based on their location before being broadcasted. Then, we introduce the prefacing scheme in LDIS for the mobile computing environment.

By using the proposed schemes, the client's access and tuning times are significantly reduced. The main contributions of our work can be summarized as follows:

- It is not necessary for the client to wait for and tune into a particular index segment, if it has already identified the nearest object before the index segment has arrived. This technique significantly reduces the access time in the broadcast based LDIS environment.
- The client simply adjusts the value of k when it performs the k-NN query processing.
- The client can also perform the a k-NN query processing without an index segment. In this case, the best access time is obtained, since no index is broadcast along with the file.

System Architecture

Client–server computing or networking is a distributed application architecture that partitions tasks or workloads between service providers (servers) and service requesters, called clients. Often clients and servers operate over a computer network on separate hardware. A server machine is a high-performance host that is running one or more server programs which share its resources with clients. A client also shares any of its resources; Clients therefore initiate communication sessions with servers which await (listen to) incoming requests (Fig. 2).

Wireless Broadcasting

The transmission schedule in a wireless broadcast system consists of a series of broadcast cycles. Within each cycle the data are organized into a number of index and data buckets [8]. A bucket (which has a constant size) corresponds to the smallest logical unit of information, similar to the page concept in conventional storage systems. A single bucket may be carried into multiple network packets (i.e., the basic unit of information that is transmitted over the air). However, they are typically assumed to be of the same size (i.e., one bucket equals one packet). Broadcast process involves the centralized server with multiple number of mobile hosts.

Data Clustering

The term Data Clustering means the process of organizing or arranging a set of objects into groups or clusters in such a manner so that objects within a cluster have the most similarity to one another and the most dissimilarity to objects in

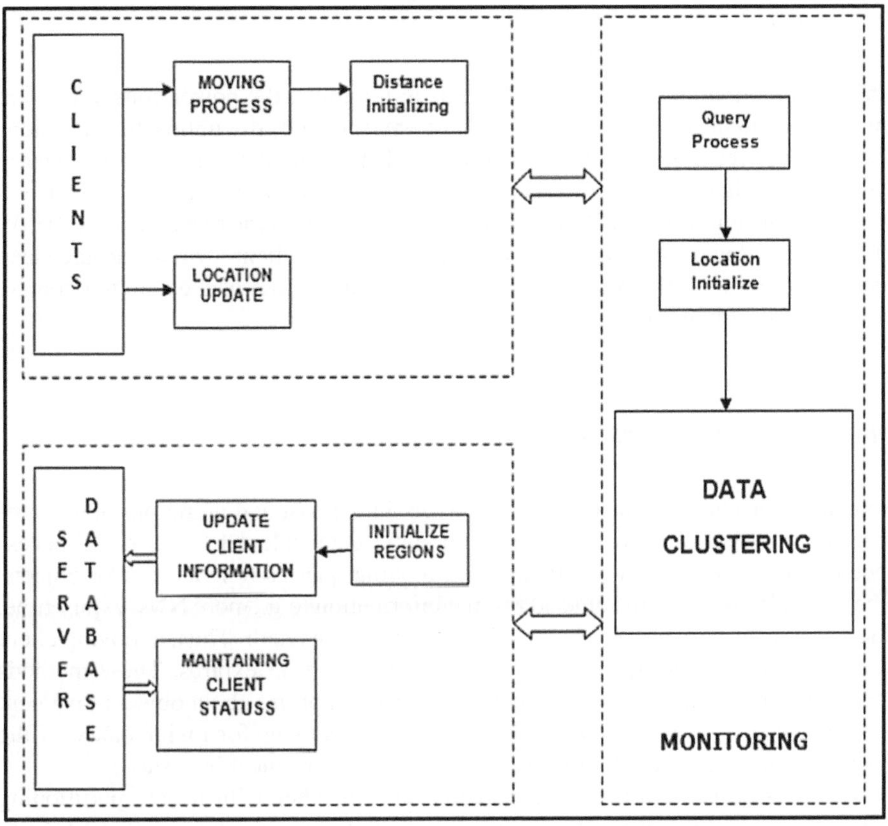

Fig. 2 System architectural diagram

other clusters. Clustering is a useful and basic task in data mining. We generally used it as initial step in data mining for future data analysis. So, here we purpose clustering of sensor data along with their sensor attributes. Here, we use the simple of K-Means clustering algorithm for making clusters of sensor node data in a wireless sensor network.

[9] Data Pre-processing involves cleaning the data by putting in missing values and removing uninteresting data. It may also include summarization and aggregation of the data. This step basically involves preparing the data for analysis. So, firstly we can detect the irregularities in the sensor data and apply pre-processing technique [10]. Pre-processing involves cleaning the data by putting in missing values and removing uninteresting data. It may also include summarization and aggregation of the data. This step basically involves preparing the data for analysis.

Air Indexing

The main motivation behind air indexes is to minimize the power consumption at the mobile client. Although in a broadcast environment, the uplink transmissions are avoided, receiving all the downlink packets from the server is not energy efficient. For instance, the Cabletron 802.11 network card (wireless LAN) was found to consume 1,400 mW in the transmit, 1,000 mW in the receive, and 130 mW in the sleep mode. Therefore, it is imperative that the client switches to the sleep mode (i.e., turns off the receiver) whenever the transmitted packets do not contain any useful information.

Spatial Query Processing

Spatial queries have been studied extensively in the past and numerous algorithms exist for processing snapshot queries on static data indexed by a spatial access method. Subsequent methods focused on moving queries (clients) and/or objects. The main idea is to return some additional information (e.g., more NNs, expiry time, and validity region) that determines the lifespan of the result. Thus, a moving client needs to issue another query only after the current result expires. These methods focus on single query processing, make certain assumptions about object movement (e.g., static in, linear in), and do not include mechanisms for maintenance of the query results (i.e., when the result expires, a new query must be issued).

The possibility of broadcasting spatial data together with a data partitioning index. They present several techniques for spatial query processing that adjust to the limited memory of the mobile device. The authors evaluate their methods experimentally for range queries (using the R-tree as the underlying index) and illustrate the feasibility of this architecture.

Continuous kNN Queries

Consider, for instance, a user (mobile client) in an unfamiliar city, who would like to know the 10 closest restaurants. This is an instance of a k nearest neighbor (kNN) query, where the query point is the current location of the client and the set of data objects contains the city restaurants. Alternatively, the user may ask for all restaurants located within a certain distance, i.e., within 200 m. This is an instance of a range query.

Conclusion

We study spatial query processing in wireless broadcast environments. A central server transmits the data along with some indexing information. The clients process their queries locally, by accessing the broadcast channel. In this setting, our target is to reduce the power consumption and access latency at the client side. We propose an on-air indexing method that uses a regular grid to store and transmit the data objects. We design algorithms for snapshot and continuous queries, over static or dynamic data. To the best of our knowledge, this is the first study on air indexing that

(1) Addresses continuous queries and
(2) Considers moving data objects.

We demonstrate the efficiency of our algorithm through an extensive experimental comparison with the current state-of-the-art frameworks for snapshot queries and with the constant re-computation technique for continuous queries. A challenging problem is to devise cost models for continuous monitoring of spatial queries in wireless broadcast environments. Such models could reveal the best technique given the problem settings, help fine-tune several system parameters (e.g., grid size), and potentially lead to better algorithms. Another interesting direction for future work is to study different types of spatial queries, such as reverse nearest neighbors, and to extend our framework to process their snapshot and continuous versions.

Future Work

Another interesting direction for future work is to study different types of spatial queries, such as reverse nearest neighbors, and to extend our framework to process the snapshots and continuous versions. Another important thing is to add more security for query processing.

References

1. Mittal, R., Bhatia, M.P.S.: Wireless sensor networks for monitoring the environmental activities (2010)
2. Hjaltason, G.R., Samet, H.: Distance browsing in spatial databases
3. Park, K.: An efficient data dissemination scheme for nearest neighbour query processing
4. Acharya, S., Alonso, R., Franklin, M.J., Zdonik, S.B.: Broadcast disks: data management for a symmetric communication environments. In: Proceedings of the ACM SIGMOD (1995)
5. Acharya, S., Franklin, M.J., Zdonik, S.B.: Disseminating updates on Broadcast disks. In: Proceedings of the International Conference on Very Large Data Bases (VLDB'96) (1996)

6. Beckmann, N., Kriegel, H.-P., Schneider, R., Seeger, B.: The R*- tree: an efficient and robust access method for points and rectangles. In: Proceedings of the ACM SIGMOD (1990)
7. Cai, Y., Hua, K.A., Cao, G.: Processing range-monitoring queries on heterogeneous mobile objects. In: Proccedings of the IEEE International Conference on Mobile Data Management (MDM '04) (2004)
8. Chen, B., Jamieson, K., Balakrishnan, H., Morris, R.: Span: an energy–efficient coordination algorithm for topology maintenance in ad hoc wireless networks. In: Proccedings of the annual international conference on mobile computing (2001)
9. Chen, M.-S., Yu, P.S., Wu, K.-L.: Indexed sequential data broadcasting in wireless mobile computing. In: Proceedings of the International Conference on Distributed Computing Systems (ICDCS '97) (1997)
10. Gedik, B., Liu, L.: MobiEyes: distributed processing of continuously moving queries on moving objects in a mobile system. In: Proceedings of the International Conference on Extending Database Technology (EDBT '04) (2004)

Enhancing Data Caching in Ad-hoc Networks Through Benefit-Based Technique

Koteswara Rao Makke, J. Thangakumar, B. V. Suresh Reddy and Yannam Somaiah

Abstract Mobile ad-hoc network (MANET) is demand-based, self configurable, network without any existing infrastructure. Data caching can significantly improve the efficiency of information access in a wireless ad-hoc network by reducing the access latency and bandwidth usage. Every Mobile Hosts (MHs) can move arbitrarily and communicate with other MHs by using multihop wireless links. However, designing efficient distributed caching algorithms is nontrivial when network nodes have limited memory. In this paper, we consider the cache placement problem of minimizing total data access cost in ad-hoc networks with multiple data items and nodes with limited memory capacity. Defining benefit as the reduction in total access cost. The approximation algorithm is amenable to localized distributed implementation, which is shown via simulations to perform close to the approximation algorithm.

Keywords MANETs · Mobile hosts

K. R. Makke (✉) · J. Thangakumar · B. V. Suresh Reddy · Y. Somaiah
School of Computer Science and Engineering, Hindustan Institute of Technology & Science, Chennai, Tamilnadu, India
e-mail: makkekoteshwar@gmail.com

J. Thangakumar
e-mail: thang.kumar@gmail.com

B. V. Suresh Reddy
e-mail: boddu.lkv@gmail.com

Y. Somaiah
e-mail: somu.yannam@gmail.com

S. Sathiakumar et al. (eds.), *Proceedings of International Conference on Internet Computing and Information Communications*, Advances in Intelligent Systems and Computing 216,
DOI: 10.1007/978-81-322-1299-7_5, © Springer India 2014

Introduction

Mobile ad-hoc networks (MANETs) are consist of Mobile Hosts (MHs) such as notebooks, Personal Digital Assistants (PDAs), cell phones, and so on. These mobile devices can create a wireless network dynamically without the aid of any network infrastructure. Every MH can move arbitrarily and communicate with other MHs by using multihop wireless links. Each MH acts as a router and forwards data packets to other neighbours in the coverage of its transmission range. This type of network can communicate with external networks such as Internet through a gateway.

A computer is envisioned to be equipped with more powerful capabilities, including the storage of a small database and the capability of data processing, which gives birth to an entirely new area of MANET. [1] MANET is demand-based, self configurable, network without any existing infrastructure. Information/ data access to mobile nodes. With the rapid development in the field of wireless technology and mobile devices, the devices come up with more capabilities like more processing power, more storage capacity, fast computation, more battery power, etc. As such there is no dearth of resources in present scenario, but at the same time it is also true that as system is growing in terms of maturity everything comes with the term "more" that is, more energy, more storage capacity, more computational power, even then there is ever increasing demand for higher bandwidth and higher processing.

In the state of sufficient cache size in mobile nodes, it is possible to hold all/ majority of data items in the cache. Neither it is possible to hold all accessed data items in the cache of each mobile node nor it is advisable to keep the data at each and every location in the network, this will create an unnecessary mess in the entire network. Moreover, the information which should be kept handy might not get a place to reside. In this paper, we proposed a novel algorithm that prefetches the data based on association among data items. Our scheme prefetches highly related data items and considers confidence of association rules. To find the relationship among data items, association rules base data mining technique [2] is used.

Since caching and prefetching are both well recognized for improving client perceived response time, the integration of both strategies may be exploited to improve the system performance. In the prefetching access to remote data is anticipated and data is fetched before it is required. If the requested data item is not prefetched earlier, the client has to send an uplink request to ask for data item when the query comes. This not only increases the query latency but also increases the uplink bandwidth requirement. The remainder of this paper is organized as follows.

In Section "Related Work", we briefly review the related studies on cache replacement and prefetching in MANETs and mobile environment. Section "System Model" gives description of system model. Section "Proposed Algorithm" describes the proposed algorithm. Section "Simulation Results" evaluates the performance. Concluding remarks are given in Section "Conclusion" and the scope of future work is discussed.

Related Work

Caching frequently accessed data on client side is an effective technique to improve performance in MANETs [3]. A lot of research has been done on cache invalidation in past few years [4, 5], with relatively little work being done on cache prefetching methods. In the following, we briefly review related studies on prefetching in mobile environments. Prefetching has been widely employed to reduce the response time in the web environment [6]. Most of these techniques concentrate on estimating the probability of each file being accessed in near future. Since these techniques are designed for point-to-point communication environment, they are not suitable for MANETs.

The client calculates the prefetch access ratio (PAR), which is the number of prefetches divided by the number of accesses for each item. If PAR is less than one, prefetching item is useful since prefetched data may be accessed multiple times. Clients prefetch aggressively when channel quality is good but reduce the prefetch rate when the channel quality becomes poor. All the previous schemes have ignored the relationship between the data items. Yin and Cao [7] realized that cache misses are not isolated events, and a cache miss is often followed by a series of cache misses. They addressed the prefetching issue among related data items by using a cache-miss initiated prefetching scheme, which is based on association rule mining technique, but the disadvantage of this scheme is that the item appearing in the consequent of mined prefetched rules cannot be prioritized.

System Model

This paper studies the prefetching of requisite data in MANETs. The client employs the rule generating engine to derive caching rules from the client's access log. The derived caching rules are stored in the caching rule depository of the client. In Global Clustering Cooperation (GCC) when a client suffers from a cache miss (called "the local cache miss"), the client looks up the required data item from the cluster members (Fig. 1).

Only when the client cannot find the data item in the cluster members' caches (called "cluster cache miss"), it will request the cache state node (CSN) which keeps the global cache state (GCS) and maintains the information about the node in the network which has copy of desired data item. If a cluster other than requesting nodes' cluster has the requested data (called "remote cache hit"), then it can serve the request without forwarding it further toward the server. Otherwise request will be satisfied by server.

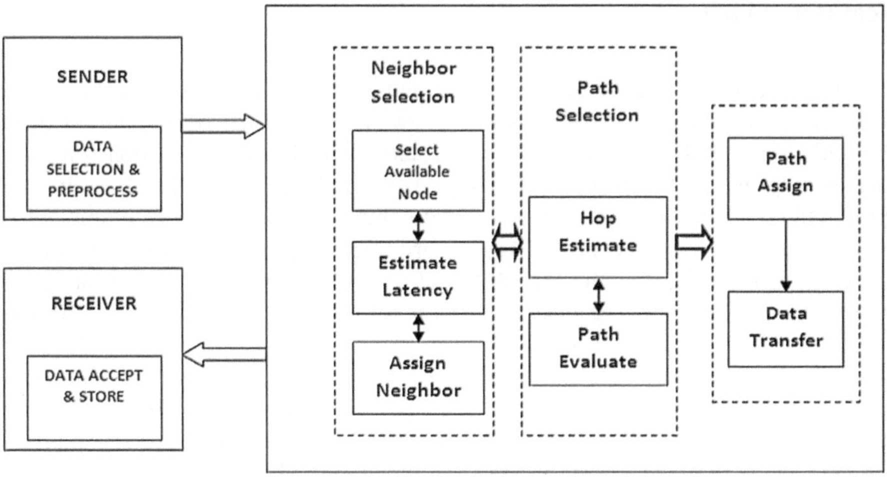

Fig. 1 System model

Prefetching Mobile Nodes

Whenever a mobile node issues a request, the cache request processing module first logs this request into record and checks whether the desired data item is available in local cache of MN or in any of the MN in the cluster. If it is a cache hit, the cache manager still needs to validate the consistency of the cached item with the copy at the original server. To validate the cached item, the cache manager checks the validation of data item from its TTL value. If the data item is verified as being up-to-date, it is returned to the MN immediately. If it is a cache hit, but the value is obsolete, the cache manager sends an uplink request to the server and waits for the data broadcast.

When the requested data item appears, the cache manager returns it to the requester and retains a copy in the cache. In the case that a cache miss occurs, the client cache manager checks the caching rule depository to derive the prefetching rules corresponding to the requested item. If this request triggers some prefetching rules, the ids of the item implied by these prefetching rules will also be piggy-backed to the server alongwith id of missed cache item.

Defining Association Rules

At a MN, data items queried during a period of time are related to each other. Observation of the history of data items queried by the client may lead to find relationship among the items. These relationships can take the form of patterns of behavior that can tell us that if the client has accessed certain items during a period of time then it is likely that one particular item will be accessed in near future [8]. For example, if a client access d_1 and d_2, then it accesses d_3 80 % of the times.

The if part of the rule is called the antecedent while the then part is called consequent. These rules are known as association rules in the data mining literature [2]. We propose to use data mining techniques to discover the association rules in the access history and apply the rules to make caching decisions (prefetching, replacement, etc.).

Let $D = \{d_1, d_2, \ldots, d_N\}$ be the set of data items at the server. Suppose a client's access trace S consists of a set of consecutive parts: $\{\text{part}_1, \text{part}_2, \ldots, \text{part}_i, \ldots, \text{part}_n\}$.

Let $A = \{d_1, d_2, \ldots, d_m\}$ denotes the set of data items accessed by the client. Let S_i denotes the data items contained in part $part_i$. S_i is called a session and $S_i \subset A$. We say a session S_i contains x if $S_i \supseteq x$, where $x \subset A$. A caching rule $r_{x,y}$ is an expression of the form $x \rightarrow d_y$, where $x \subset A$, $d_y \in A$, and $x \cap \{d_y\} = \varphi$. x is called antecedent of the rule and d_y is called consequent. In general, a set of data items is called an itemset. The number of data items in an itemset is called the size of the itemset and an itemset of k size is called k-itemset. The support of an itemset x, support(x), is defined as the percentage of sessions that contains x in the client's access trace S [7, 9]. The support of a caching rule $r_{x,y}$, support($r_{x,y}$), is defined as the support of the itemset that consists data items in both the antecedent and the consequent of the rule, that is, support($r_{x,y}$) = support($x \cup \{d_y\}$). The confidence of a caching rule $r_{x,y}$, confidence($r_{x,y}$) is defined as the support of the rule divided by the support of the antecedent [7, 10].

Given an access trace S, the problem of mining caching rules is to find all the association rules that have support and confidence greater than the user defined minimum support (*minsup*) and minimum confidence (*minconf*), respectively. The problem of mining caching rules can be decomposed into the following subproblems [2]:

1. Find all the itemsets x such that support(x) $\geq$ minsup. An itemset x that satisfies this condition is called frequent itemset.
2. Use the frequent itemsets to generate association rules with minimum confidence.

Proposed Algorithm

Generating Frequent Itemsets

In this section, we present an algorithm to generate frequent itemsets from the client's access trace. Table 1 shows the notations used in the algorithm. The main steps of the algorithm are given. It accepts an access trace S, a minimum support (minsup), and the maximum number of items NR to be used in a rule as

Table 1 Notations

NR	Maximum number of items in a rule
k-itemset	An itemset with k items
F_k	The set of frequent k-itemsets (those with minimum support)
f_i, f	Any of the frequent (k-1)-itemsets within F_{k-1}
$f_i[m]$	m-th item in itemset f_i
f	A new frequent k-itemset obtained by combining a frequent (k-1)- itemset with one item
Z	The set of caching rules

parameters. In line 1, S is analyzed to generate the frequent 1-itemset. This is done by calculating the support of each data item and comparing it to the minimum support. Every data item that has minimum support forms one frequent 1-itemset.

Loop from lines 3–20 is used to generate all the frequent 2-, 3-,…, k-itemsets. Each iteration of the loop, say iteration k, generates frequent k-itemsets based on the (k-1)-itemsets generated in the previous iteration. This loop continues until it is not possible to generate new itemsets or the number of items in an itemset exceeds the predefined maximum NR. Lines 3–14 generate all the new candidate frequent k-itemsets out of the frequent (k-1)-itemsets. Lines 15–18 remove those candidate frequent k-itemsets that do not fulfill the minimum support requirement. In line 21, the algorithm returns all the frequent itemsets generated.

1) $F_1 = \{\text{frequent 1-itemsets}\}$
2) $k = 2$
3) **while** $F_{k-1} \neq \varphi \wedge k \leq NR$ **do**
4) $F_k = \varphi$
5) **for each** itemset $f_i \in F_{k-1}$ do
6) **for** each itemset $f_j \in F_{k-1}$
7) **if** $f_i[3] = f_j[3] \wedge \ldots \wedge f_i[k\text{-}2] = f_j[k\text{-}2] \wedge f\,[k\text{-}1] < f\,[k\text{-}1]$
8) **then** $f = \{f_i\} \cup \{f_j[k\text{-}1]\}$; $F_k = F_k \cup \{f\}$
9) **for each** (k-1)-subsets $s \in f$ do
10) **if** $s \notin F_{k-1}$
11) **then** $F_k = F_k - \{f\}$; break
end
end
end
15) **for each** itemset $f_i \in F_k$ do
if support(f_i) < minsup
then $F_k = F_k - \{f]$
18) **end**
19) $k = k + 1$
20) **end**
21) **return** $\cup_k F_k$

Table 2 Sample trace in sessions

Itemset	Frequent itemsets
1-itemset	$\{d_1\},\{d_2\},\{d_3\},\{d_5\}$
2-itemset	$\{d_1,d_2\},\{d_1,d_3\},\{d_1,d_5\}$, $\{d_2,d_3\},\{d_2,d_5\}$
3-itemset	$\{d_1,d_2,d_3\},\{d_1,d_2,d_5\}$

Algorithm to generate frequent itemsets. For example, Table 2 shows a sample trace for a client divided into five sessions. Table 3 shows the frequent itemsets generated from the sample trace by applying frequent itemsets generating algorithm. Here, minsup = 60 %, and NR = 3.

Generating Algorithm for Caching Rules

We are interested in generating, from a frequent *k*-itemset f_i, rules of the form $x \rightarrow d_y$, where x is a (*k-1*)-itemset, $d_y$is a 1-itemset and $f_i = x \cup \{d_y\}$. Table 2 shows the notations used in the algorithm.

Algorithm that generate the caching rules, which, illustrates the main idea of the algorithm. The algorithm accepts the frequent itemsets and a minimum confidence (minconf) as parameters. For each frequent itemset, the rules are generated as follows. Of all the data items within frequent itemset, one item becomes the consequent of the rule, and all other items become the antecedent. Thus, a frequent *k*-itemset can generate at most *k* rules. For example, suppose $\{d_1, d_2, d_3\}$ is a frequent 3-itemset. It can generate at most three rules: $\{d_1, d_2\} \rightarrow d_3$, $\{d_1, d_3\} \rightarrow d_2$, and $\{d_2, d_3\} \rightarrow d_1$. After the rules have been generated, their confidences are calculated to determine if they have the minimum confidence (Fig. 2).

Only the rules with at least the minimum confidence are kept in the rule set Z. For example, for the rule $\{d_1, d_2\} \rightarrow d_3$, confidence $conf = support(\{d_1, d_2, d_3\})$/support($\{d_1, d_2\}$). If $conf \geq minconf$, the rule holds and it will be added to the rule set Z.

Simulation Results

The simulation area is assumed of size 1,500 m × 1,500 m The clients move according to the random waypoint model. The time interval between two consecutive queries generated from each client follows an exponential distribution

Table 3 Frequent item sets

Session #	Data item requests
1	$d_1d_2d_3d_4d_5$
2	$d_1d_2d_3d_4$
3	$d_1d_2d_3$
4	$d_1d_2d_5$
5	$d_1d_2d_3d_5$

Fig. 2 Algorithm to generate the cachingrules

```
k = 2
while F_k ≠ ϕ do
    for each itemset f_i ∈ F_k do
        for each item fi[j] ∈ f_i
            if support(f_i) / support(f_i − f_i[j]) ≥ minconf
            then Z = Z ∪ {{f_i-f_i[j]}→f_i[j]}
        end
    end
k = k+1
end
return Z
```

with mean T_q. Each client generates accesses to the data items following Zipf distribution with a skewness parameter θ. There are N data items at the server (Table 4).

Figure 3 shows the effects of mean query generate time on average query latency on NC, CacheData, GCC, and GCC with prefetching schemes [11]. For all the algorithms, the average query latency drops when T_q increases since fewer queries are generated and the server can server queries more quickely.

Results for Cache Size

We investigated the performance of various algorithms under different cache sizes. The simulation results are shown in Figure 3. Our algorithm outperforms the other ones in terms of average query latency [12] (Fig. 4).

Table 4 Simulation parameters

Parameter	Default value	Range
Database size (N)	1,000 items	
s_{min}	10 KB	
s_{max}	100 KB	
Number of clients (M)	70	50–100
Client cache size (C)	800 MB	200–1,400 KB
Client speed ($v_{min} \sim v_{max}$)	2 m/s	2–20 m/s
Bandwidth (b)	2 Mbps	
TTL	5,000 s	200–10,000 s
Pause time	300 s	
Mean query generate time (T_q)	5 s	2–100 s
Transmission range®	25 m	25–250 m
Skewness parameter (θ)	0.8	0–1

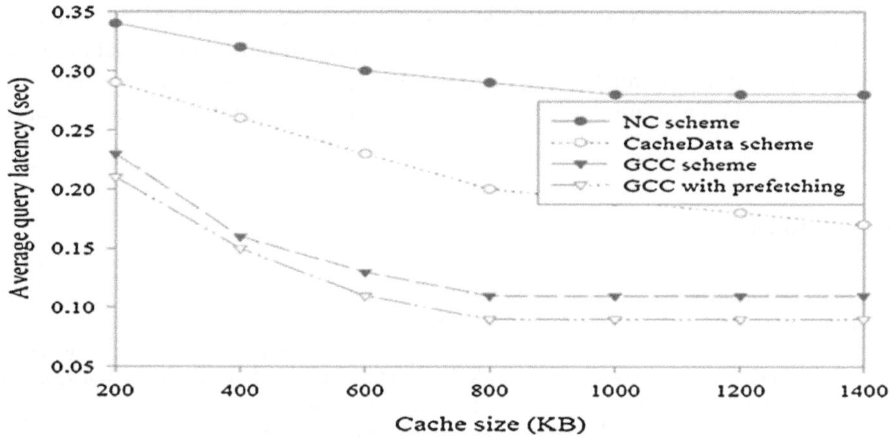

Fig. 3 Effects of cache size on average query latency

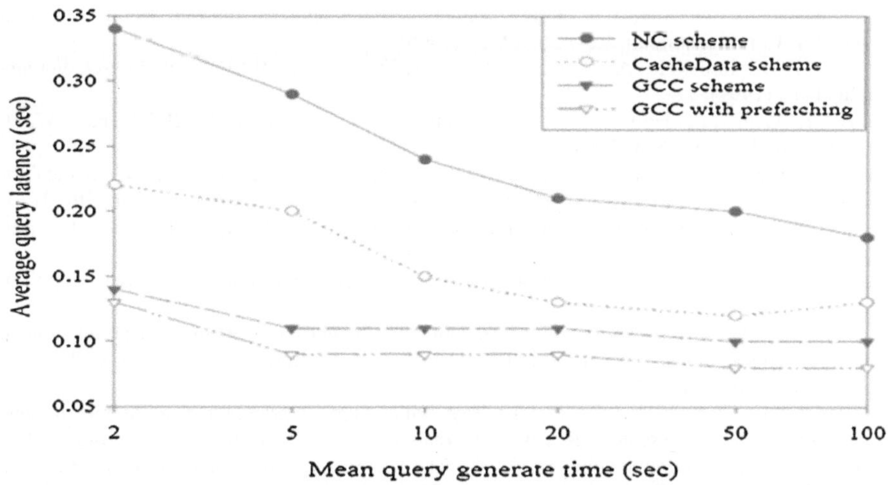

Fig. 4 The effects of mean query generate time on average query latency

Conclusion

Prefteching technique can be used to improve the system performance in MANETs. However, prefetching also consumes a large amount of system resources such as computation power and energy. Thus, it is very important to only prefetch the right data. In this paper, we have proposed an algorithm which maximizes the performance improvement due to caching in MANETs.

An algorithm not only considers the caching parameters of a data item, but also the relationship of this item with the cache set. Association rule based data mining is applied to find the relationship among data items. To enhance the caching performance, the generated association rules are used to prefetch the data item(s).

References

1. Cao, G.: A scalable low latency cache invalidation strategy for mobile environments. IEEE Trans. Knowl. Data Eng. **15**, 1251–1265 (2006)
2. Agarwal, R., Imielinski, T., Swami, A.: Mining association rules between sets of items in large database. Paper presented at ACM SIGMOD International Conference on Management of Data, Washington, D.C., May 26-28, pp. 207–216, May 1993
3. Acharya, S., Alonso, R., Frankli, M., Zdonik, S.: Broadcast disk: data management for a symmetric communication environments. Paper presented at ACM SIGMOD International Conference on Management of Data, San Jose, California, May 22-25, pp. 199–210, 1995.
4. Chand, N., Joshi, R.C., Misra, M.: Energy efficient cache invalidation in a disconnected wireless mobile environment. Int. J. Ad Hoc Ubiq. Comp. **1**, 119–125 (2006)
5. Padmanabhan, V., Mogul, J.: Using predictive prefetching to improve world wide web latency. Comp. Commun. Rev. **26**, 22–26 (1996)
6. Jiang, Z., Kleinrock, L.: An adaptive network prefetch scheme. IEEE J. Sel. Areas Commun. **16**, 1–11 (1998)
7. Yin, L., Cao, G.: Adaptive power aware prefetch in wireless networks. IEEE Trans. Wirel. Commun. **3**, 1648–1658 (2004)
8. Cheluvaraju, B., Kousik, A.S.R., Rao, S.: Anticipatory retrieval and caching of data for mobile devices in variable-bandwidth environments. Paper presented at 5th IEEE International Systems Conference, Montreal, QC, 4–7, 531–537 April 2011. doi: 10.1109/SYSCON.2011.5929049
9. Song, H., Cao, G.: Cache-miss-initiated prefetch in mobile environments. Comp. Commun. J. **28**, 741–753 (2005)
10. Shen, H., Kumar, M., Das, S., Wang, Z.: Energy efficient data caching and prefetching of mobile devices based on utility. ACM/Kluwer Journal of Mobile Networks and Applications (MONET) **10**(4), 475–486 (2005)
11. Denko, M., Tian, J.: Cooperative caching with adaptive prefetching in mobile ad hoc networks. Paper presented at IEEE international conference on wireless and mobile computing and communication, pp. 38–44. The Guangdong HotelShanghai, China (2007)
12. Bae, I.H., Olariu, S.: Design and evaluation of fuzzy ching for mobile media streaming. Paper presented at 6th international workshop on data engineering for wireless and mobile access, pp. 67–74, 2007

Intrusion Detection in Cloud Computing Implementation of (SAAS & IAAS) Using Grid Environment

S. Manthira Moorthy and M. Roberts Masillamani

Abstract Security requires user authentication with password digital certificates and confidentiality for transmission of data in a distributed system. The Intrusion Detection System (IDS) detect intrusions by means of knowledge and behavior analysis. In this paper, we introduce concept called cloud computing to increase data efficiency and satisfies user request. We also include grid computing to make cloud computing more efficient, reliable, and increase the performance of the systems that are accessing server. This is because of more user logins at the same time and the server is not able to provide equal performance to all other system. We can achieve performance by getting performance from the system that are connected to server and providing it to system that accessing the server.

Keywords Cloud computing · Grid computing · Data efficiency · Data security · Intrusion detection system · Knowledge analysis · Behavior analysis

Introduction

In Information Security, intrusion detection is the act of detecting actions that attempt to compromise the confidentiality, integrity, or availability of a resource. When intrusion detection takes a preventive measure without direct human intervention, then it becomes an intrusion–prevention system.

Intrusion detection can be performed manually or automatically. Manual intrusion detection might take place by examining log files or other evidence for signs of intrusions, including network traffic. A system that performs automated

S. M. Moorthy (✉) · M. R. Masillamani
School of Computing Sciences, Hindustan University, Chennai, Tamil Nadu, India
e-mail: moorthysgm@gmail.com

M. R. Masillamani
e-mail: deancs@hindustanuniv.ac.in

S. Sathiakumar et al. (eds.), *Proceedings of International Conference on Internet Computing and Information Communications*, Advances in Intelligent Systems and Computing 216,
DOI: 10.1007/978-81-322-1299-7_6,

intrusion detection is called an Intrusion Detection System (IDS). An IDS can be either host-based, if it monitors system calls or logs, or network-based if it monitors the flow of network packets. Modern IDSs are usually a combination of these two approaches. Another important distinction is between systems that identify patterns of traffic or application data presumed to be malicious (misuse detection systems), and systems that compare activities against a 'normal' baseline (anomaly detection systems) [1].

When a probable intrusion is discovered by IDS, typical actions to perform would be logging relevant information to a file or database, generating an email alert, or generating a message to a pager or mobile phone.

Determining what the probable intrusion actually is and taking some form of action to stop it or prevent it from happening again are usually outside the scope of intrusion detection. However, some forms of automatic reaction can be implemented through the interaction of IDSs and access control systems such as firewalls. Intrusion prevention is an evolution of intrusion detection.

Literature Survey

Toward Taxonomy of IDSs

To combat attackers, intrusion-detection systems (IDSs) can offer additional security measures for grid and cloud computing environment by investigating configurations, logs, network traffic, and user actions to identify typical attack behavior. However, the IDS must be distributed to work in a grid and cloud computing environment. It must monitor each node and, when an attack occurs, alert other nodes in the environment [2].

Security Architecture for Computational Grids

This requires compatibility between heterogeneous hosts, various communication mechanisms, and permission control over system maintenance and updates typical features in grid and cloud environments. Cloud middleware usually provides these features, so we propose an IDS service offered at the middleware layer [3].

Intrusion Detection for Computational Grids

We use a feed-forward artificial neural network, because in contrast to traditional methods this type of network can quickly process information, has self-learning capabilities, and can tolerate small behavior deviations. These features help overcome some IDS limitation [4].

Grid M: Middleware to Integrate Mobile Devices, Sensors and Grid Computing

We developed a prototype to evaluate the propose architecture using Grid-M. We created data tables to perform the experiments with audit elements coming from both the log system and from data captured during node communications [5].

Artificial Intelligence Techniques Applied to Intrusion Detection

We evaluated the behavior-based technique using artificial intelligence enabled by a feed forward neural network. In the simulation environment, we monitored five intruders and five legitimate users [6].

Improvements in the Model for Interoperability of Intrusion Detection Responses Compatible with the IDWG Model

In testing our prototype, we learned that it has a low processing cost while still providing satisfactory performance for real-time implementation. Sending data to other nodes for processing did not seem necessary. The individual analysis performed in each node reduces the complexity and the volume of data in comparison to previous solutions, where the audit data is concentrated in single points [7].

Existing System

In these days, a single server handles the multiple requests from the user. Here the server has to process the all the requests from the users simultaneously, so the processing time will be high. This may lead to loss of data and packets may be delayed and corrupted. On doing this, the server cannot process the query from the user in a proper manner.

So the processing time gets increased. It may lead to traffic and congestion. To overcome these problems, we are going for the concept called "cloud computing". In this cloud computing, we are going to implement the Proxy server to avoid these problems.

But in this system, data efficiency is improved but not the data security. Whenever we speak all about data efficiency we should speak about data security also, because in the cloud computing we don't know from which cloud the data is coming, so in the existing system there is no system to find the data security.

Proposed System

The system based on the new architecture has better scalability and fault tolerance. A cluster consists of a single server and multiple proxy servers and is accessed by multiple clients. Proxy servers stores data on local disks and read or write data specified by a server. The server maintains the index for all file stored in different proxies. When a client wants to download some data, it will first send a request to the Server and the Server then redirect the request to a corresponding proxy that have the required data and hence the data will be sent to the client. With the combination of cloud and grid computing concepts, the data request can be efficiently serviced in a timely manner. The major part of the project is security, so above-mentioned phase speaks all about cloud and grid technology, but not about security. The security implementation in this project is achieved by two phases, namely, behavioral and knowledge.

Behavior Analysis

Using this method, we need to recognize expected behavior or a severe behavior deviation. The network must be correctly trained to efficiently detect intrusions. For a given intrusion sample set, the network learns to identify the intrusions. However, we focus on identifying user behavioral patterns and deviations from such patterns. With this strategy, we can cover a wider range of unknown attacks [8].

Knowledge Analysis

Using an expert system, we can describe a malicious behavior with a rule. One advantage of using this kind of intrusion detection is that we can add new rules without modifying existing ones [9].

IDS System

IDS Architecture

Fig. 1, represent a model for response interoperability between IDS, compatible with the model for alert interoperability developed by the IDWG group. Alterations in the IDS IDWG architecture are proposed in order to provide for response

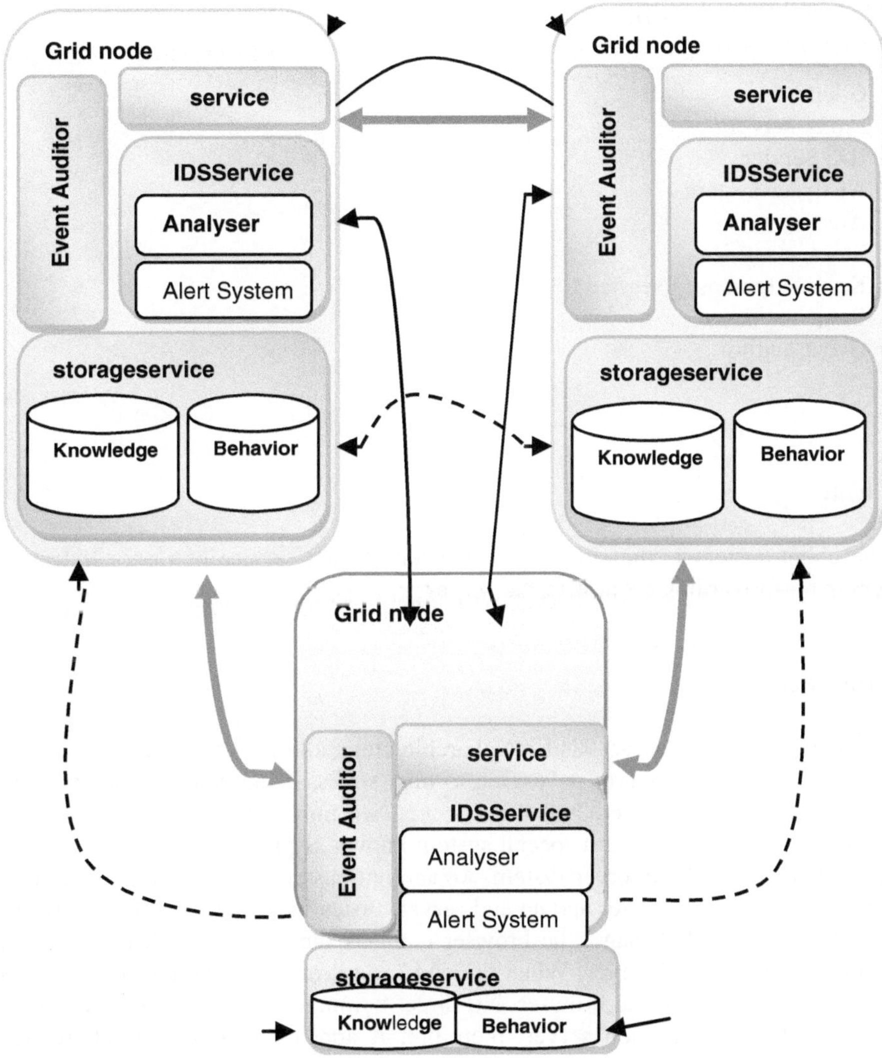

Fig. 1 IDS architecture

interoperability support. The development and testing of the proposed model and its components are also presented. IDSs can offer additional security measures for these environments by investigating configurations, logs, network traffic, and user actions to identify typical attack behavior. However, IDS must be distributed to work in a grid and cloud computing environment [10].

Module Description

- Client
- Proxy server
- IDS service
- Analyzer system
- Alert system
- Storage service
- Knowledge-based service
- Behavior based
- Event auditor

Client

The client system is the system which wants to get service or response from a server by forwarding request to the server.

Proxy Server

In the existing client server/distributed architecture, all requests from all clients are accessed by a single server only. Due to high stress, the system may hang or the data may be lost. To overcome this concept, we moved for new cloud and grid computing which provides a special system known as proxy server which extends the functionality of the server system. An anonymous proxy serves as a middleman between your web browser and an end server. Instead of contacting the end server directly to get a Web page, the browser contacts the proxy, which forwards the request on to the end server. When the end server replies to the proxy, the proxy sends the reply on to the browser. No direct communication occurs between the client and the destination server; therefore, it appears as if the HTTP request originated from the intermediate proxy server [11].

IDS Service

The IDS service increases a cloud's security level by providing two methods of intrusion detection. First method is behavior-based method which dictates how to compare recent user actions to the usual behavior. The second approach is knowledge-based method that detects known trails left by attacks or certain sequences of actions from a user who might represent an attack. The audited data is sent to the IDS service core, which analyzes the behavior using artificial intelligence to detect deviations. This has two subsystems, namely, analyzer system and alert system [12].

Analyzer System

The analyzer uses a profile history database to determine the distance between a typical user behavior and the suspect behavior and communicates this to the IDS service. The rules analyzer receives audit packages and determines whether a rule in the database is being broken. It returns the result to the IDS service core. With these responses, the IDS calculate the probability that the action represents an attack and alerts the other nodes if the probability is sufficiently high.

Alert System

This subsystem will work when intrusion is detected. If any node among the cloud system is affected by intrusion then this alert system will alert the remaining nodes about thc intrusion.

Storage Service

The storage service is a database system which contains two types of services namely knowledge based service and behavior based service. Whenever a node gets requests or responses, the analyzer system compares the node information in the storage service.

Knowledge-based Service

We used audit data from both a log system and the communication system to evaluate the knowledge based system. We created a series of rules to illustrate security policies that the IDS should monitor.

The knowledge service is nothing but set of rules which is formed from previous attacks. Following things that come under this category are:

- Password cracking and access violation,
- Trojan horses,
- Interceptions, most frequently associated with TCP/IP stealing and interceptions that often employ additional mechanisms to compromise operation of attacked systems man in the middle attacks.
- If any packets come with .exe extension
- Packets containing worms

Behavior-based Service

Here we have to classify the behavior in two types as follows:

User Behavior

For user's behavior, we have to analyze the users' behavior. Using this method, we need to recognize expected behavior (legitimate use) or a severe behavior deviation.

Examples: Ex 1 -IP spoofing

There is a range of attacks that take advantage of the ability to forge (or 'spoof') your IP address. While a source address is sent along with every IP packet, it is not actually used for routing. This means an intruder can pretend like you when talking to a server. The intruder never sees the response packets (although your machine does, but throws them away because they don't match any requests you've sent). The intruder won't get data back this way, but can still send commands to the server pretending to be you.

Ex 2- DNS poisoning through sequence prediction

DNS servers will "recursively" resolve DNS names. Thus, the DNS server that satisfies a client request will become itself a client to the next server in the recursive chain. The sequence numbers it uses are predictable. Thus, an intruder can send a request to the DNS server and a response to the server forged to be from the next server in the chain. It will then believe the forged response, and use that to satisfy other clients.

Node Behavior

The most common way people approach network intrusion detection is to detect statistical anomalies. The idea behind this approach is to measure a "baseline" of such states as CPU utilization, disk activity, user logins, file activity, and so forth. Then, the system can trigger when there is a deviation from this baseline. The benefit of this approach is that it can detect the anomalies without having to understand the underlying cause behind the anomalies.

Event Auditor

To detect an intrusion, we need audit data describing the environment's state and the messages being exchanged. The event auditor can monitor the data that the analyzers are accessing. The first component monitors message exchange between nodes. Although audit information about the communication between nodes is being captured, no network data is taken into account only node information. The second component monitors the middleware logging system. For each action occurring in a node, a log entry is created containing the action's type (such as error, alert, or warning), the event that generated it, and the message. With this kind of data, it's possible to identify an ongoing intrusion.

Implementation Steps

- User sends the query to the main Cloud Server, to which all the Cloud Servers (proxy servers) are connected.
- The main Cloud Server will have the index information of the data information by all the Cloud Servers (proxy servers).
- After receiving the request given by the client, the main Cloud Server will verify the exact proxy server which can carry the work more efficiently. By this process, we are implementing both cloud and grid computing together.
- But in the cloud computing, plenty of Hackers may intrude the data, so we are working on IDS.
- IDS is carried by two process, namely, knowledge and behavior.
- Knowledge is done by violating the set of rules.
- Behavior is done by comparing the previous behavior with the present behavior.
- If the system finds that there is Intrusion attack has occurred, it should alert other Cloud Server, so that data access is prevented.

Activity Diagram

Figure 2 represents the activity diagram of IDS. Once the user logs in the system, the client sends request to the server. The server corresponding proxy and forward the requested service. The proxy processes the request and sends response to the client. Here each node monitors the other nodes based on knowledge and behavior. if any intrusion is detected, then it alerts all other nodes.

Result Analysis

In testing our prototype, we learned that it has a low processing cost while still providing a satisfactory performance for real-time implementation. Sending data to other nodes for processing did not seem necessary. The individual analysis performed in each node reduces the complexity and the volume of data in comparison to previous solutions, where the audit data is concentrated in single points

From the above Figs. 3 and 4, we conclude that the latency of cloud computing is higher than that of cloud and grid computing. From this, we can conclude that the use of grid computing in cloud environment increases data efficiency and provide reliability. Thus increasing the performance of the network.

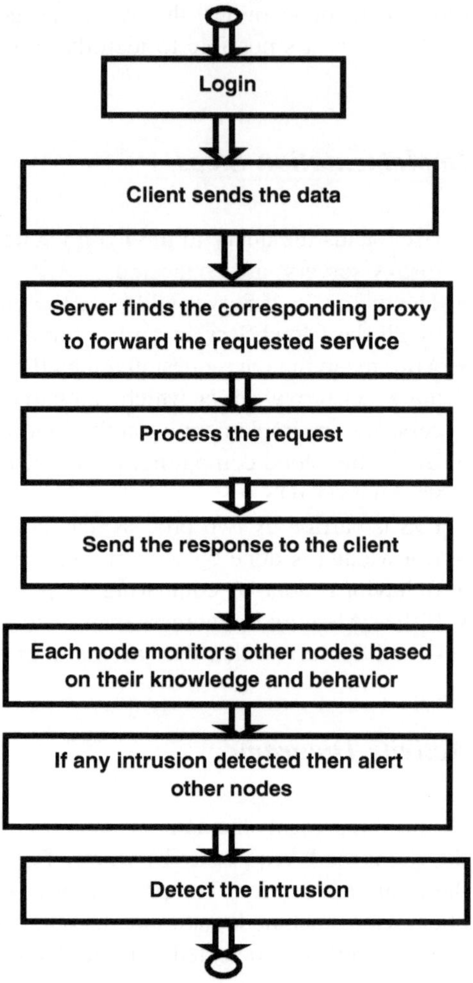

Fig. 2 Activity diagram

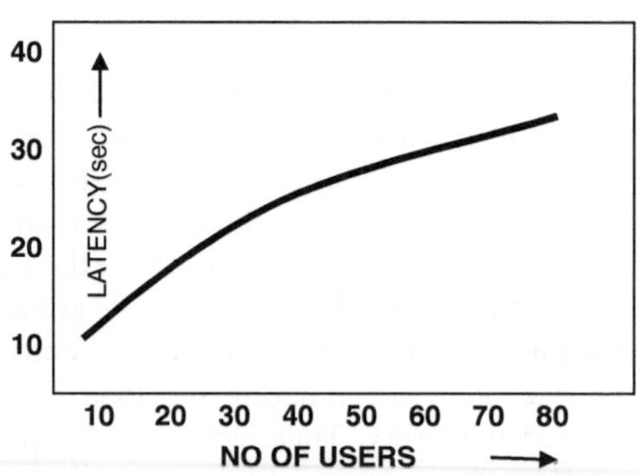

Fig. 3 Response time of cloud computing

Fig. 4 Response time of cloud and grid computing

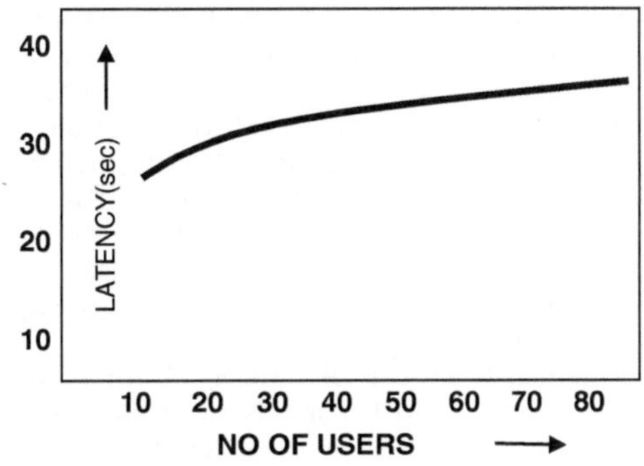

Conclusion and Future Enhancement

Conclusion

The contribution of this paper that the system provides security in a distributed system rather than user authentication with passwords or digital certificates and confidentiality in data transmission. The grid and cloud computing IDS integrates knowledge and behavior analysis to detect intrusions. In addition provides (SAAS & IAAS) as service to the installed network which provides a common platform for nodes in the networks. The major limitation is performance, a noticeable overhead at run time.

Future Enhancement

In future, we will implement our IDS, helping to improve green (energy-efficient), white (using wireless networks), and cognitive (using cognitive networks) cloud computing environments. We also intend to research and improve cloud computing security.

References

1. Chaboya, D.J., Raines, R.A., Baldwin, R.O., Mullins, B.E.: Network intrusion detection: automated and manual method prone to attack and evaton. IEEE Secur. Priv. **4**, 36–43 (2006)
2. Overill, R.E.: Intrusion detection systems: threats, taxonomy, tuning. J. Finan. Crim. **6**, 49–51 (1998)
3. Smarr, L., Catlett, C.E.: Meta computing. Commun. ACM **35**, 44–52 (1992)
4. Schulter, A.: Intrusion detection for computational grids. Proceedings of 2nd International 2nd Conference on new technologies, mobility, and security, Tangier - Morocco, pp. 1–5. IEEE Press, (2008)

5. Bolosky, W.J., Douceur, J.R., Ely, D., Theimer, M.: Feasibility of a serverless distributed file 284 system deployed on an existing set of desktop PCs, SIGMETRICS, ACM, Performance Evaluation 285 Review, 34–43 (2000). http://dl.acm.org/citation.cfm?doid=339331.339345
6. Idris, N.B., Shanmugam, B.: Indicon. Annual IEEE, pp. 52–55, 2005. http://ieeexplore.ieee.org/xpl/articleDetails.jsp?arnumber=1590122&punumber%3D10601
7. Cuppens, F., Gombault, S.: Thierry Sans: selecting appropriate counter-measures in an intrusion detection framework. In: Proceedings of the 17th IEEE workshop on computer security foundations, pp. 78–80, 2004. http://ieeexplore.ieee.org/xpl/articleDetails.jsp?arnumber=1310733&sortType%3Dasc_p_Sequence%26filter%3DAND%28p_IS_Number%3A29101%29
8. Zhang, G.Y., Lee, W., Huang, Y.A.: Intrusion detection techniques for mobile wireless networks. ACM J. Wirel. Netw. **9**, 545–546 (2003)
9. Denning, D.E.: An intrusion-detection model. IEEE Trans. Softw. Eng. **13**, 222–232 (1987)
10. Li, T., Song, M., Alam, M.: Compromized sensor node detection: a quantitative approach. In: Proceedings of the IEEE international conference on distributed computing systems, pp. 352–357, 2008.
11. Shapiro, M.: Structure and encapsulation in distributed systems: the proxy principle, Proc. 6th Int. Conf. on Distributed Computing Systems(ICDCS), Cambridge MA (USA), pp. 198–204, 1986
12. Lunt, Teresa F.: "IDES: An intelligent system for detecting intruders," Proceedings of the symposium on computer security; threats, and counter measures; Rome, Italy, pp. 110–121, 22–23 Nov 1990

Inter Departure Time Analysis Over Heterogeneous Platforms Using Distributed IPv6 Traffic Generator Tool

S. P. Meenakshi, S. V. Raghavan and S. M. Bhaskar

Abstract IPv6 networks are at present under vast deployment in production networks as well as in the Internet. Tasks such as resource reservation, capacity planning, and effective security deployments necessitate the understanding of IPv6 flow behavior in the nodes as well as in the network. To accomplish the understanding of IPv6 flow behavior, a tool that generates application and attack flows based on sound mathematical models is essential. Since different flows have different characteristics, the model parameters for a particular flow feature are to be assigned with appropriate values. In our tool, we have identified five flow features for characterizing the flows. They are, namely, inter departure time (IDT), packet size (PS), flow count (FC), flow volume (FV), and flow duration (FD). Random sampling from distributions such as Exponential, Pareto, Poisson, Cauchy, Gamma, Student, Wei-bull, and Log Normal are considered. IPv6 TCP and UDP packets are constructed and transmitted for the flows according to the feature values received from the assumed distribution model. The direct model parameter specification and trace-based model parameter learning are incorporated in our tool. Using our tool, we have analyzed the node behavior for IDT feature over different hardware platforms to IPv6 flows. We have presented the IDT analysis and modeled the bit rate using a three parameter function from the experimental measurements. The parameters estimated for different platforms to this model are also reported.

S. P. Meenakshi (✉) · S. V. Raghavan
Department of CSE, Indian Institute of Technology Madras,
Chennai 600036, India
e-mail: meena@cs.iitm.ernet.in

S. V. Raghavan
e-mail: svr@cs.iitm.ernet.in

S. M. Bhaskar
Society for Electronic Transactions and Security, CIT Campus,
Taramani, Chennai 600113, India
e-mail: smb@nic.in

S. Sathiakumar et al. (eds.), *Proceedings of International Conference on Internet Computing and Information Communications*, Advances in Intelligent Systems and Computing 216,
DOI: 10.1007/978-81-322-1299-7_7, © Springer India 2014

Keywords IPv6 · Inter departure time · Packet size · Flow count · Flow volume · Flow duration

Introduction

IPv6 protocol is at present deployed vastly and rapidly in the Internet as well as in the production networks. Understanding the application and also the attack flow behavior in the end nodes and the networks for this protocol is very much required. Tasks such as resource reservation, capacity planning and effective security deployments necessitate this understanding. To accomplish these tasks efficiently, experimental measurements and analysis are to be conducted in a controlled and distributed environment such as a testbed. An IPv6 flow generator tool that generates traffic according to user specification is a quintessential requirement in this scenario. In this work, we have presented the IPv6 flow generator tool developed by us and the measurement analysis performed on IDT that supports the manager component of the tool.

Flow characteristics differ with respect to applications and attacks. Popular applications such as FTP, WWW, and streaming media are well studied in the literature [1–4] for their statistical behavior. To capture the respective flow behavior, it has been found that the flow features are to be modeled appropriately using distributions. In our IPv6 traffic generator tool, five flow features represented by the tuples $< IDT, PS, FC, FV, FD >$ are identified to generate the required user specified traffic. These five features are modeled as random variables and they take values from different probability distributions. The distributions such as Constant, Random, Exponential, Pareto, Poisson, Cauchy, gamma, Student, Wei-bull, and Log normal are considered in our tool. These distributions are chosen based on the empirical study of the traffic reported in [1, 2, 4].

IPv6 protocol provides many additional features. They are optional headers such as authentication header (AH) and hop by hop header and fields such as flow label. To have control over the features during flow generation process, we have constructed IPv6 TCP and UDP packets in our tool. This packet construction feature provides the advantage of including QoS and network management capability to flows. Attack flow generation is also feasible due to this feature.

The flow scalability which is a feature of large networks is supported in our tool.

The tool is deployed in a distributed environment and controlled by a manager. The manager estimates and controls the traffic generators. A web-based GUI is used to input the user requirement to the manager.

In a node, different hardware platforms, operating systems and system load play a crucial role in flow generation. We address the problem of how reliably the software-based traffic generation tools perform for user supplied traffic profiles in different hardware platforms [5]. In this work, we study the flow generation behavior in nodes with different CPU architectures namely Intel-based and AMD-based architectures. The unix-based operating system and minimum system load are considered. Our goal is to identify precisely the determiners of bit rate generation in the nodes. The bit rate generation performance is measured and reported for flow features like IDT and PS. IPv6 Traffic generator tool for generating synthesized application/attack traffic and experimental analysis to determine the bit rate generation influencers in a heterogeneous node environment are our contributions in this paper.

The rest of the paper is organized as follows. The related work is presented in Section "Related Work". The architecture and implementation details of the tool are discussed in Section "The Architecture". Section "Experiments and Analysis" presents experiments, flow measurements and analysis that are performed using our tool. Section "Summary" concludes with summary of the work.

Related Work

At present there are many free and nonfree tools for IPv6 traffic generation enhanced from IPv4 protocol exist. iPerf [6] is one such tool which supports packet level traffic generation. It works by sending as many packets as possible to measure the throughput but cannot generate according to the specific traffic profile requested by the user.

In [7], focus is on synthetic traffic generation of realistic traffic over real networks. The Distributed Internet Traffic generator (D-ITG) tool discussed in [7] provides support to evaluate performance metrics related to throughput, loss, delay, and jitter. It provides setting the Type of Service field (TOS) and Time To Live Field (TTL). Different protocols such as UDP, TCP, DCCP, SCTP, and ICMP can be tested and their performance can be analyzed.

Traffic generators Rude/Crude [8], Mgen [9] and TG [10] support only IPv4 traffic generation. They work on operating systems such as Linux, FreeBSD and Solaris. The distributions such as Constant, Uniform, On/Off, and Exponential are supported. TG and MGEN after reaching their maximum values, saturate to a fixed packet rate at about 70 kilo packets per second as the requested rate increases [5]. RUDE/CRUDE shows decreasing trend when maximum packet generation limit is reached. D-ITG starts to deviate from its behavior when 130 kilo packets per second is reached. In [5], the architecture of the generators were shown to have significant impact on the ability of the generators in generating the requested packet rate. When the requested packet rate increases, the generator's IDT distribution deviates more from the expected distribution.

Our tool considers packet level and flow level features in traffic generation. Also learning components are incorporated in our tool to learn the bit rate generation capabilities on various hardware platforms and on different system loads. We consider only linux-based operating systems for our tool.

The Architecture

The architecture plays a critical role in traffic generation accuracy of the software-based tool. There are six important functional components in our tool. They are namely the web-based GUI interface, the manager, Trace Learner, the core traffic generators, the traffic receivers, and the traffic logger. The manager is a central entity that issues commands to the traffic generators. The traffic generators and receivers are deployed in a distributed fashion. The traffic flow generated measurements are first stored locally in the generators and receivers. Then, it is sent to the centralized traffic logger for plotting. The relationship between these functional components is shown in Fig. 1.

Implementation and Functional Details

The implementation and functionality details of each functional module are discussed in this section. We have used programming languages such as C and html, and scripting languages such as php, java script and bash in coding the functional modules. Client server model is employed to enable the communication between the distributed components.

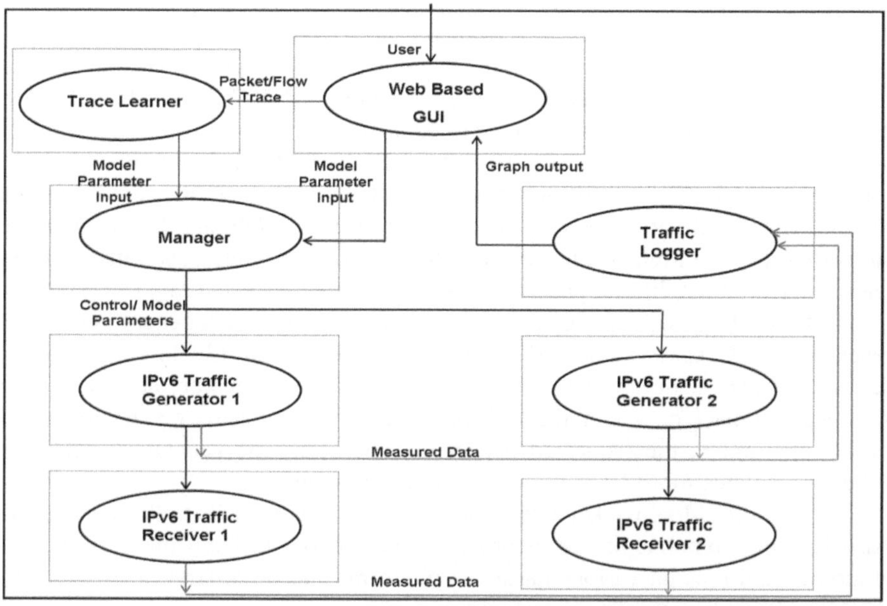

Fig. 1 Tool architecture

Web-based GUI

The flow generation parameters are required to generate the IPv6 flows. These parameters are provided to the manager through the GUI component. There are two ways the user can give the flow parameters.

1. Direct parameter specification
2. Trace-based parameter specification

Direct Parameter Specification

In the direct model parameter specification, the user needs to have prior knowledge with respect to the features for the type of flows one intends to generate. In a given duration, user can choose single flow or multiflow option. In case of single flow option, the user can specify source address, destination address, source port, destination port, bit rate, flow duration, average packet size, IDT distribution and PS distribution. In case of multiflow option (FC), the user has to provide additional input on the distributions of multiflow generation namely flow volume and flow durations. The GUI interface for single flow option is given in Fig. 2. To generate flows for specific applications, the user has to choose a set of distributions for the flow features. The applications and the associated distributions are given as help to guide the users.

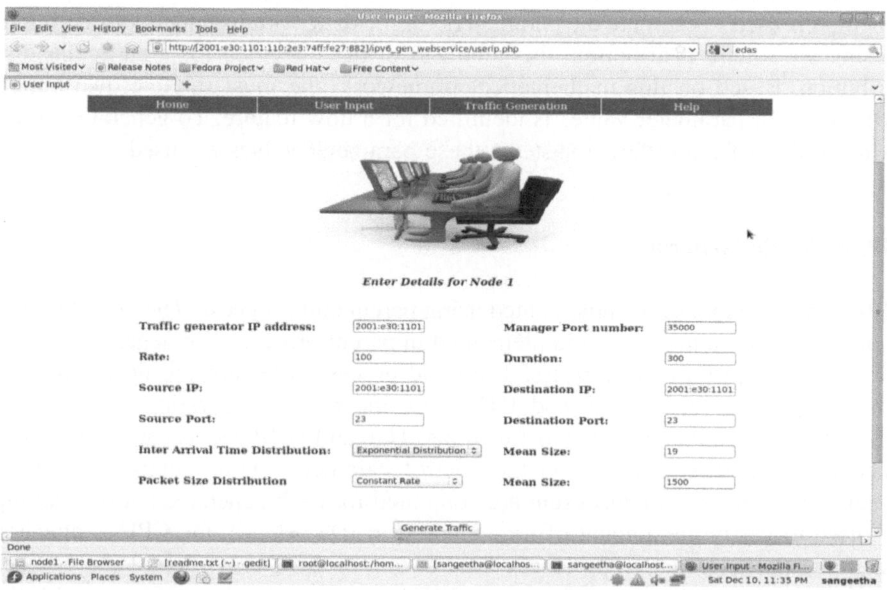

Fig. 2 Web-based GUI

Trace-based Parameter Learning

Instead of giving the model parameters, a user can give a packet level trace or flow level trace as input. The flow features are learned by the Trace Learner component. For the learned features appropriate model with parameters are fitted based on chi-square goodness-of-fit method. Then the parameter is given as input to the manager for flow generation. The web-based GUI is developed using html, javascript, and php. It is deployed in apache webserver on linux platform. A php UDP client program is implemented to send the user input to the manager and the trace learner.

Manager

Manager is implemented as server program using C. It has a list of traffic generator IPv6 addresses and flow generation capacity. Based on user requested bit rate, the manager chooses one or more number of traffic generators. Then the control parameters are issued to the generators for the required flow generation.

Trace Learner

This module takes the input from the real traffic data traces and extract the flow features from the given trace, that is, PS, IDT, and FD. From the extracted features it estimates the parameter for the appropriate distribution. Maximum likely hood method is used to estimate the parameters for the distributions under consideration. Using chi-square distribution, the goodness of fit is decided for a particular distribution. Based on this mathematical framework, the most suitable distribution with optimal parameter values is identified for a flow feature. To generate similar kind of flows from different systems these parameter values are used.

IPv6 Traffic Generator

IPv6 traffic generator is implemented using parent child process. The flow duration and flow volume models are implemented in parent process. The generated values are passed on to a child process. The child process generates the flow using the models for the features PS and IDT using the provided duration or size. The distributions are implemented as functions. TCP and UDP packets are constructed and transmitted as specified in the control parameter. Timestamps, TCP flags, sequence numbers and checksum are computed for each generated packet. Polling is used to verify the amount of time elapsed in IDT. Hence, the CPU is attached with the packet generation process even during the IDT execution. This provides the tool with an advantage of accurate IDT realization that is concurrent with the imposed value. Generated flow traffic is logged per second and send to the traffic logger at the end of the flow duration. Since the process of writing the log to the disk has significant performance overhead, we store the log data in a memory array of size 50. As soon as it gets filled, it is written to the disk.

IPv6 Traffic Receiver

IPv6 Traffic receiver is a simple tcp dump tool in the edge nodes. The received packet data is filtered for flow details and logged. The logged data is sent to the traffic logger.

Traffic Logger

Graphs are generated using the received log data from the sender and the receiver. We use gnuplot [11] and R statistical package [12] for plotting and analyzing the data. The graphs are stored in a directory which can be accessed using the web-based GUI.

Experiments and Analysis

Experiments are conducted using our tool to understand the hardware dependency of IDT during flow generation process. This understanding can be incorporated in the functional modules of the tool to enhance the accuracy of traffic generation rate. Here IDT is defined as the time delay between two consecutive packet generations. This includes the packet generation time (PGT) and a delay. Experiments are conducted in an IPv6 testbed environment to measure the flow generation rates by varying the flow feature values. To estimate the PGT for different packet sizes experiment 1 is conducted. Experiment 2 is conducted to understand the flow rate when the IDT takes values from exponential or constant distribution. Experiment 3 is conducted to model the bit rate based on the empirical measurements. To analyze the relationship between imposed IDT and achieved IDT under various load conditions experiment 4 is performed.

Experiment 1

In the experiment 1, we are measuring the IDT for different packet sizes over differ- ent hardware platforms. We are considering PGT in this experiment. The flow rate is measured for a single flow in flooding mode without varying any flow features to measurement intervals of 1 min. In flooding mode, the IDT assumes only the generation time and the delay is assumed to be zero. The packet generation time IDT is estimated from the flow rate measurement using the Eq. 1.

$$\text{IDT PGT} = \frac{60}{\text{Packet count}} \tag{1}$$

For comparison over different hardware platforms, we have generated the flows in three different servers (Server1, Server2, and Server3) and in a user laptop. The servers are built with Intel Xeon Quadcore processor 2.4 GHz speed, AMD Opteron processor 2.8 GHz speed and Intel Xeon Quadcore processor 2.6-GHz speed, respectively. The laptop has the built-in CPU of Intel Core 2 Duo 1.66 Ghz processor. The primary RAM available in servers(1&2) is 14 GB and in server3 is 32 GB. The laptop has the memory capacity of 2 GB. The IDT delay time is assumed as zero throughout the measurement time. Incrementing the packet size in terms of 100 bytes, the flow packet count, and rate have been measured. The measurements are given in Tables 1, 2, 3 and 4.

We can discern from the tabulated data of the servers that up to 800 bytes of packet size, the IDT (contributed only by PGT) differences are not much significant in the microsecond clock resolution range. In 801–1,500 bytes packet range, for every 100 bytes, the IDT increases in an average of 8.11 % in Server1, 9.48 % in Server2 and 8.45 % in Server3. In the laptop measurements, we can observe that the IDTs increase linearly when the packet size increases. For every 100 byte increase, the IDT increases in an average of 14.95 % from 100 bytes onward.

The IDT comparison for different packet sizes in the flooding mode on the servers and the laptop is given in Fig. 3. From Fig. 3, it can be observed that the packet size falls into two classes, namely, A and B based upon the packet generation time in the servers. Class A assumes the range of packet sizes from 1 to 800

Table 1 Flooding mode traffic measurements in Server1

Packet size	Packet count	Inter departure time (sec)	Rate (Mbps)
100	7876789	7.62	102
200	7991941	7.51	208
300	7836962	7.66	306
400	7946926	7.55	413
500	7799518	7.69	507
600	7848875	7.64	613
700	7751825	7.74	706
800	7749814	7.74	807
900	7567716	7.93	886
1000	7182656	8.35	935
1100	6557572	9.15	939
1200	6036028	9.94	943
1300	5586980	10.74	945
1400	5201740	11.53	948
1500	4864354	12.33	950

Table 2 Flooding mode traffic measurements in Server2

Packet size	Packet count	Inter departure time (sec)	Rate (Mbps)
100	9147610	6.56	119
200	9160054	6.55	238
300	9098351	6.59	355
400	9156717	6.55	476
500	9098850	6.59	592
600	9144177	6.56	714
700	9131667	6.57	832
800	8799935	6.82	916
900	7865603	7.63	921
1000	7121447	8.43	927
1100	6488222	9.25	929
1200	5990624	10.02	936
1300	5535497	10.84	936
1400	5151648	11.65	939
1500	4528025	13.25	884

Table 3 Flooding mode traffic measurements in Server3

Packet size	Packet count	Inter departure time (sec)	Rate (Mbps)
100	8343085	7.19	108
200	8251710	7.27	214
300	8181211	7.33	319
400	8316751	7.21	433
500	8081246	7.42	526
600	8131297	7.38	635
700	8003141	7.50	729
800	8177850	7.34	851
900	7931685	7.56	929
1000	7218907	8.31	939
1100	6587461	9.11	943
1200	6058063	9.90	946
1300	5605321	10.70	948
1400	5215503	11.50	950
1500	4876694	12.30	952

bytes and class B assumes the range from 801 to 1,500 bytes. When compared to class B, in class A the servers have significant difference in packet generation time. But in the laptop, the IDT increases when the packet size increases. From this we can infer that, the system hardware platform plays a significant role in PGT and flowrate. The software-based flow generators should learn this platform dependent information for generating flows according to the user specified flow rates.

Table 4 Flooding mode traffic measurements in Laptop 1

Packet size	Packet count	Inter departure time (sec)	Rate (Mbps)
100	3657489	16.4	47.62
200	2492602	24.07	64.91
300	1870971	32.06	73.08
400	1870971	40.06	77.99
500	1248421	48.06	81.27
600	1070323	56.06	83.62
700	936711	64.05	85.37
800	832755	72.05	86.75
900	749541	80.05	87.83
1000	681465	88.05	88.73
1100	624778	96.03	89.48
1200	576725	104.03	90.11
1300	535565	112.03	90.65
1400	499880	120.02	91.12
1500	486900	123.22	95.10

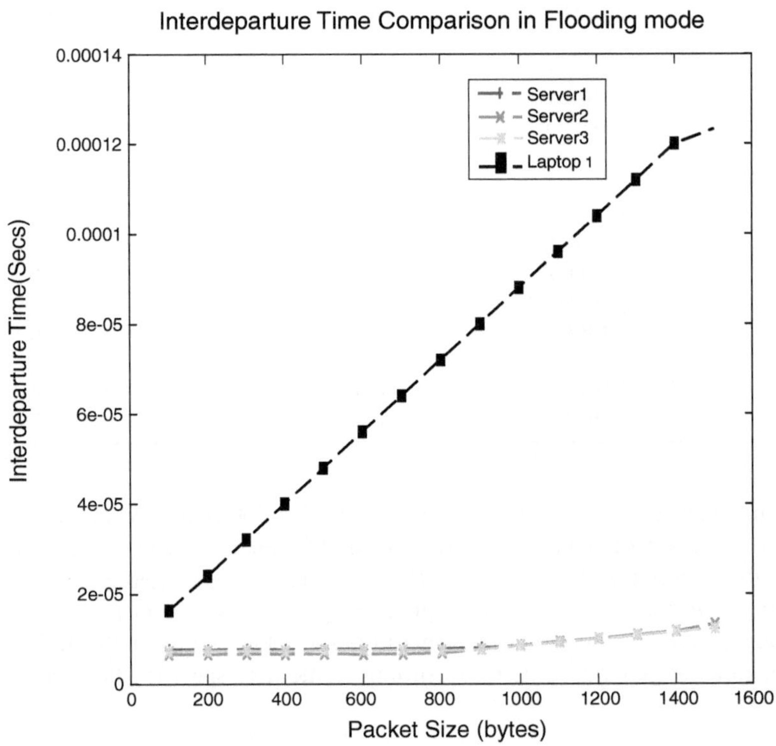

Fig. 3 Comparison of inter departure time versus packet size in flooding mode

Experiment 2

The influence of IDT delay component variations on bit rate is studied in this experiment. From the definition, IDT is the sum of the two components, namely packet generation time and delay. The packet generation time is estimated from experiment 1 for packet sizes from 100 bytes to 1,500 bytes. In this experiment the delay component is considered for two cases. Case 1 assumes constant IDT and constant packet size. Case 2 assumes exponential distribution for IDT and constant packet size. By incrementing the delay in terms of microseconds and keeping the packet size as 100 bytes we have measured the bit rate. The experiments have been conducted in Server1.

The measurements taken are given in Table 5. The comparison of generated bit rates when delay assumes constant or exponentially distributed values is given in Fig. 4. We can observe the following from the aforesaid figure. There is no significant difference between the bit rates for constant and exponential delay distributions. But in both the cases, the bit rate decreases nonlinearly when delay increases.

Table 5 Generated bit rate for constant, exponent IDT and constant packet size

IDT (sec) Case 1	Case 2 (Delay)	Bit rate (Mbps)
1	98.30	96.21
4	81.91	76.31
7	62.10	59.98
10	50.15	49.81
13	42.04	41.44
16	36.03	35.71
19	31.78	31.49
22	28.34	28.12
25	25.51	25.34
28	23.25	23.13
31	21.34	21.47
34	21.33	21.22
37	21.53	21.20
40	21.35	21.21
43	21.30	21.18
46	15.21	15.00
49	14.25	14.18
51	13.72	13.62
54	13.07	13.15
57	12.44	12.47
60	11.86	11.77
63	11.34	11.21
66	10.87	10.81
69	10.42	10.37
72	10.03	9.88
75	9.47	9.60

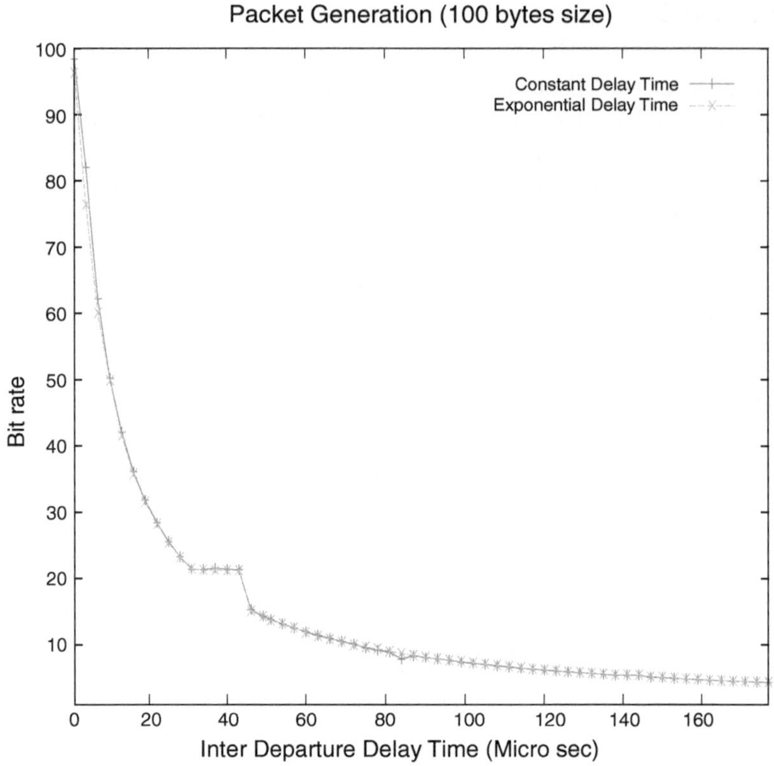

Fig. 4 Bit rate comparison for exponential and constant IDT

Experiment 3

In this experiment, we study the impact of delay component over bit rate in the server and the laptop. From the empirical measurement, we model and predict the bit rate for the corresponding IDT to a specific hardware platform. The packet size is kept as 100 bytes. The delay value is considered as constant from 1 s and incremented in terms of 3 s. The comparison of bit rates from the measurements taken in Server1 and Laptop1 is given in Fig. 5. Both the measurements exhibit heavy tailed nonlinear nature. So we have used nonlinear regression curve fitting to identify the appropriate function for bit rate modeling. The function given in Eq. 2 having three parameters predicts the bit rate values. Table 6 provides the estimated parameter values for bit rate generation in Server1 and the Laptop1. The parameter estimation error for the function is measured using red-Chi square value. For the Server1 the parameter estimation error rate is 1.00698 and for the laptop it is 0.0103. This function can be incorporated in the manager module of our tool to decide the IDT delay for the user specified traffic rate.

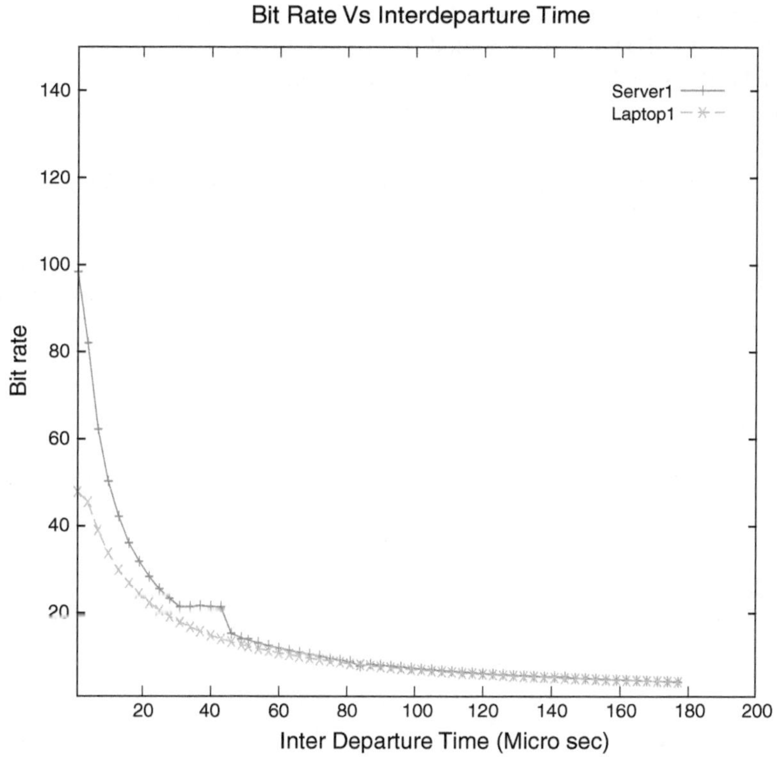

Fig. 5 Bit rate comparison in server and laptop for constant IDT

Table 6 Estimated model parameters for bit rate generation in Server1 and Laptop1

Node	Parameters		
Type	*A*	*B*	*C*
Server1	0.006801	0.001261	0.002101
Laptop11	0.0162	0.00128	0.00338

$$Y = 1/(A + B * X + C_x) \tag{2}$$

Experiment 4

In this experiment, we are testing the accuracy of bit rate generation using our tool by measuring the difference between the achieved IDT and the imposed IDT. From the experiment 3 Laptop1 measurements, we have taken the mean IDT as 22 s that can generate 22 Mbps rate. Using exponential distribution, the imposed IDT is generated for a duration of 1 min. The achieved IDT is measured in the generator at the Laptop1. We have used tcp dump timestamp to measure the achieved mean IDT. The mean IDT of the sampled data(imposed IDT) is 22.03 s whereas the mean IDT measured when the packets are generated is 22.31 s. From the measurements taken under default system load conditions, we can see that there is no

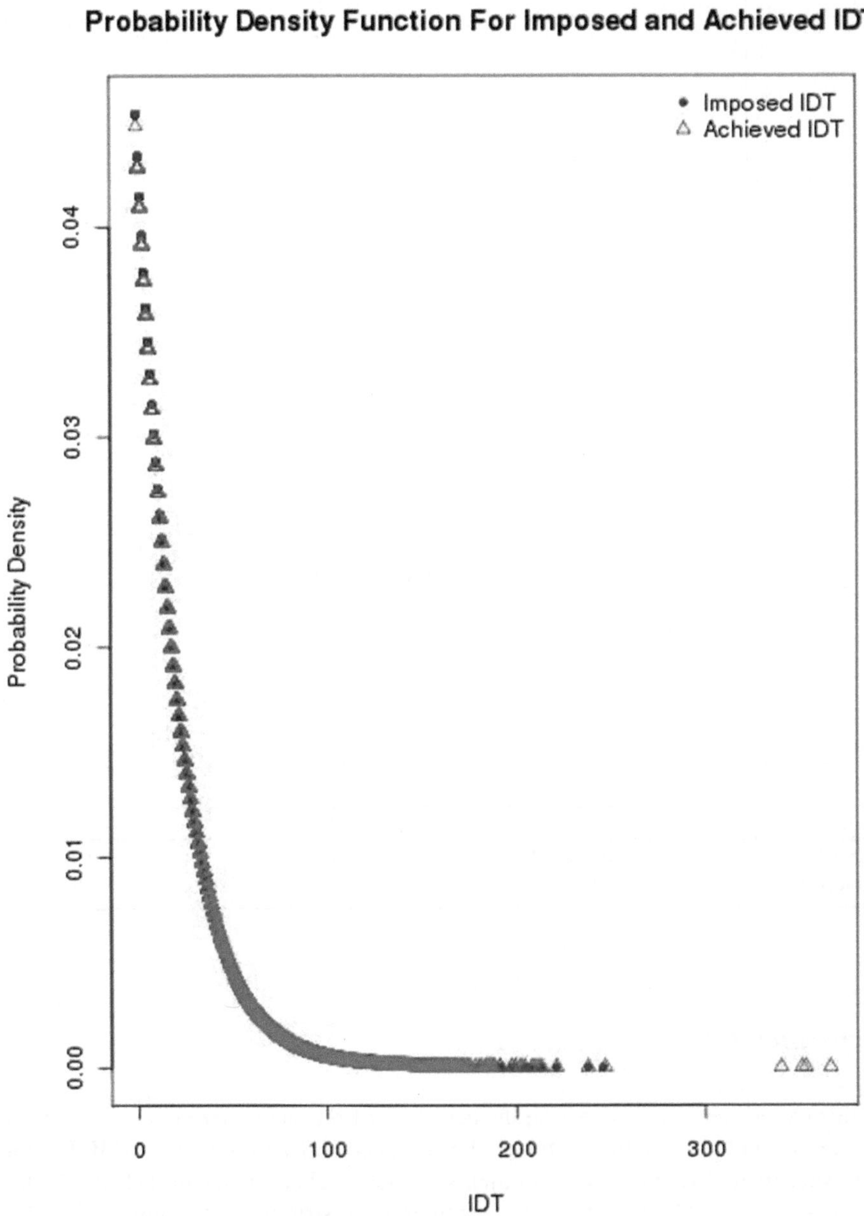

Fig. 6 Probability density function for imposed and achieved IDT

significant difference between the achieved and the imposed mean IDT. Hence we can conclude that our tool is capable of generating the expected bit rate for the imposed mean IDT under this load condition. The probability density function of the imposed IDT and the achieved IDT is given in the Fig. 6.

Summary

We have presented a distributed IPv6 traffic generator tool that generates IPv6 traffic flows based on 5 flow features. The flows are characterized by the five tuples $<$ IDT, PS, FC, FV, FD $>$. Various distributions are used to model the flow features. By selecting different combinations of distributions for the flow features, various types of traffic can generated. Web-based GUI is provided for submitting the traffic profiles by the users. The exponential and constant distribution IDTs are analyzed for different hardware platforms using the tool. We have observed that the packet generation component of the IDT doesn't have significant increase up to 800 bytes in the server hardware platforms. But in the case of laptop hardware platform it in- creases lincarly in the second clock resolution range. The packet generation rate is maximum when the delay component of IDT is 0 in all packet sizes. When delay component of IDT follows constant or exponential distribution, it does not have significant difference in traffic generation bit rates while packet sizes are kept constant.

But when the delay component increases, the traffic generation rate decreases nonlinearly. Using bit rate empirical measurement data for different IDT delay, we have arrived at a three-parameter equation for bit rate prediction. This function can be used in traffic generators to generate traffic for different hardware platforms. Under default load conditions, it is shown from the probability density graph that the generated IDT and achieved IDT are almost similar. This can be interpreted as under certain load conditions, for a user specified IDT delay, the corresponding bit rate can be accurately generated using software-based traffic generators.

References

1. Barford, P., Crovella, M.: Generating representative web workloads for network and server performance evaluation. ACM SIGMET- RICS Perform. Eval. Rev. **26**, 151–160 (1998) http://portal.acm.org/citation.cfm?doid=277858.277897
2. Deng, S.: Empirical model of www document arrivals at access link. In: Proceedings of the 1996 IEEE international conference on communication, Dallas, pp. 1797–1802, 1996
3. Paxson, V., Floyd, S.: Wide-area traffic: the failure of poisson modeling. IEEE/ACM Trans. Netw. **3**, 226–244 (1995)
4. Tang, W., Fu, Y., Cherkasova, L., Vahdat, A.: Modeling and generating realistic streaming media server workloads. Comp. Netw. **51**, 336–356 (2007)
5. Botta, A., Dainotti, A., Pescap, A.: Do you trust your software-based traffic generator? IEEE Commun. Mag. **48**, 158–165 (2010)

6. Iperf, http://dast.nlanr.net/Projects/Iperf/
7. Avallone, S., Emma, D., Pescap, A., Ventre, G.: A distributed multiplatform architecture for traffic generation. In: International symposium on performance evaluation of computer and telecommunication systems, California, 2004
8. Rudecrude, http://rude.sourceforge.net/
9. Mgen, http://cs.itd.nrl.navy.mil/work/mgen/index.php
10. Traffic Generator, http://www.postel.org/tg/tg.htm
11. Gnuplot, http://www.gnuplot.info/
12. R Project, http://www.r-project.org/

Modeling-Simulation of an Underground Wireless Communication Channel

M. N. Jayaram and C. R. Venugopal

Abstract Wireless communication inside mines and tunnels is very different from that in terrestrial environment because of the strong attenuation of signals. Here, we have tried to develop an empirical model for the underground wireless communication channel based on experimental data. The model developed is based on the available outdoor and indoor propagation models such as Okumara-Hata, COST231, ITU indoor propagation models. Modeling is done by super position method. Choosing the most appropriate model among the available ones for the given data and performing regression methods to do curve curve fitting. Correction factors are then added based on two parameters, namely, diffraction and low frequency interference losses. Losses due to penetration and multi path loss are assumed to be constants. MAT LAB is used for curve fitting.

Keywords Model · Diffraction · Low frequency Interference

Introduction

To study the behavior of wireless signals in the underground environment, we have done some measurements at a gold mine. Simple modulation techniques, namely, BPSK, BFSK, and 0.3GMSK are considered [1]. A very sensitive receiver with good fidelity was used for signal measurements. Measurements were possible up to a depth of 30 m only.

M. N. Jayaram (✉) · C. R. Venugopal
Department of E&C, SJCE, Mysore 570006, Karnataka, India
e-mail: jayaram_mn2002@yahoo.com

C. R. Venugopal
e-mail: venu713@gmail.com

S. Sathiakumar et al. (eds.), *Proceedings of International Conference on Internet Computing and Information Communications*, Advances in Intelligent Systems and Computing 216,
DOI: 10.1007/978-81-322-1299-7_8,

Equipments and Specifications

To generate 30 W of power with 20 % efficiency in the UHF band MOSFET hybrid modules with proper heat sink is used.

Transmitted power = P_T = 30 Wt = 44.77 dBm

Operating frequency is from UHF band (GSM 900),

Transmitting antenna outside tunnel is a parabolic dish which is horn fed, antenna gain is 35 dBi or 32.26 dBd or 316.28 (ratio), VSWR <1.4, 3 dB beam width is 3.05°.

Height of transmitting antenna w.r.t. ground plane is = 50 m.

Receiving antenna is a loop antenna (magnetic dipole). Height of the receiving antenna varies w.r.t reference ground plane, height of the person holding the antenna, whether the person is standing on the ground or standing on the trolley, etc.

Receiver Specifications

Receiver sensitivity

For BPSK= −106 dBm (minimum power detectable = 1.99 × 10exp − 14 W)
For BFSK= −107 dBm (minimum power detectable = 2.512 × 10exp − 14 W)
For 0.3 GMSK = −105 dBm (minimum power detectable = 0.0316 × 10exp − 12 W)
Rx bandwidth = 280 kHz

Three propagation models are used as base models. They are:

Cost 231 model. (COperation européenne dans le domaine de la recherche Scientifique et Technique) is a European Union Forum for cooperative scientific research which has developed this model accordingly to various experiments and researches. It is an extension to the Hata model and holds good for the following range of parameters:

Frequency: 500–2,000 MHz
Mobile station antenna height: 1–10 m
Base station antenna height: 30–200 m

The path loss equation is given by:

$$L = 46.3 + 33.9\log f - 13.82\log hB - a\,(hR) + [44.9 - 6.55\log hB]\log d \quad (1)$$

where a (h_R) = Mobile station antenna height correction factor
Given by (1.1logf − 0.7) h_R − (1.56logf − 0.8)

L Median path loss. Unit: Decibel (dB)
f Frequency of transmission. Unit: Mega Hertz (MHz)
h_B Base station antenna effective height. Unit: Meter (m)
d Link distance. Unit: Kilometer (Km)
h_R Mobile station antenna effective height. Unit: Meter (m)

Log distance path loss model. It predicts the path loss of signal inside a building or densely populated areas over a distance.

The path loss equation is given by:

$$L = 20\log(4 * \pi * \mathrm{d}/\lambda) + 10 * \gamma * \log(\mathrm{d}/\mathrm{d0}) + \mathrm{Xg} \tag{2}$$

where

L is the total path loss measured in Decibel (dB)
d is the length of the path
d_0 is the reference distance (1 m)
γ is the path loss exponent
Xg 0(fading not considered).

ITU model for indoor propagation. It estimates the path loss inside a room or a closed area inside a building delimited by walls of any form and is restricted to following range of parameters:

Frequency: 900 MHz to 5.2 GHz

Proposed model for path loss:

$$L = 20\log\mathrm{f} + \mathrm{N}\log\mathrm{d} + \mathrm{Pf}(\mathrm{n}) - 28 \tag{3}$$

where,

L the total path loss. Unit: decibel (dB)
f Frequency of transmission. Unit: Mega Hertz (MHz)
d Distance. Unit: meter (m)
N the distance power loss coefficient
n Number of floors between the transmitter and receiver.

The regression method used throughout the paper is a simple polynomial regression (first and second order). For a model which has the closest match; we incorporate correction factors to fit two parameter variations. Here we have assumed penetration and multi path scattering losses as constants. Former varies with frequency, type of material used in the wall, etc. For indoor propagation, we have assumed a multipath loss (including other losses) of 40 dB . The two variable parameters considered are:

Fig. 1 Knife edge diffraction model

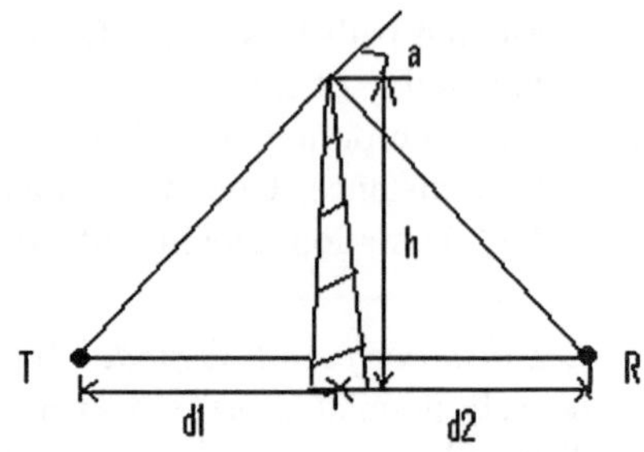

Diffraction Loss (Bending Loss)

Diffraction is the bending of electromagnetic wave (EMW) along the corners and it occurs when the size of the obstacle is comparable to the wavelength of the signal. The diffraction considered here is knife edge diffraction. Diffraction loss occurs from the blockage of secondary waves such that only a portion of the energy is diffracted around an obstacle. That is an obstruction causes a blockage of energy from some of the Fresnel's zones, thus allowing only some of the transmitted energy to reach the receiver. For the knife edge diffraction model shown in Fig. 1, *v* is the diffraction factor, *h* is the height of knife edge, d1 distance to knife edge from transmitter (T), d2 distance to knife edge from receiver (R) .

$$v = h\surd[2(\mathrm{d1} + \mathrm{d2})/(\lambda \mathrm{d1d2})] \tag{4}$$

Diffraction gain/loss is given by,

$$\mathrm{G_d(dB)} = 20\log(0.225/\mathrm{v}). \tag{5}$$

Low Frequency Interference Loss (Interference Loss due to Power Cables)

Low frequency noise arises within the spectral range from the fundamental generation frequency either 50 or 60 Hz into the UHF range. Power line noise depends on KVA of cable, distance at which it is located from measuring equipment, height of receiving antenna, frequency at which measurement is done, gain of receiving antenna through which it is measured. First electric field strength μV/m/MHz or dBμV /m/MHz is converted to dBm noise power and then noise power can be converted to dB loss. Half wave dipole is used in the measurement.

For a measuring instrument having Rt = 50 ohm (terminating resistance), 0 dB μV corresponds to −107 dBm .

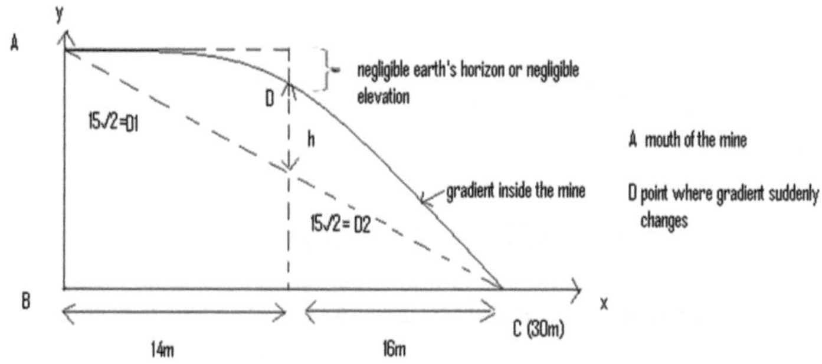

Fig. 2 Terrain profile inside the mine

Calculations

1. Modeling

The approximate terrain geometry inside the main is as shown in Fig. 2. Point C is the place inside the mine up to which measurement was possible.

The various losses considered are penetration loss at the entrance, obstruction loss due to machinery, interference loss due to power cables and the bending loss. Interference loss due to power cables and the bending loss are already discussed above.

Penetration Loss at the Entrance

There was a 35 cm thick concrete brick wall at the entrance which results in penetration loss.

For lossy dielectric medium

$$\gamma^2 = (\alpha + j\beta)^2 = (\sigma + j\omega\epsilon)(j\omega\mu) \tag{6}$$

Equating real and imaginary parts we have,

$$\alpha^2 - \beta^2 = -\omega^2\mu\epsilon \tag{7}$$

And

$$2\alpha\beta = \omega\mu\sigma \tag{8}$$

Substituting for β from Eq. (8) in Eq. (7) gives,

$$\alpha^2 - (\omega\mu\sigma/2\alpha)^2 = -\omega^2\mu\epsilon \tag{9}$$

Or

$$\alpha \exp 4 + \omega^2 \mu \epsilon \alpha^2 - (\omega \mu \sigma / 2)^2 = 0 \tag{10}$$

On solving,

$$\alpha = \omega \sqrt{\left(\mu \epsilon / 2 * \{ \surd \left(1 + (\sigma / \omega \epsilon)^2 \right) - 1 \} \right)} \tag{11}$$

For $\sigma/\omega\epsilon << 1$& using Binomial expansion for

$$\sqrt{\left(1 + (\sigma / \omega \epsilon)^2 \right)} \approx 1 + 1/2 (\sigma / \omega \epsilon)^2 \tag{12}$$

So Attenuation constant

$$\alpha \approx \sigma / 2 \surd (\mu / \epsilon) \tag{13}$$

For concrete ϵr = 8.9 or 9 F/m, μr = 1 H/m, σ = 0.1 A/m^2 in GSM900 band With ϵo = 8.854 × 10exp-12 F/m, $\mu o = 4\pi \times$ 10exp-7 H/m, from Eq. (13),

$\alpha = 60\pi\sigma / (\surd\, \epsilon r)$ or penetration loss in dB = (20α) dB = 15.96 = 16 dB.

As the EMW is propagating perpendicular to the earth's magnetic field with E field vector along the earth's magnetic field, the magnetic field of earth does not affect the propagation characteristics of the wave. In other words propagation constant of the wave is same as that in the absence of the earth's magnetic field (i.e., ordinary wave propagation).

Obstruction Loss due to Machinery

Machines are used to move trolleys inside the mine. For a machinery of medium area < 15 sqm area loss will be around 4 dB in GSM900 band. For two machinery (one is used as back up or redundancy) the total machinery obstruction loss is = 8 dB.

Normal distance between the successive points for the reception of radio wave signal in rough mining is 20–35 m. In the present gold mine, this distance is only 12 to 14 m and beyond 30 m there was 75% loss in signal reception. Hence the losses are calculated at 3 points A (at entrance), D (intermediate point) and C (at a depth of 30 m) along the mine terrain. These are critical points along the terrain where appreciable change in signal strength are noticed.

Signal loss at a given point varies. At point A (Mouth of the mine) we have to consider penetration loss, machinery loss, and cable interference loss. At point D (intermediate point inside the mine), we have to consider additional loss like scattering. At point C (at a depth of 30 m), we have to consider additionally bending loss of microwave signal.

Table 1 Total losses. **a** Path loss of models

POINT	BPSK(dB)	BFSK(dB)	0.3-GMSK(dB)
A	119.7185	119.7185	119.7185
D	137.4461	134.7918	132.5873
C	150.2984	147.9036	146.1967

Model	A(d = 1)	D(d = 14)	C(d = 30)
Okumara	22.3	69.4	75.27
Hata	21.65	68.75	74.62
Cost 231	21.2	42.62	41.04
ITU model	-70.34	-49.52	-51.33

Table 2 Path loss for COST231 model

POINT	LOSS(dB)
A (d = 1 m)	124.877
D (d = 14 m)	163.58
C (d = 30 m)	174.762

Table 3 List of correction factors

Modulation format	Correction factor c1
BPSK	$5.2 + 36 \log_{10} d - 16(\log_{10} d)^2$
BFSK	$5.2 + 41 \log_{10} d - 18(\log_{10} d)^2$
0.3-GMSK	$5.2 + 46 \log_{10} d - 20(\log_{10} d)^2$

The total losses are tabulated as in Table 1:

By substituting the value of d in the respective model we get path loss (Table 1A),

The path loss calculated for COST231 model is least compared to other models. Hence, this model is used. Path loss for COST231 model is tabulated as in Table 2.

The differences between Table 1 and Table 2 for COST231 model are fitted using second order linear regression with logarithm of distance (log d) as independent variable. The correction factors obtained are tabulated in Table 3: The corrected COST231 model is shown in Fig. 3 .

Parameter Analysis

The parameters considered as correction factors are

- *Bending loss.*

Knife edge diffraction model is used to account for bending loss here. Bending loss is considered only at point C which varies with frequency.

Fig. 3 Plot of the corrected COST231 model for all three modulation formats with respect to frequency

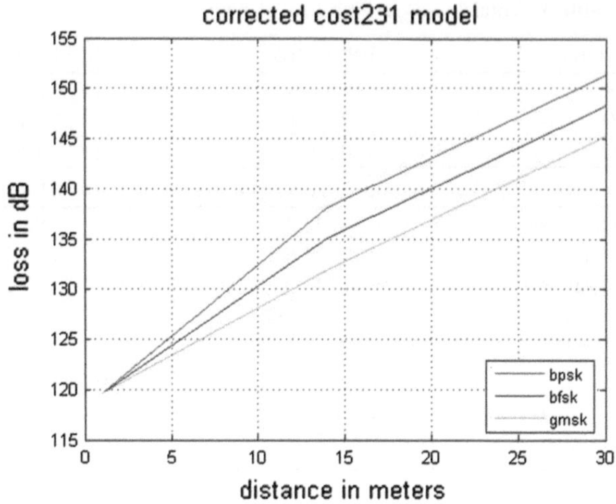

Considering all other losses to be constant,

$$\begin{aligned}\text{Total loss at C} &= \text{penetration loss} + \text{multipath loss} + \text{power line loss} \\ &\quad + \text{bending loss (Gd)} \\ &= 16 + 40 + 57.9 + G_d + e\end{aligned}$$

Since the loss calculated above will differ from measured data values, we add an error term e = 1.6766

For the range of GSM9000 with v > 2.4, G_d = 20log (.225/v).

The differences in the total loss and model loss for various frequencies are now corrected by using a regression method. Since the difference plot is a straight line we go for the linear regression method. Thus, we obtain second correction factor C1 = $24\log_{10}$f-71

Adding it to the already corrected cost231 model we get the cost231 model loss equation as Eq. (14),

$$L = 46.3 + 33.9\log f - 13.82\log(h_B) - a(h_R) + (44.9 - 6.55\log h_B)\log d - C1 - C2 \tag{14}$$

The plot of the model loss of Eq. (14) versus frequency is given in Fig. 4.

- *Low frequency interference loss.*

At point C, considering all other losses as a constant and varying only low frequency (power line) interference loss, we get,

$$\begin{aligned}\text{Total loss at C} &= \text{penetration loss} + \text{multipath loss} \\ &\quad + \text{power line loss} + \text{bending loss} \\ &= 16 + 40 + G_{rf} + 34.63 + e\end{aligned}$$

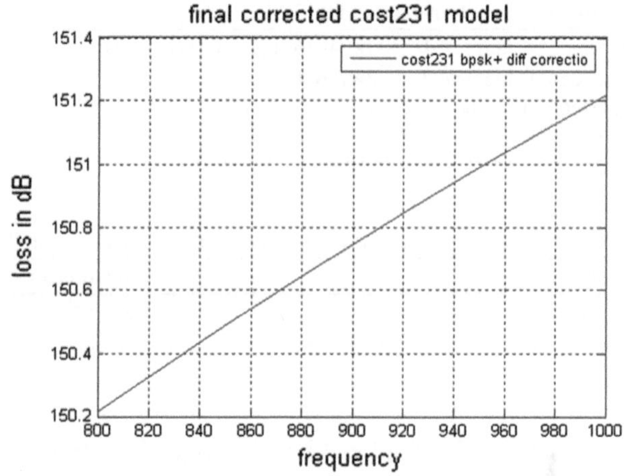

Fig. 4 Plot of the final corrected COST231 model with respect to frequency

where, G_{rf} (dB) = E-107 = −0.094f-12

Since the loss calculated above will differ from measured data values, we add an error term e = 1.7208.

The differences in the total loss and model loss for various frequencies are now corrected by using a regression method. Since the difference plot is a curve we go for the polynomial regression method. Thus, we obtain a third correction factor $C3 = 185.93*(\log f)^2 - 1079*\log f + 1566$.

Adding it to the already corrected cost231 model we get the final cost231 model as,

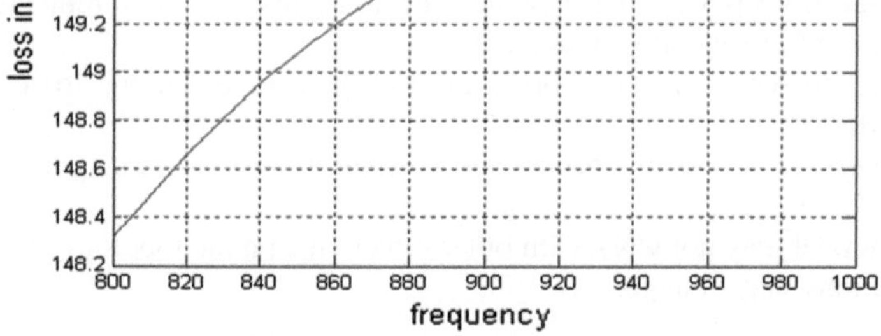

Fig. 5 Plot of the final corrected COST231 model with respect to frequency

Table 4 Correction factors for BPSK

Correction factor C1	$5.2 + 36 \log_{10} d\text{-}16(\log_{10}d)^2$	
Correction factor C2	$24\log_{10}f\text{-}71$	(considering diffraction loss)
Correction factor C3	$185.93*(\log f)^2 - 1079* \log f + 1566$	(considering low frequency interference loss)

Table 5 Mean and standard deviations for the corrected model

Considering	Mean error	Standard deviation
Diffraction loss	0.1954	0.5342
Low frequency interference loss	0.9708	0.5342

$$L = 46.3 + 33.9\log f - 13.82\log(h_B) - a(h_R) + (44.9 - 6.55\log h_B)\log d - C1 - C2 - C3$$

The plot of the final model versus frequency is shown in Fig. 5.

Conclusions

1. The final cost231 model which can be applied to an underground wireless channel is as follows:

   $$L = 46.3 + 33.9\log f - 13.82\log(h_B) - a(h_R) + (44.9 - 6.55\log h_B)\log d - C1 - C2 - C3$$

   where modulation format is BPSK. This can be repeated for BFSK and 0.3 GMSK modulation formats. The correction factors for BPSK is given in Table 4.

2. The mean and standard deviations for the corrected model are given in Table 5.
3. The error occurs because of various factors such as:

   - Measurements for the low frequency interference loss are done at a particular distance from power line. For more accurate results, measurements must be carried out at several distances.
   - Fading losses such as short-term and long-term fading losses are not considered.
   - RF interference loss that occurs due to exhaust fans, blowers, lights etc are not considered.
   - This model may not work with other mines since mine topology, environment conditions may change.

References

1. Rapport, T.S.: Wireless communications: principle & practice. In: Eastern Economy, 2nd ed. Pearson (2010)
2. Lee, W.C.Y.: Mobile communications engineering: theory and applications. Macgraw Hill (1998)
3. Delisle, G.Y.: School of information technology & engg. University of Ottawa, Canada Underground mine wireless propagation modeling. NRC Publications Archive (NPArC)Archives des publications du CNRC (NPArC) (2004)
4. Signal propagation techniques in underground mines. http://www.ece.gatech.edu/research/labs/bwn/surveys/underground09.pdfhttp://www.ece.gatech.edu/research/labs/bwn/surveys/underground09.pdf
5. Helchel. Comparison of 900 MHz band and 800 MHz band indoor propagation. IEEE Trans. Antenna Propag. **54**, (2006)
6. Shi, L., Xu, Q.: Bin, C., Cheng, G.: Measurement of the frequency dependent dielectric constant of concrete materials by TDR & wavelet modeling method. Paper presented at Asia pacific conference on environmental electromagnetic. CEEM 2003, pp. 626–629, 4–7 Nov 2003
7. Hindawi Publishing Corporation.: Int. J. Antennas Propag. **2008**, Article ID 806326, 10 pp., doi:10.1155/2008/806326
8. Jain, R.: Channel models a tutorial. http://www.cse.wustl.edu/ ~ jain/cse574-08/ftp/channel_model_tutorial.pdf. Accessed 1 Feb 2007
9. About Modulation Format. Http://en.wikipedia.org/wiki/Modulation
10. Nerguizian, C., Despins, C.L.: Aff'es, S., Djadel, M.: Radio channel, characterization of an under ground mine @2.4GHz. IEEE Trans. Wirel. Commun. **4**, 2441–2453 (2005)
11. Zhi, S., Akyildiz, I.F.: Channel modeling of wireless net-works in tunnels. Paper presented at IEEE global telecommunications conference, (GLOBECOM'08), New Orleans, LA, November 30–December 4, 2008.
12. Empirical formula for propagation loss in land mobile radio service. IEEE Trans. Vehicul. Technol. **VT-29**, 1980
13. About Radio Propagation Models .http://en.wikipedia.org/wiki/Radio_propagation model
14. About Path Loss .http://en.wikipedia.org/wiki/Path_loss
15. Bertoni, H.L.: Radio Propagation for Modern Wireless Systems. Prentice Hall, Englewood Cli®s, New Jersey, (2000).
16. Boutin, M., Benzakour, A., Despinsand, C., Affes, S.: Radio wave characterization and modeling in underground mine tunnels. IEEE Trans. Antennas Propag. **2**, 540–549 (2008)
17. Lienard, M., Degauque, P.: Natural wave propagation in mine environments. IEEE Trans. Antennas Propag. **48**(9), 1326–1339 (2000)
18. Pakala, W.E., Chartier, V.L.: Radio noise measurements on overhead power lines from 2.4 to 800 kv. IEEE Trans. **PAS-90**, 1155–1165 (1971)
19. Ndoh, M., Delisle, G.Y.: Underground mines wireless propagation modeling. Paper presented at 60th IEEE vehicular technology conference, Los Angeles, CA, pp. 3584–3588, 2004.
20. Kjeldsen, E., Hopkins, M.: An experimental look at RF propagation in narrow tunnels. Proceedings of IEEE military communications conference (MILCOM'06), Washington, D.C., October 23–25, 2006.
21. Wenfeng, L., Yingli, L.V., Baiping, L.: Experiment on characteristic of radio propagation in mine. J. Xi an Univ. Sci. Technol. **2,** 327–330 (2008) (in Chinese).

References

Review on Heart Sound Analysis Technique

U. More Monali and R. Shastri Aparana

Abstract Heart auscultation (the interpretation by a physician of heart sounds) is a fundamental activity of cardiac diagnosis. It is, however, a difficult skill to acquire. So it would be convenient to diagnosis the failure using some monitoring techniques. This paper reviews different signal processing technique for analyzing Heart Sound (HS) Vibration signals which is mainly used to diagnose these diseases. Conventional methods for fault diagnosis are mainly based on observing the amplitude differences in time or frequency domain such as Fourier Transform (FT), Short Time Fourier Transform (STFT), and Wavelet transform. This paper includes Spectral analysis method of heart sound by using autoregressive power spectral density (AR-PSD) for discriminating normal and abnormal HS, another method to diagnose heart sound such as Wavelet packet analysis and classifiers like Hidden Markov Model (HMM), Artificial Neural Network (ANN).

Keywords Fault diagnosis · Heart sound · STFT · Wavelet transform · Autoregressive power spectral density · HMM · ANN

Introduction

Auscultation, listening to sounds emanating from human organs, is a primary routine for screening and diagnosing many pathological conditions of the heart. Heart sound signal contains physiological and pathological information, which is

U. More Monali (✉) · R. Shastri Aparana
Department of Electronics and Telecommunication, Vidya Pratishtan's COE, Baramati, India
e-mail: monalimore12@yahoo.co.in

R. Shastri Aparana
e-mail: aparnadjoshi@yahoo.com

S. Sathiakumar et al. (eds.), *Proceedings of International Conference on Internet Computing and Information Communications*, Advances in Intelligent Systems and Computing 216,
DOI: 10.1007/978-81-322-1299-7_9,

related to each part of heart (atrial, ventricular, cardiovascular, and valvular function). With the improvement of domestic living standard, numbers of patients with heart disease are increasing rapidly. The heart disease is associated with living conditions, such as coronary heart disease (angina pectoris, myocardial infarction) and hypertension [1]. It becomes more and more difficult to recognize and diagnose heart sound in traditional auscultation way, because of limitation of human hearing sensitivity and auscultator's clinical experience. Now, heart disease is not the "patent" of elder, it also threatens many striplings' health. How to find symptom and know the incidence status in advance, is necessary to prevent and diagnose heart disease [1]. Heart sound and murmurs are of relatively low intensity and is band limited to 10–1000 Hz. The mechanical activities of the heart during each cardiac cycle produce the sounds, which are called heart sounds. The factors involved in the production of heart sounds are as follow:

(1) The movement of the blood through the chambers of the heart.
(2) The movements of cardiac muscle.
(3) The movement of the valves of the heart [2].

Human heart generates four sounds during its activity for one cardiac cycle. These sounds identified as S1, S2, S3, and S4 are not all audible. Figure 1 shows four heart sound S1, S2, S3, and S4 with systole and diastole.

S1 is generated at the end of atrial contraction, just at the onset of ventricular contraction. S1 can be heard obviously in the interval of the fifth rib which lies in the midline of left clavicle. The main feature of S1 is low tone and long time lasted. S2 occurs during ventricular diastole and can be heard clearly at the auscultation region between aortic valve and pulmonary valve. In contrast with S1, it has characteristics of high tune and short time lasted. S1 and S2 contain important information of cardiac sounds auscultation [3].

In recent ten years, with the rapid development of computer hardware and digital signals processing techniques, heart sound could be easily recorded and analyzed, the research [2] on automatic heart sound analysis showed its new tendency. Most of these researches were concerning on the characteristic extraction by frequency analysis method includes FT Fourier transform), Short time Fourier transform (STFT), Continuous wavelet transform (CWT), etc. Some other researchers were on how to extract of the heart beat from weeping noises of a baby and noise cancellation by an adaptive filtering method.

Fig. 1 Location of heart sound in cardiac cycle

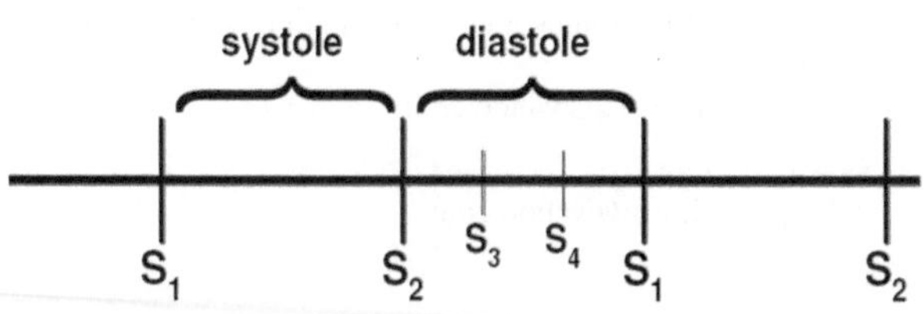

Analysis Techniques

STFT

(1) Fan and Brooks [4], proposed *Detection of Hypovolemia Using Short-Time Fourier Transform Analysis of S1 Heart Sound.* The goal is to detect hypovolemia as revealed by changes in the S1 and S4 heart sound. Heart sounds early in systole (S1 and S4) are sensitive to changes in blood volume in the heart chambers. Thus, the purpose of this study is to investigate whether changes in acoustically monitored heart sounds during anesthesia can be used to detect hypovolemia. In this paper author describe the use of the STFT to represent S1 sounds, and the use of signal processing methods to extract detection statistics from the modulus of the STFT. This method can be summarized in four steps: Waveform, STFT, Band-limited energy signal (BES), and Quantized pulse train signal (QPT) Classification.

Band-Limited Energy Signal (BES). The short-time nature of our STFT (small *L)* implies poor spectral resolution, so only frequency-averaged quantities can be reliably estimated. Since S1 and S4 normally have most of their energy below 200 HZ, we next calculate a band-limited energy signal, denoted BES, at short-time interval *l*, as

$$BES(l) = \sum_{k=i1}^{i2} |\mathrm{X}(\mathrm{l},\mathrm{k})|^2$$

where, the frequency indices corresponding to 20 and 200 Hz, respectively.

Thresholding in amplitude. Again based upon the observed audible changes, our major goal is to detect the extent of a "split" sound in *SI.* Thus we next generate a quantized pulse train signal, QPT (l), from BES (l), by means of a threshold.

Classification statistic based upon interpulse interval distribution. After obtaining the QPT signal for each heart sound, we have transformed a noisy S1 waveform to an energy-based signal which estimates when the sound is "on" and when it is "off".

The Fig. 2 shows Histogram of the interpulse interval distribution of all expiration S1 sounds during 20 s epochs before, during, and after the hypovolemic episode. It seems that interpulse interval statistic seems promising as a discriminator of the hypovolemic state. In addition, we noticed the clear presence of a small but perhaps significant number of S4 sounds during expiration during normovolemia but not during hypovolemia. The presence of the *S4* sound can easily be seen in the *QPT* signal.

Advantage. STFT may provide us with a tool to objectively observe the extent of splitting of S1 heart sounds.

Disadvantage. STFT has problem of fixed resolution. Also it is not good to analysis second heart sound S2 as good as S1.

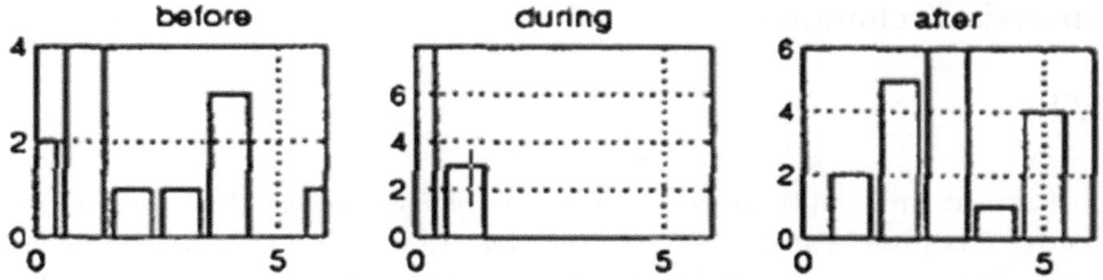

Fig. 2 Histogram of the interpulse interval distribution of all expiration S1 sounds during 20 s epochs before, during, and after the hypovolemic episode

(2) Vikhe and Nehe [2], proposed “Heart Sound Abnormality Detection Using Short Time Fourier Transform and Continuous Wavelet Transform” in which analysis of the first (S1) and second (S2) heart sound of the Phonocardiogram signal (PCG) using STFT and CWT. The STFT combines traditional time domain and frequency domain concepts into a single time–frequency framework. The frequency content and time duration of S1 and S2 can be determined by the STFT without difficulties. The second heart sound S2 consists of two major components A2 and P2. The time delay between them is very important for the medical diagnosis. With STFT it is impossible to determine the time duration between A2 and P2 which plays vital role in diagnosis of the PCG signal. This drawback of the STFT is over come using CWT. The time delays between A2 and P2 have been measured using CWT.

Experiments are performed on normal and pathological PCG signals. Frequency contents of S1 and S2 of PCG as well as time duration of them have been measured using STFT. Split between A2 and P2 have been measured using CWT.

This research shows for the normal heart, S1 includes a single frequency spectral component of energy and the duration of the sound is less in the range of 0.04–0.15 s. For the normal heart, S2 represents a uniform frequency spectral component of energy and the duration of the sound is less in the range of 0.03–0.12 s The magnitude of the energy components of normal heart S1 is higher than that of S2. But in case of pathological, there are more chances of S2 sound energy components to have larger magnitude than that of S1. Similarly, CWT is applied to normal heart sound and abnormal heart sound and result in splits is as follows (Table 1).

Advantage. Frequency content of S1, S2 can be easily measured using STFT, CWT.

Disadvantage. Split between A2 and P2 is not measured with STFT accurately.

Table 1 Split time for normal and pathological conditions

Types of signals	Normal	Aortic stenosis	Pulmonic stenosis	Atrial septal defect
Split (ms)	24	43	48	50

Wavelet Transforms

(1) Vikhe and Hamde [5], proposed "*Wavelet Transform Based Abnormality Analysis of Heart Sound*". This paper is concerned with the analysis of the first (S1) and second (S2) heart sound of the (PCG) using Discrete Wavelet Transform (DWT) and (CWT). The second heart sound S2 consists of two major components A2 and P2. The time delay between them plays very vital role in medical diagnosis. DWT has been used to determine the best split between A2 and P2 of the second heart sound. Frequency component of each heart sound is determined using DWT. Normal heart sound S1 and S2 has frequency range 0–125 Hz and 125–150 Hz, respectively, but in pathological cases such as Aortic Stenosis, Pulmonic Stenosis, Atrial Septal Defect frequency range exceed to 250–500 Hz.

Using DWT it is impossible to determine the time split between A2 and P2 which plays vital role in diagnosis of the PCG signal.

This drawback of the DWT is over come using CWT. The time delays between A2 and P2 have been measured using CWT. It is observed that the time delay between A2 and P2 is less than 30 ms for normal case and it is greater than 30 ms for pathological cases.

Advantage

(i) Window size is variable
(ii) It can handle the point discontinuity.

Disadvantage

(i) It does nnot handle curve discontinuities
(ii) Wavelets do not supply good direction selectivity

2) *Using Wavelet and Hidden Markov Model*

Lima and Barbosa [6], proposed "Automatic Segmentation of the Second Cardiac Sound by Using Wavelets and Hidden Markov Models". This work is concerned with the segmentation of the second heart sound (S2) of the (PCG), in its two acoustic events, aortic (A2), and pulmonary (P2) components. An automatic technique, based on discrete wavelet transform and hidden Markov models, is proposed in this paper to segment S2, to estimate de order of occurrence of A2 and P2 and finally to estimate the delay between these two components (split). A discrete density hidden Markov model (DDHMM) is used for phonocardiogram segmentation while embedded continuous density hidden Markov models are used for acoustic models, which allows segmenting S2. The main objective of the work described in this paper was to develop a robust segmentation technique for segmenting the phonocardiogram into its main components.

Advantage. HMM can perform well not only in segmenting the PCG in its main components but also in segmenting the components of the main components allowing automatic diagnosis related with abnormal order of appearance of the components of the main PCG sounds.

Disadvantage. The performance of the method degrades significantly in severe murmurs, especially in aortic and mitral regurgitation.

DWPA: Discrete Wavelet Packet Analysis

A-Naami et al. [7], proposed Identification of Aortic Stenosis Disease using Discrete Wavelet Packet Analysis. This work includes wavelet packet transforms in detection of an Aortic Stenosis (AS) using heart sound data. The discrete wavelet packet analysis utilizes both the low frequency components, and the high frequency components. From these frequency components and using entropy based criterion, a method for choosing the optimum scheme for the identification of AS Disease is developed.

Entropy is a common concept in signal processing. Classical entropy-based criteria describe information-related properties for an accurate representation of a given signal. There are many entropy criteria among them: Shannon entropy, energy entropy, norm entropy, and threshold entropy. In this study, norm entropy is used to extract some features from the PCG signals.

The procedure followed in the identification of aortic stenosis disease can be divided into three processes described as,

(1) A PCG signal is to clean it from noise associated with PCG systems. Noise is caused by breast sounds; contact of the stethoscope with skin, ambient noise that may corrupt the heart sounds. The data is filtered with high-pass Butterworth filter to eliminate noise. The Butterworth filter is selected because it has the least steepness of the amplitude response in the transition region.
(2) The DWPT is used to extract features that can be useful in the classification stage. The wavelet base Daubechies 'db4' is used since it has oscillations very similar to those of a PCG signals.
(3) The norm entropy-based criterion is used for the classification of PCG signal.

In this method, Frequency is divided into different subbands and for each subband entropy is calculated. If E1 is larger than E2 and both are larger than E12 and E21, then the heart sound signal is normal. If E1 is larger than E12 and both are larger than E2 and E21, then the heart sound signal has the symptom of aortic stenosis disease.

Advantages. The DWPA utilizes both the low frequency components, and the high frequency components.

Limitation. The number of data is limited and more is needed to validate the proposed criteria.

Using Autoregressive Power Spectral Density

Wang and Zhang [1], proposed heart sound analysis technique based on Autoregressive Power Spectral Density for discriminating normal and abnormal Heart Sound (HS). Digital stethoscope was used to collect HS and transmitted into computer by USB interface, so as to store, display and analysis. Practical cases of normal/abnormal HS analysis are demonstrated to validate the usefulness and efficiency of the proposed method.

(1) Initially, using digital stethoscope to collect high-quality sounds, and then proprocessing them preliminarily;
(2) Second, AR-PSD analysis method was used for discriminating normal and abnormal HS.

The corresponding frequency range of the normal and abnormal AR-PSD are significant difference, representing them by simple curves and digital parameters, it provides a new analytical method to a variety of quantitative evaluation of heart murmurs. The analyzing of normal/abnormal heart sound case verifies the validity of the method.

Figure 3 shows a scatter gram 20 normal and 20 abnormal HS, their parameters distributed in different obvious areas, So, it may provides a novel idea to recognize normal and abnormal HS.

Advantage. Simple to analyze and capable of representing stationary as well as nonstationary signals.

Disadvantage. Difficult to choose appropriate model order.

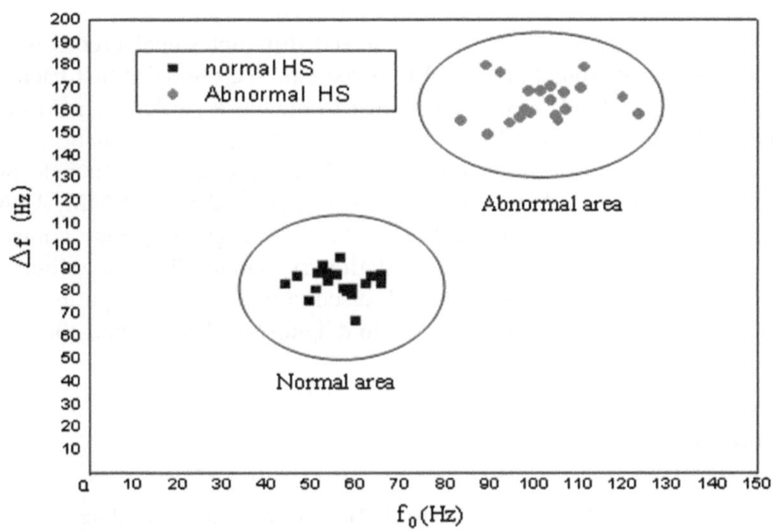

Fig. 3 Scatter gram of f0 and Δf

Artificial Neural Network

Omid mokhlessi, Naser mehrshad, Hojat Moayedi Rad [8], proposed "Utilization of 4 types of Artificial Neural Network on the diagnosis of valve-physiological heart disease from heart sounds". This work includes sound heart recognition for diagnosing heart disease with four type of Artificial Neural Network (ANN). Here they develop a simple model for the recognition of heart sounds, and demonstrate its utility in identifying features useful in diagnosis. They then present a prototype system intended to aid in heart sound analysis. Based on a wavelet decomposition of the sounds, feature vectors are formed and ANNs finds use in classification of Heart valve diseases for its discriminative training ability and easy implementation. four type of ANN which used for this approach are Multilayer perception (MLP), Back Propagation Algorithm (BPA), Elman Neural Network (ENN), and Radial Basis Function (RBF) Network. Using these ANN classifiers would appeared ability of classifying heart disease and will be shown an accuracy of 81.25 % for MLP, 87.17 % for BPA, 91.59 % for ENN, and 96.42 % for RBF was achieved.

The results demonstrate that the RBF can detect heart diseases with higher Diagnosing accuracy and have best.

Training performance than other ANNs, While, ENN extracts temporal contents from the whole signal epoch after epoch and may support for productive behavior or support inference. Besides, BPA can announce heart diseases with lower Neurons numbers; have best MSE quantity than others.

Conclusion

In this paper, we have reviewed and analyzed different signal processing techniques used for heart sound abnormality detection. We observed that each of the signal processing methods has an advantage to diagnose particular heart disease. Initially, STFT has been used for diagnosis but because of limitation of fixed resolution wavelet transform is used later. Other methods-mentioned such as Wavelet Packet Analysis, Autoregressive power spectral density, HMM, ANN also has own advantages such as utilization of low and high frequency component, capable of analyzing stationary and nonstationary signals. Because of some limitation of human hearing sensitivity and auscultator's clinical experience signal processing technique plays important role in diagnosis of heart sound abnormality.

References

1. Wang, H., Hu, Y., Liu, L., Wang Y., Zhang, J.: Heart sound analysis based on autoregressive power spectral density. ICSPS, 978-1-4244-6893-5© 2010 IEEE

2. Vikhe, P.S., Nehe, N.S., Thool, V.R.: Heart sound abnormality detection using short time fourier Transform and continuous wavelet transform ICETET. 978-0-7695-3884-6/09 © 2009 IEEE
3. Johnston, M., Collins, S.P., Storrow, A.B.: The third heart sound for diagnosis of acute heart failure. Curr. Heart Fail Rep. **4**(3), 164–168 (2007)
4. Fan, W., Brooks, D.H., Mandel, J., Calalang, I., Philip, J.H.: Detection of hypovolemia using short-time fourier transform analysis of S1 Heart Sound. 0-7803-2693-8195, © 1995 IEEE
5. Vikhe, P.S., Hamde, S.T., Nehe, N.S.: Wavelet transform based abnormality analysis of heart sound. ICSPS 978-0-7695-3915-7/09 ©2009 IEEE
6. Lima, C.S., Barbosa, D.: Automatic segmentation of the second cardiac sound by using wavelets and hidden markov models. EMBS conference, 2008 IEEE
7. Al-Naami, B., Chebil, J., Torry, J.N.: Identification of aortic stenosis disease using discrete wavelet packet analysis. 0276-6547/05© 2005 IEEE
8. Mokhlessi, O., Mehrshad, N., Rad, H.M.: Utilization of four types of artificial neural network on the diagnosis of valve-physiological heart disease from heart sounds, ICBME, 2010

Solutions for Security in Mobile Agent System

Neelam Dayal and Lalit Kumar Aswathi

Abstract For getting instant access to the data at one place, the software called 'Mobile Agent' is used, which moves from host to host and brings back the information required. With advancement in technology, access to information has become more and more easy, but it also brings certain concerns about the security of systems involved in the process and the data being used or taken. This paper is concerned with the security threats that a mobile agent system could experience and solutions that have been proposed for securing the mobile agent system. There are many solutions proposed for security; some are simulation based while some are mathematical derivation based. Simulation-based models are more practical for solving the real-world scenarios.

Keywords Mobile agent · Extended elementary object system model · Police office model · Mobile agent security

Introduction

Mobile agents (MA) are autonomous software that move from one host to another for retrieving certain information. Using the itinerary table, mobile agent makes decisions about where to move and finally brings back the information to the home agent. MAs are goal directed and have the capability to suspend their execution on the current platform and move to the next platform where it resumes its execution. The final data retrieved and brought back to the home agent by the MA is called

N. Dayal (✉) · L. K. Aswathi
National Institute of Technology, Hamirpur, India
e-mail: neelu.dayal@yahoo.co.in

L. K. Aswathi
e-mail: lalitdec@yahoo.com

S. Sathiakumar et al. (eds.), *Proceedings of International Conference on Internet Computing and Information Communications*, Advances in Intelligent Systems and Computing 216,
DOI: 10.1007/978-81-322-1299-7_10,

execution result. MA are autonomous, have learning ability, and most importantly, they are mobile. They have made retrieval of data much easier. They find application in various fields such as ecommerce, maintenance of data, and parallel processing. The final data retrieved by mobile agent through various nodes (hosts) is termed as execution results. These execution results are expected to be true and unaltered. However, if the data are altered or false, or the agent is faulty, it will be a security threat to the system. To make it possible for the host to have faith in the agent, and vice versa, it is important to assure both of them of having a secure medium of interaction.

Mobile Agent

A mobile agent (MA) is a process consisting of software and data. It has two components:

1. Program code: program code and static data
2. Program execution state: variable data such as the mobile agent itinerary data and execution results.

The Mobile Agent system comprises the User, Home Agent (HA), Mobile Agent, and Platforms. User asks a particular site (Platform) for certain information that it requires. That site known as Home Agent creates software called Mobile Agent, which is further dispatched for collecting information from various sites. MA using its itinerary table moves from site to site to gather the required information. Generally, user is responsible for creating the itinerary table for the mobile agent being created, but in some cases MA is expected to make its decision itself as it is considered to be intelligent enough to choose its route by itself. When MA reaches a Platform, it requests the platform for execution; when the host platform allows the MA, MA executes itself on that platform and appends the result retrieved to its actual data. Finally, the MA takes back the result to its Home Agent. HA provides this execution result to the user who has asked for the data.

Security Threats

As MA has to visit throughout the network and also has to execute itself on various platforms, it becomes exposed to various threats. In addition, if MA arriving at a platform is malicious it is a threat to that platform also. Based on these two aspects security is concerned with securing both MA and the platforms on which it is executing.

Hence there are two categories of concern for securing the MA system:

A. *Mobile agent security*. Mobile agent security is concerned with securing the MA that has to travel through a vast insecure network and has to execute on some other platform. There might be various possible attacks on this MA such as:

   (i) *Traveling from one platform to other*. While traveling through one platform to another there are possibilities that the MA could be attacked by eavesdrops. The data being carried by the MA and also the execution code are at risk during this movement.
   (ii) *Threat by host*. When MA is executing on a Host platform that is malicious, it could harm the executing MA. Host can harm the data carried by the MA of other sites. In order to dominate in business the host can alter the information provided by other sites that is carried by MA to it. Host can also change certain execution codes so that when MA executes on other sites, it damages those sites and their data. There could be possibilities that the host on which MA is executing is forged, hence it is dangerous for the MA.
   (iii) *Threats by other agents*. There are possibilities that when MA is executing on a platform, there are several other MAs being executed on that platform. If any of these MAs are insecure it could create a security threat for other MAs and harm the data carried by them.

B. *Host security*. A malicious or forged MA could be a threat to the host on which it is being executed. MA could ask for critical information about the host that could be used to harm it. If malicious MA gains access to the sensitive data, it can use this information against the host. Also, denial of service attacks is possible on the host by malicious mobile agent.

Security Approaches

Many researches have tried to solve issues of providing a secure and reliable interaction between mobile agent and host. These approaches can either be implemented for securing MA or the Platform or a model could be derived for securing both the systems. Some approaches are:

For Securing Platform

Authentication and authorization [1, 2]. Platform must allow only authentic MAs to be executed; this will protect host from forged MAs.

Sand boxing [3]. The main concept behind sand boxing is to isolate the MA command lines into different parts; safe command line codes and unsafe command line codes. The safe codes are directly allowed to be executed, while the code that seems to be unsafe is either made safe or denied for execution.

Software-based fault isolation [4]. The entire software is isolated into different modules for identifying unsafe (fault domains) modules. The unsafe modules are allowed to execute under separate allocated space so that it may not affect other modules. This method is useful for identifying the software

Signed code [5, 6]. This technique makes use of digital signature. Authenticity of the MA can be ensured to the host if the originator digitally signs the MA. Digitally signed MA ensures that it is a genuine agent originated from a genuine platform. Hence its authenticity and integrity are ensured.

State appraisal [7]. Appraisal functions are used to determine the privileges that are granted to the MA, so that access to the MA could be controlled by the host platform. Hence, it is a type of access control function used by the platform. These appraisal functions are determined by the platform that creates the MA. This scheme has to do with the privileges to an MA, but has no specification about attacks on MA code that has privilege to access the data.

Path history [8, 9]. In this scheme the MA carries the records of all the platforms it has visited, i.e., the history of the path followed by MA is carried with it. When MA arrives at a new platform, the platform checks for its path history. If all the platforms visited previously are trusted, MA is allowed to execute. If any of the previously visited sites is untrusted, platform will not allow the MA to execute. This scheme only takes care of possible attacks by a platform; however, if the MA has been attacked while on the move in the network, there is no solution for such situation.

Proof carrying code [10]. In this scheme the creator of the MA has to provide the proof of MA being secure and authentic. The originator attaches the proof that the MA fulfills the security requirements needed with it. It allows hosting platform to verify the identity of MA, hence making it easy for host to decide whether to allow the MA to execute or not. All this is done by HA, hence it reduces the burden of the hosting platform, as it now has to check the authenticity of MA based on the details provided.

For Securing Mobile Agent

Trusted platform[11]. Security of the MA can be guaranteed if the MA is executing itself on a trusted platform. The platform on which the MA has arrived must provide the proof that it satisfies all the security requirements and is a secure platform, only then will the MA start executing itself on it. Hence, a mutual authentication can provide assurance for a secure MA system.

Encrypted algorithms[12, 13]. MA is transmitted from the HA to various platforms in encrypted format. When it reaches the host platform, if the host is authentic and has the key to decrypt MA, it can decrypt the MA and only then it is executed. Security of this scheme is completely dependent on the Key shared between the platforms. If this shared secret (key) is retrieved or hacked by an intruder the whole security system fails. The intruder can easily access the MA and misuse its information, and may also alter its code to harm other platforms. Here, this security scheme lacks in guaranteeing the security of the system, mainly for the MA.

Model for Secure Mobile Agent System

Many scientists have also tried to guarantee the security of the complete MA system that includes security of the MA, platforms, and the underlying network. This paper is mainly concerned with schemes that have provided simulation-based models for providing the secure system. These models represent the whole system in a graphical manner so that it is easy to understand the system and analyze the security of the system in spite of using mathematical derivation-based proofs. Some of the important graphical models for securing mobile agent system are underlined below.

Extended Elementary Object System Model [14, 15]

EEOS model is based on the elementary object system (EOS) that was proposed by Valk[16] for modeling workflow and flexible manufacturing system. It is based on Object Petri Net (OPN) system of representation of transaction processing. Since MA system is concerned with the mobility of the MA throughout the network, EOS is one of the most suitable methods for representing this mobility. MA in the MA system is represented using the object nets in the EOS, platforms are represented by system nets, and the movement of the MAs and various tokens are represented by the transaction arcs. EOS was not as it is used in the EEOS model, there were few additions in the existing system for representing various features of Mobile Agent system. The additions made to the EOS system were: Multiple System Nets for representing multiple platforms, multiple layers for representing multiple layered architecture, Token pool for representing complex network environment, two new arcs for representing transactions in a better way, and extended interaction relation.

This model defines three layers of the Mobile agent system. The first layer is a Platform layer; it contains mobile agent and token pool, and is a system net having Mobile Agent as object net. It defines the migration of mobile agent to various platforms. Various places and transitions are defined in the system for representing how the movement should take place and what the result will be of each transition. It helps in identifying the abnormal behavior of MA if it is attacked. Hence, the system security can be ensured. The second layer is the mobile agent layer that contains security mechanisms; it is system net for the security mechanism layer.

This layer represents the processes taking place in mobile agent and also defines the security mechanism for the MA execution. The third layer is the security mechanism layer that defines how the MA has to behave. If the MA has been attacked it will behave in an abnormal manner, this abnormal behavior would be tracked by this security mechanism, and hence attack is detected.

This model is suitable for the detection and avoidance of attacks on the mobile agent system. It provides the mutual authentication and dynamic tracking of the movement of mobile agent and could prove an effective solution for security of MA system. However, its shortcomings are that it is really a complex model to be implemented and could prove costly.

Police Office Model [17]

POM is based on the concept of the Police Office system. All the hosts in the system are divided into different groups, with a particular host in each group allotted the duty of Police Office (PO). This PO has the same duty as a police office in the real-world scenario to ensure the security of the system. Whenever any Mobile agent tries to enter the group to interact with any of the hosts in the group, MA has to contact the PO. If PO authenticates the MA to proceed further, only then can it interact with the desired host, else it will not be allowed to move further. The responsibility of security comes to the host that works as PO. It can lead to the bottleneck problem, but for networks with less traffic it is an appropriate model.

Novel MA Security Mechanism [18]

This method integrates the trusted platform module into the platform for determining the security of the system. Integrity of data contained by the MA is the prime concern of the novel security mechanism. The model is based on the interaction between two agents; task agent (TA) and security agent (SA). When an MA is dispatched, SA is sent to a trusted anonymous third party (ATP). With the movement of the MA the TA also moves to various hosts. When a TA moves to a new host, SA sends an integrity report to it and checks for the integrity of the data. The MA can freely move in the system anywhere it wants to and its integrity is assured by the SA assisted with the ATP, a trusted third party. This approach is appropriate when MA has to determine its movement itself without the help of itinerary table.

Conclusion

As the mobile agent has to move in the whole network whether secure or insecure, it also has to interact with different hosts, hence its security becomes of utmost importance. Various researches have provided different solutions to the security issues discussed in this paper. All these solutions provide security to the mobile agent to some extent. Simulation-based security systems are more appropriate to the real-world scenario. Recently, researchers have developed simulation-based models for solving the overall security issues in the mobile agent system. The security aspect covers securing the mobile agent, the host, and the underlying network. A security model is complete when it covers all these aspects. Simulation-based methods can easily represent the overall mobility and can help in detecting and avoiding attacks on the system. These approaches have solved a number of issues related to MA system security. Yet, flaws remain that need to be overcome in the future to provide a better security system, to utilize the Mobile agent system in a better way.

References

1. Gray, R. S.: Agent TCL: a flexible and secure mobile-agent system. In: Proceedings of the Fourth Annual Tcl/Tk Workshop (TCL) 96, pp. 9–23 Monterey, California, (1996)
2. Farmer, W.M., Guttman, J.D., Swarup, V. (1966) Security for mobile agents: Authentication and state appraisal. Proceedings of th 4th Europian Symposium on Research in Computer Security (ESORICS'96), Rome, Italy pp.118–130
3. Gong, L.: New security architectural directions for java extended abstract. In: Proceedings of IEEE COMPCON, pp. 97–102 San Jose, California, February (1997)
4. Wahbe, R., Lucco, S., Anderson, T.E., Graham, S.L.: Efficient software-based fault isolation. In: SOSP'93: Proceedings of the fourteenth ACM symposium on operating systems principles. ACM, New York, pp. 203–216 (1993)
5. Jansen, W., Karygiannis, T.: Mobile agent security. In: Nist special publication 800-19 - (2000)
6. Wang, C., Zhang, F., Wang, Y.: Secure web transaction with anonymous mobile agent over internet. J. Comp.Sci. Technol. **18**(1), 84–89 (2003)
7. W. M. Farmer, J. D. Guttman, V. Swarup, S. Wakid, J. Davis, (eds.), In: Security for mobile agents: Authentication and state appraisal, Proceedings of the 19th National Information Systems Security Conference, Vol. 2, pp. 591–597, Baltimore Convention Center, Baltimore, Maryland, Oct 22–25, (1996)
8. Chess, D., Grosof, B., Harrison, C., Levine, D., Parris, C., Tsudik, G.: Itinerant agents for mobile computing. IEEE Pers. Commun. **2**(5), 34–49 (1995)
9. Bellavista, P., Corradi, A., Montanari, R., Stefanelli, C.: A mobile computing middleware for location and context- aware internet data services. ACM Trans. Int. Technol. **6**(4), 356–380 (2006)
10. Necula, G.C., Lee, P.: Safe kernel extensions without runtime checking. SIGOPS Oper. Syst. Rev. **30**, 229–243 (1996)
11. Peine , H., Stolpmann, T.: The architecture of the platform for mobile agents. In: Proceedings of the First International Workshop on Mobile Agents. Springer-Verlag, London, pp. 50–61 (1997)

12. Hohl, F.: Time limited blackbox security: protecting mobile agents from malicious hosts. In: G.Vigna (ed.) Mobile Agents and Security, pp. 92–113. Springer-Verlag, London (1998)
13. Sander, T.: On cryptographic protection of mobile agents. In: Proceedings of the 1997 Workshop on Mobile Agents and Security. University of Maryland, Baltimore, USA Oct (1997)
14. Ma, L., Tsai, J.J.P., Deng, Y., Murata, T.: Extended elementary object system model for mobile agent security. In: Proceedings of World Congress on Integrated Design and Process Technology, pp. 169–178, (2003)
15. Ma, L., Tsai, J.J.P.: Formal modeling and analysis of a secure mobile-agent system. In: IEEE Trans. Systems, Man Cybern.- A: Syst. Humans. 38(1), pp.180–196 (2008)
16. Valk, R.: Petri Nets as token objects—an introduction to elementary object nets. In: Proceedings of 19th International Conference on Application and Theory of Petri Nets, vol. 1420, pp. 1–25. (1998)
17. Xudong Guan, Yiling Yang, Jinyuan You. POM – A mobile agent security model against malicious hosts. In Proc. HPC-Asia 2000, pp.1165-1166, Beijing, China, May 2000
18. Lei, S., Liu, J., Xiao, J.: A novel free-roaming mobile agent security mechanism by trusted computing technology. In: 2008 International Conference on Computer Science and Software Engineering 12–14 Dec Wuhan, China
19. Borselius, N.: Mobile agent security. Electron. Commun. Eng. J. **14**(5), 211–218 (2002)
20. Dewi Agushinta, R., Suhendra, A.: Secure mobile agent system in peer-to-peer networks: a review of security mechanisms based on several security issues. In: Proceedings of IEEE World Congress on Software Engineering Xiamen, China, 19–21 May (2009)
21. Hasan, M.B., Prasad, P.W. C.: A review of security implications and possible solutions for mobile agents in e-commerce. In: 2009 Conference on Innovative Technologies in Intelligent Systems and Industrial Applications (CITISIA 2009), 25–26 July 2009, Monash
22. Vigna, G.: Cryptographic traces for mobile agents. In: Mobile Agents and Security, pp. 137–153. Springer-Verlag (1998)
23. Riordan, J., Schneier, B.: Environmental key generation towards clueless agents. In: G., Vigna (ed.) Mobile Agents and Security. LNCS 1419, pp. 15–24. Springer-Verlag, London April 1998
24. Balfe, S., Lakhani, A.D., Paterson, K.G.: Trusted computing: providing security for peer-to-peer networks. In: P2P'05: Proceedings of the Fifth IEEE International Conference on Peer-to-Peer Computing, pp. 117–124. IEEE Computer Society, Washington, DC (2005)
25. Sander, T., Tschudin, C. F.: Protecting mobile agents against malicious hosts. In: Mobile Agents and Security, pp. 44–60. Springer-Verlag, London (1998)
26. Silva, L.M., Soares, G., Martins, P., Renato, C., Almeida, L., Stohr, N.: JAMES: a platform of mobile agents for the management of telecommunication networks. In: G. Vigna (ed.) Proceedings of IATA99 Stockholm, Sweden, Aug 1999
27. Binder, W., Roth, V.: Secure mobile agent system using java: where are we heading? In: Proceedings of Symposium of Applied Computing, pp. 115–119 Madrid, Spain (2002)

An Automatic MRI Brain Segmentation by Using Adaptive Mean-Shift Clustering Framework

J. Bethanney Janney, A. Aarthi and S. Rajesh Kumar Reddy

Abstract A novel fully, automatic, adaptive, robust procedure for brain tissue classification from three-dimensional (3D) magnetic resonance head images (MRI) is described in this paper. We propose an automated scheme for magnetic resonance imaging (MRI) brain segmentation. An adaptive mean-shift methodology is utilized in order to categorize brain voxels into one of three main tissue types: gray matter, white matter, and cerebro spinal fluid. The MRI image space is characterized by a high dimensional feature space that includes multimodal intensity features in addition to spatial features. An adaptive mean-shift algorithm clusters the joint spatial-intensity feature space, thus extracting a representative set of high-density points within the feature space, otherwise known as modes. Tissue segmentation is obtained by a follow-up phase of intensity-based mode clustering into the three tissue categories. By its nonparametric nature, adaptive mean-shift can deal successfully with nonconvex clusters and produce convergence modes that are better applicant for intensity based categorization than the initial voxels. The performance of this brain tissue classification procedure is demonstrated through quantitative and qualitative validation experiments on both simulated MRI data (10 subjects) and real MRI data (43 subjects). The proposed method is validated on 3-D single and multimodal datasets, for both simulated and real MRI data. It is shown to perform well in comparison to other state-of-the-art methods without the use of a preregistered statistical brain atlas.

Keywords Adaptive mean-shift · Brain magnetic resonance imaging · Segmentation

---

J. Bethanney Janney (✉) · A. Aarthi · S. Rajesh Kumar Reddy
Department of Biomedical Engineering, Sathyabama University,
Chennai, Tamilnadu, India
e-mail: jannydoll@gmail.com

S. Rajesh Kumar Reddy
e-mail: rajeshsr86@gmail.com

S. Sathiakumar et al. (eds.), *Proceedings of International Conference on Internet Computing and Information Communications*, Advances in Intelligent Systems and Computing 216,
DOI: 10.1007/978-81-322-1299-7_11,

Introduction

A magnetic resonance imaging instrument (MRI scanner) uses powerful magnets to polarize and excite hydrogen nuclei (single proton)) in water molecules in human tissue, producing a detectable signal which is spatially encoded resulting in images of the body. In brief, MRI involves the use of three kinds of electromagnetic field: a very strong (of the order of units of Tesla) static magnetic field to polarize the hydrogen nuclei, called the static field; a weaker time-varying (of the order of 1 kHz) for spatial encoding, called the gradient field(s); and a weak radio frequency (RF) field for manipulation of the hydrogen nuclei to produce measurable signals, collected through an RF antenna [1–3]. Like CT, MRI traditionally creates a two-dimensional image of a thin "slice" of the body and is therefore considered a tomographic imaging technique. Modern MRI instruments are capable of producing images in the form of three-dimensional (3D) blocks, which may be considered a generalization of the single-slice tomographic concept. Unlike CT, MRI does not involve the use of ionizing radiation and is therefore not associated with the same health hazards.

Fully automatic brain tissue classification from magnetic resonance images (MRI) is of great importance for research and clinical studies of the normal and diseased human brain. Operator assisted classification methods are non reproducible, and also are impractical for the large amounts of data **A**utomated MRI segmentation systems classify brain voxels into one of three main tissue types: gray matter (Gm), white matter (Wm), and Cerebro-spinal fluid (Csf). Volumetric analysis of different parts of the brain is useful in assessing the progress or remission of various diseases, such as Alzheimer's disease, epilepsy, multiple sclerosis, and schizophrenia. Both supervised and unsupervised approaches have been used for this task. In the supervised approach, intensity values of labeled voxel samples from each tissue (prototypes) must be provided during the learning phase. In a subsequent classification phase, unlabeled voxels are classified using a selected classifier. This method requires human interaction to select the prototypes and is therefore semi-automatic. To avoid re-training the classifier for each new scan, methods are required to normalize the intensity between MRI scans This allows selecting the prototypes and training the supervised classifier on a reference scan, following which voxels of any other scan, previously normalized with regard to the reference scan, can be classified using the same classifier without further human intervention. Unsupervised approaches often rely on a Gaussian approximation of the voxel intensity distribution for each tissue type. This is justified due to the Rician behavior of the noise present in the MRI intensity signal [4]. A Gaussian mixture model (GMM) is fitted to the voxels intensity using the expectation–maximization (EM) algorithm following which every voxel is assigned to the tissue class for which it gives the highest probability. The GMM–EM intensity-based framework has been refined to account for partial volume effects and blood vessel signals that may alter Csf segmentation. However, using intensity information alone has proven to be insufficient for a reliable automated segmentation of the brain tissues. Local signal perturbations caused by additive noise and multiplicative bias-fields are responsible for cluster overlaps in the intensity feature space, resulting in poor tissue-class separability. Due to the artifacts present, voxel-wise intensity-based classification methods, such as *K*-means modeling and GMMs may give unrealistic results, with tissue class regions appearing granular, fragmented, or violating anatomical constraints. Incorporating spatial information via a statistical atlas provides a means for improving the segmentation results. The statistical atlas provides the prior probability for each pixel to originate from a particular tissue class [5–7]. Each tissue's intensity is modeled by a Parzen density fitted to voxels selected from an affine-registered atlas. Co-registration of the input image and the atlas is critical in this scenario. It is important to stress that an appropriate atlas does not always exist for the data at hand. Two such cases are brain data with pathologies or brain data obtained from young infants. An additional conventional method to improve segmentation smoothness is to model neighboring voxels interactions using a Markov random field (MRF) statistical spatial model. MRF-based algorithms are computationally intensive, requiring critical parameter settings. It is possible to use MRF with predefined settings, which is faster but possibly less accurate. Statistical modeling frameworks can include the spatial information by augmenting the input feature space. By appending the spatial position coordinates [x, y, z] to the intensity features, a higher dimensional feature space is obtained where clusters represent both voxels' intensity and spatial position distribution. Clusters in the augmented feature space are more closely related to the brain anatomy. This observation may prove problematic for parametric models such as GMMs that implicitly assume cluster convexity, as the brain anatomy cannot be decomposed into a small number of convex regions in the joint spatial-intensity feature space. A recently published solution includes using a large number of Gaussians per brain tissue, in order to capture the complicated spatial layout of the individual tissues. Supervised classification by the *K*-nearest-neighbors algorithm in an augmented position-intensity feature space has also been proposed In this case a manually segmented prototype atlas or an anatomical template is aligned with the considered brain in order to obtain the segmentation. An alternative to statistical parametric approaches is the use of unsupervised nonparametric schemes. One such approach is the mean-shift algorithm. Here, adaptive gradient ascent is used to detect local maxima of data density in feature space. Data points are associated with local maxima, or modes, thereby defining the clusters. Key characteristics of the mean-shift algorithm include the fact that no initial cluster positions are required, as well as the fact that the final number of extracted clusters is a result of the algorithm. A description of the mean shift algorithm is provided. In recent years, mean-shift has been used for image segmentation, object tracking, and medical image analysis applications in the current work; the objective is to utilize the mean-shift formalism to provide a robust segmentation framework for brain tissues in MRI that integrates spatial information in a simple way without requiring the use of a statistical brain atlas or HMRF modeling [8]. The developed framework, initially proposed is based on a variation of the mean-shift algorithm, termed the adaptive mean-shift algorithm by assigning a distinct bandwidth to every data point, the adaptive mean-shift allows for increased sensitivity to local data structure even in a higher dimensional feature space corresponding to multimodal MRI.

The issues in the existing work are (1) A GMM is fitted to the voxels intensity using the EM algorithm following which every voxel is assigned to the tissue class for which it gives the highest probability. The GMM-EM intensity based framework has been refined to account for partial volume effects and blood vessel signals that may alter Csf segmentation. (2) A Markov random field (MRF) based algorithms are computationally intensive, requiring critical parameter settings. It is possible to use MRF with predefined settings, which is faster but possibly less accurate. Statistical modeling frameworks can include the spatial information by augmenting the input feature space [9, 10].

Our proposed work is that the developed framework, initially proposed is based on a variation of the mean-shift algorithm, termed the adaptive mean-shift algorithm By assigning a distinct bandwidth to every data point, the adaptive mean-shift allows for increased sensitivity to local data structure even in a higher dimensional feature space corresponding to multimodal MRI. We present an automated segmentation framework for brain MRI volumes based on adaptive mean-shift clustering in the joint spatial and intensity feature space. The method was validated both on simulated and real brain datasets, and the results were compared with state-of-the-art algorithms. The advantage over intensity-based GMM EM schemes as well as additional state-of-the-art methods was demonstrated. We will show that using the AMS framework, segmentation of the normal tissues is not degraded by the presence of abnormal tissues.

Methodology

Proposed Algorithm

The fast adaptive mean-shift algorithm (FAMS) is utilized to analyze 3-D multimodal MRI data and provide segmentation maps of the three main tissue types (Gm, Wm, Csf). One to four MRI modalities are available per segmentation task. Standard preprocessing steps include: (1) brain parenchyma extraction using the brain extraction tool (BET). The obtained brain masks were visually inspected and corrected for outliers when needed. When a binary mask was available from the dataset, it was used instead of applying BET. (2) Intensity in-homogeneities (bias) correction by homomorphic low pass filtering and (3) Intensity values normalization across input channels (input modalities) via linear histogram stretching based on the darkest and brightest percentage points. The normalization sets the darkest percent of voxels to zero and rescales the brightest percent to 4095. The purpose is to obtain similar dynamic ranges for all the considered modalities. Following the initial data processing, feature-vectors are extracted per input voxel.

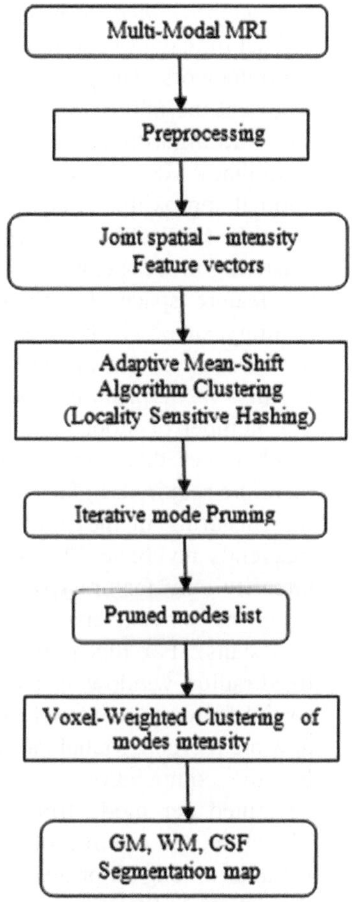

Fig. 1 Block diagram for the proposed algorithm

The set of feature-vectors is input to the adaptive mean-shift clustering stage of the framework. The output of the clustering step is a set of modes which provides a compact representation of the data. A follow-up merging stage is proposed to further prune the initial set of modes. Finally the categorization of the resultant modes into three categories, as defined in the brain segmentation task, is achieved via an intensity-based clustering stage. Figure 1 shows a summarizing block diagram for the proposed algorithm.

A. *Pre processing module.*
The initial data processing, feature-vectors are extracted per input voxel. Intensity as well as spatial features (x, y, z voxel coordinates) is used for an overall dimensionality of $3 + n$, where n is the number of input intensity channels (modalities). The set of feature vectors is input to the adaptive mean-shift clustering stage of the framework.

B. *Fast adaptive mean-shift clustering.*
The set of feature vectors is input to the adaptive mean-shift clustering stage of the framework. The process starts by clustering the input feature vectors, which represent the multimodal MRI brain data using the FAMS implementation of the AMS algorithm.
In that AMS, we can include Locality-Sensitive Hashing. That can produce optimal approximate neighborhood with radius. This stage, each feature vector bears the label of its convergence mode, or cluster. Each mode obtained by the clustering process expresses the local structure of the data in a given region of the feature space. It should be emphasized that modes define clusters of arbitrary shape, without any convexity constraints. The number of obtained modes is an output of the fast adaptive mean-shift algorithm.

C. *Iterative mode pruning module.*
The number of modes is a large compression of the initial data but it is still much larger than the targeted number of classes. A mode pruning step is therefore required. In fact, we have used the nonparametric adaptive mean-shift for clustering in the joint spatial-intensity feature space as the clusters are inherently no convex. For the pruning of the modes, however, we switch to an intensity-only feature space for which clusters can be conveniently approximated as convex, enabling the use of parametric models (i.e., multivariate Gaussians). For this purpose, a pruning mechanism is added as follows. A fixed-radius window is shifted across the intensity feature space (ignoring spatial features), centered on each mode. Modes that co-exist within the window are merged. Mahalanobis distance is utilized for the distance computation. For the computation of the Mahalanobis distance, a covariance matrix is computed per mode from the intensity values of its corresponding voxels. Therefore, the pruning of the initial modes list is performed until the largest variance among all pruned modes exceeds a preset threshold. That can produce Pruned mode list.

D. *Voxel-weighted clustering module.*
That results from the preceding clustering and pruning steps, is indicated by an arrow in the color of the corresponding segmentation map. It can be observed that the intensity value for each sample mode is closer to a peak of the whole brain intensity distribution than for most of the voxels it represents. Therefore, the modes intensity provides a higher probability classification into one of the three tissue types than the intensities of the voxels they represent. The effect observed at a single mode level is the sharpening of the intensity distribution peaks (similar to the effect of a bias correction algorithm) which results in a stronger intensity separation between the different tissue types. It can produce the Final segmentation results.

Results and Discussion

In this section, we present the performance of the proposed segmentation framework on 3-D simulated and real datasets. Simulated data was downloaded from the Brain web Simulated Brain Database (SBD) repository. Real data was downloaded from the center for Morph metric Analysis. Mass-achusetts General Hospital Repository, which is a standard repository for algorithm comparison (hereon termed IBSR). Both qualitative and quantitative validation is conducted, with a comparison to additional state of-the-art segmentation algorithms and the ground truth data when available. The AMS framework performance on real data sets is demonstrated. A set of 20 normal T1-weighted real brain data was downloaded from the IBSR repository. Each volume consists of around 60 coronal T1 slices. Segmentation overlap index values obtained with several segmentation algorithms are available for comparison in the IBSR site. Three Sample slices are shown in Fig. 2. The original data; the ground truth and the AMS segmentation are presented in Fig. 2 (a)–(c), respectively.

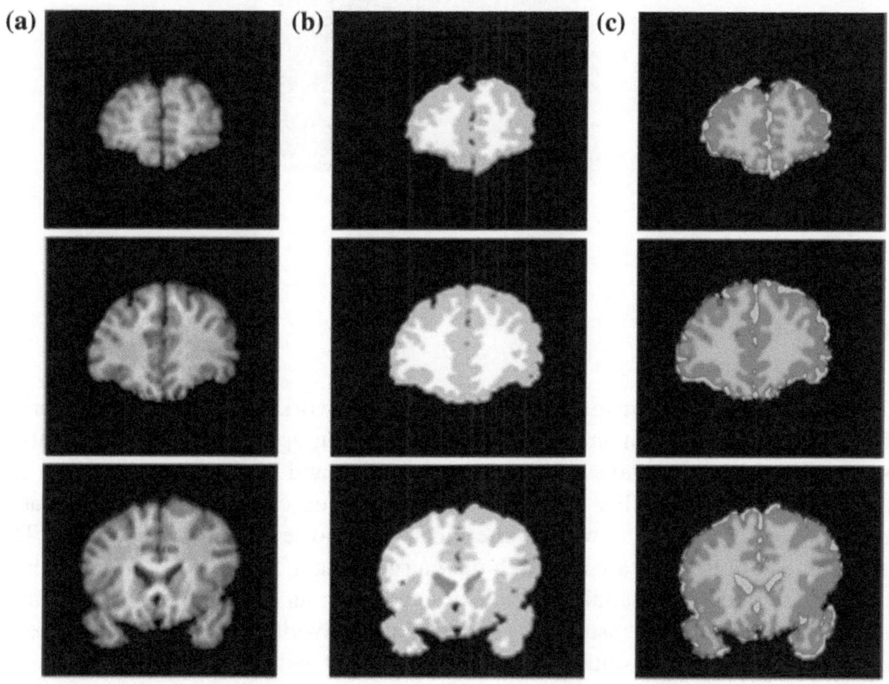

Fig. 2 Three sample slices from IBSR. **a** Input slices. **b** Ground truth. **c** AMS segmentation (Wm in *cyan*, Gm in *yellow*, Csf in *blue*)

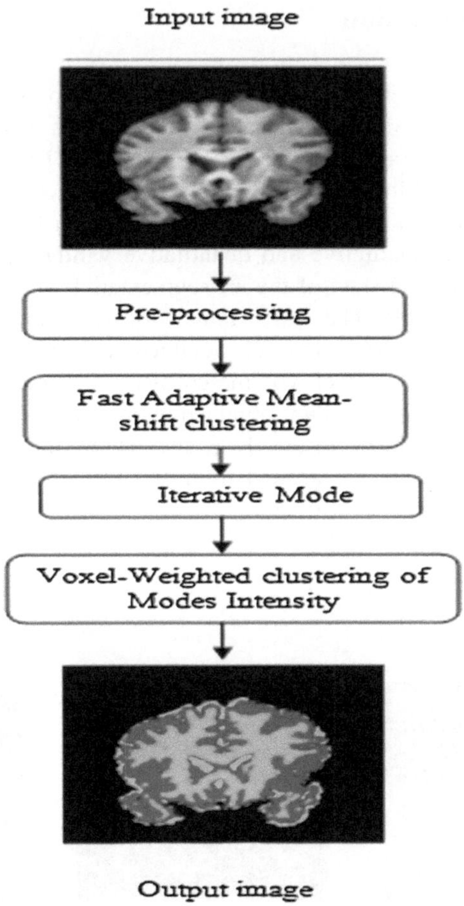

Conclusion

We presented an automated segmentation framework for brain MRI volumes based on adaptive mean-shift clustering in the joint spatial and intensity feature space. The algorithm gave good results on noisy and biased data thanks to the adaptive mean-shift ability to work with non-convex clusters in the joint spatial intensity feature space as well as the mean-shift noise smoothing behavior. The advantages of our work are although only a rudimental bias field correction step is implemented and no spatial prior is extracted from an atlas, Moreover, by using the adaptive mean-shift instead of the constant bandwidth algorithm, we ensure an appropriate bandwidth value for each feature point without requiring per-dataset manual tuning In the current implementation (Matlab and C), the typical algorithm execution time is about 30 min for a four-modal $256 \times 256 \times 46$ brain volume. As mean-shift runs a large loop on the whole feature vectors set, we believe that a full C/C++ multithreaded implementation on a multicore PC can reduce by more than half the running time. In future research, we will examine ways to improve the current algorithm's limitations. In particular, the current bandwidth selection algorithm based on the k-nearest neighbor makes no use of application specific information. Edge information, for instance, could help define the region of influence of a kernel by a given point since edges generally delimit regions corresponding to different tissue types. Another important issue regards the final mode merging step. Currently, it is based on the intensity clustering with k-means. This provides a robust and straightforward way of getting the desired number of classes tissue types but at the cost of losing some local spatial information contained in the modes found with mean-shift. The proposed framework will be extended to incorporate the detection of abnormal tissues such as sclerotic lesions and tumors.

References

1. Pham, D.L., Xu, C.Y., Prince, J.L.: A survey of current methods in medical image segmentation. Annu. Rev. Biomed. Eng. **2**, 315–337 (2000)
2. Rouaïnia, M., Medjram, M.S., Doghmane, N.: Brain MRI segmentation and lesions detection by EM algorithm. World Academy of Science, Engineering and Technology 24 2006
3. Cocosco, C., Zijdenbos, A., Evans, A.: A fully automatic and robust brain MRI tissue classification method. Med. Image Anal. **7**(4), 513–527 (2003)
4. Georgescu, B., Shimshoni, I., Meer, P.: Mean-shift based clustering in high dimensions: a texture classification example. In Proceedings of IEEE Conference on Computer Vision (ICCV), France, 2003, pp. 456–463
5. Derpanis, K.G.: Mean shift clustering. Shape Modeling and Applications, 2005 International Conference, Germany. August 15, 2005
6. Van Leemput, K., Maes, F., Vandeurmeulen, D., Suetens, P.: Auto-mated model-based tissue classification of MR images of the brain. IEEE Trans. Med. Imag. **18**(10), 897–908 (1999)
7. Dugas-Phocion, G., González Ballester, M.Á., Malandain, G., Le-brun, C., Ayache, N.: Improved EM-based tissue segmentation and partial volume effect quantification in multi-sequence brain MRI," in partial volume effect quantification in multi-sequence brain MRI. Paper presented at International Conference on Medical Image Computing and Computer Assisted Intervention (MICCAI), France, pp. 26–33, 2004
8. Gudbjartsson, H., Patz, S.: The Rician distribution of noisy MRIdata. Magn. Reson. Med. **34**, 910–914 (1995)
9. Dempster, A., Laird, N., Rubin, D.: Maximum likelihood from in-complete data via the EM algorithm. J. Roy. Stat. Soc. B **39**, 1–38 (1977)
10. Van Leemput, K., Maes, F., Vandermeulen, D., Suetens, P.: A uni-fying framework for partial volume segmentation of brain MR images. IEEE Trans. Med. Image. **22**(1), 105–119 (2003)

Automatic Silkworm Egg Counting Mechanism for Sericulture

Rupali Kawade, Jyoti Sadalage, Rajveer Shastri and S. B. Deosarkar

Abstract Sericulture is an art of rearing silkworm for the production of cocoons, which is the raw material for the production of silk. The silkworm seed production is one of the important activities of sericulture in which the silkworm seed known as Disease Free Layings (DFLs) are prepared in their centers and supplied to the farmers for rearing. It is very important to count the number of silkworm eggs accurately so that farmers can pay accordingly and they should not suffer a loss. In order to generate some statistics, the fecundity and hatching percentage is measured by counting silkworm eggs. This counting is usually performed in a manual, visual, and non-automatic form, which is erroneous and time-consuming. This work approaches the development of automatic methods to count the number of silkworm eggs using image processing, particularly color segmentation and mathematical morphology.

Keywords Sericulture · DFLs · Image processing · Mathematical morphology

Introduction

No other fabric has fascinated man so continuously over millennia as silk. It is royal in its splendor, exotic, and sensuous in its radiance. Sericulture is essentially a village-based industry in which silkworms are reared for the production of

R. Kawade (✉) · J. Sadalage · R. Shastri · S. B. Deosarkar
Vidya Pratishthan's College of Engineering, Baramati, Pune, Maharashtra, India
e-mail: rupali2118@gmail.com

J. Sadalage
e-mail: jyotika2k1@gmail.com

R. Shastri
e-mail: rajveer_shastri@yahoo.com

S. B. Deosarkar
e-mail: sbdeosarkar@yahoo.com
URL: http://www.vpcoe.org

S. Sathiakumar et al. (eds.), *Proceedings of International Conference on Internet Computing and Information Communications*, Advances in Intelligent Systems and Computing 216,
DOI: 10.1007/978-81-322-1299-7_12,

cocoons. The cocoon is the raw material for the production of silk. India has the unique distinction of being the only country producing different kinds of silk. The larva of mulberry silk moth is a domesticated form which grows feeding mulberry leaves. The silkworm seed production is one of the important activities of sericulture in which DFLs are prepared in their centers known as grainages and supplied to the farmers for rearing. In sericulture, the demand of eggs for rearing silkworms is not uniform throughout the year. When seed cocoons are available in plenty of favorable seasons, surplus quantity of eggs are prepared and stored in cold storage to release at the time of demand [1].

In silk production, the number of silkworms required for a particular plantation of mulberry trees, should be approximate one for good yield of silk, so that mulberry leaves will not be wasted. For this farmers must purchase approximate number of eggs from grainages. Variability of egg quantities laid on sheets during production can cause economic losses. In addition quality control measures to monitor the egg numbers is tedious and laborious. While selling the silkworm eggs for rearing it is necessary to count the silkworm eggs accurately. This counting also determines the fecundity and hatching percentage of silkworm rearing.

The conventional method for the silkworm egg counting is by using ink/sketch pen. The transparent paper is put on the egg sheet and eggs from one DFL are counted by marking it using sketch pen. Each DFL approximately contains 450–550 eggs. Eggs from all DFLs on sheet are calculated by multiplying number of eggs from one DFL by total number of DFLs on sheet. This is usually performed manually which can be erroneous and time consuming. Karnataka State Sericulture Research and Development Institute, Bangalore (India) has implemented a calculator for egg counting. A simple pocket calculator is modified into a egg counter using small probe to count the silkworm eggs. The probe is attached to a pen, which will be pressed gently against each egg. As a result, the counted number will be marked and at the same time calculator will record the number. This avoids remembering of the counted egg number and concentrate only on marking. This improves the accuracy of counting and efficiency. This technology can be adopted in grainages, CRCs and seed a multiplication center which reduce the error during the egg counting and increases the quality parameters. The main disadvantage is that it puts the pressure on eggs because of which the embryo in an egg may get harm, and because of it the hatching efficiency decreases. This is also time-consuming [2].

Image processing techniques are being used frequently to count objects, orient pieces, or discriminate between objects with different visual characteristics. Most automated image systems perform counting by segmenting the object to be counted from background by applying a threshold based on the pixel intensity and/or intensity slope (or rate of change of intensities). Using this methodology, automated imaging systems have been developed to count mosquito eggs [3], [4], to count the objects in a video [5], to count number of feeder fish [6]. In addition, there are several commercial software programs that use this methodology to count objects given a digitized image. Pearson et al. have proposed method for counting insect eggs by image analysis [7]. Haouari and Chassery proposes a method to detect schistosome eggs in a microscopic environment connected to a computer. The advantage of this method is the use of a simple two pass algorithm. The first pass detects and analyses the elementary cases of isolated objects. The second pass introduces an enhancement process to improve the classification ratio [8]. Liao et al. have presented pelagic egg identification system by image analysis. Some efficient techniques have been developed such as the edge detection method using human visual characteristics, the improved Hough method for circle extraction and the embryo segmentation method [9]. These object counting systems can work well if the objects are only one layer thick and have good contrast from background. These algorithms are experimented for counting the silkworm eggs for sericulture.

This paper is organized as follows: next section describes the images acquired and the algorithms developed to perform automatic counting of silkworm eggs. In Section "Discussion and Result," the results are related and analyzed. Section "Conclusion and Future Scope" concludes and describes the future scope.

Materials and Methods

A digital camera with 12 megapixels resolution and 4.5 times optical zoom is used. The camera is connected to personal computer. A digital image captured from camera of size 1600 versus 1,200 pixels is processed. The amount of eggs is counted from each DFL by visual inspection to compare with count obtained from image processing algorithms experimented. Image processing algorithms and GUI implemented to display the results to user in MATLAB.

The explored algorithm evaluated is based on color segmentation and mathematical morphology. Figure 1 shows the sample image of one DFL which actually consist of 585 eggs. A color digital image of size 1,600 × 1,200 is converted to gray image as RGB color models are not well suited for describing the colors it terms that are not well suited for human interpretation[10].

This conversion of RGB image to grayscale is carried out by eliminating the hue and saturation information while retaining the luminance. The applied transformation is as per equation (1)

$$\text{Grayscale} = (0.2989 \times \text{R}) + (0.5870 \times \text{G}) + (0.1140 \times \text{B}) \tag{1}$$

where R, G, and B are the values of the red, green, and blue components.

The image shown in Fig. 2 has a uniform background but is now a bit too bright. The contrast of gray image is enhanced by mapping the values in intensity image to new values such that 1 % of data is saturated at low and high intensities of original image. This increases the contrast as shown in Fig. 3 .

Figure 3 is binarized using Otsu's method which uses a global threshold to convert the intensity image to binary image[10]. Otsu's method is used to find out

Fig. 1 Sample of DFL

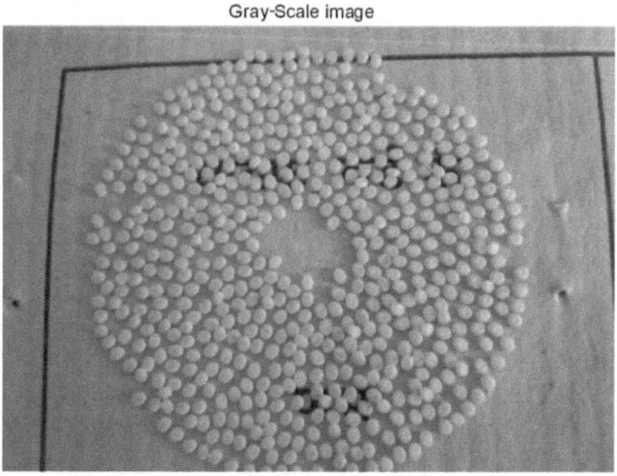

Fig. 2 Gray scale image

the global threshold level to minimize the intraclass variance of thresholded black and white pixels. The applied transformation is as per equation (2)

$$\begin{aligned} S(x,y) &= 1\ldots\text{if } r(x,y) > \text{ threshold} \\ &= 0\ldots\text{if } r(x,y) < \text{ threshold} \end{aligned} \tag{2}$$

where $r(x, y)$ represents the pixels of original image and $S(x, y)$ represents the pixels of thresholded image. Figure 4 shows the result of thresholding. In this image, the background illumination is brighter in the top right corner of the image than at the bottom left corner. This is the effect of nonuniform illumination. Hence, this gray image shown in Fig. 3 is filtered using TopHat Filtering. An important use of top-hat transformation is in correcting the effects of nonuniform illumination. It performs morphological opening on grayscale or binary image with structuring element. Morphological opening operation is erosion followed by dilation using same structuring element for both operation.

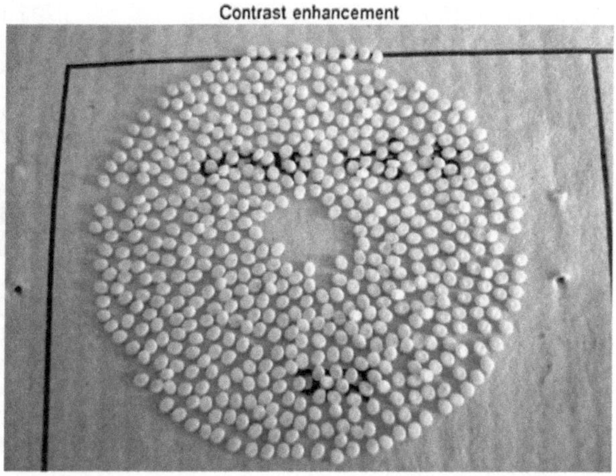

Fig. 3 Contrast enhancement

Fig. 4 Thresholding before applying top-hat filtering

A morphological opening operation is used to estimate the background illumination.

$$\text{TopHat filtering} = \text{Image} - \text{Opening of image}$$

The opening operation has the effect of removing objects that cannot completely contain the structuring element. After applying top-hat filtering, we get the better results for thresholding as shown in Fig. 5.

Small areas can be deleted as it could not contain an egg. This experiment has defined that every area with less than 100 pixels should be deleted. Figure 6 shows the result of removing background noise due to which unwanted small objects from background are removed. Figure 6 shows there are some eggs which overlap with one another. To separate them, erosion is applied which reduces the size of each egg. Erosion is performed with the disk shaped structuring element so that

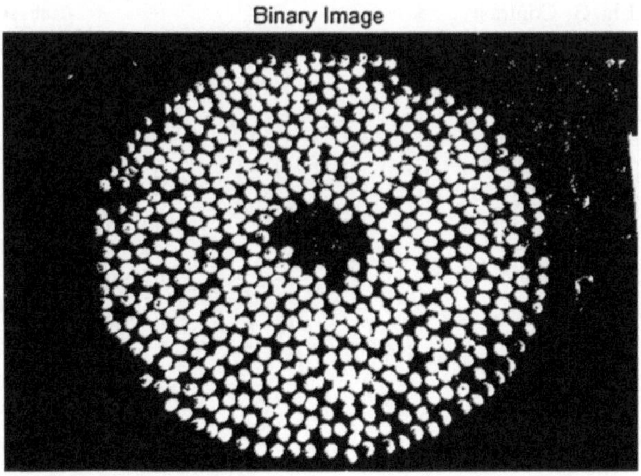

Fig. 5 Thresholding after applying top-hat filtering

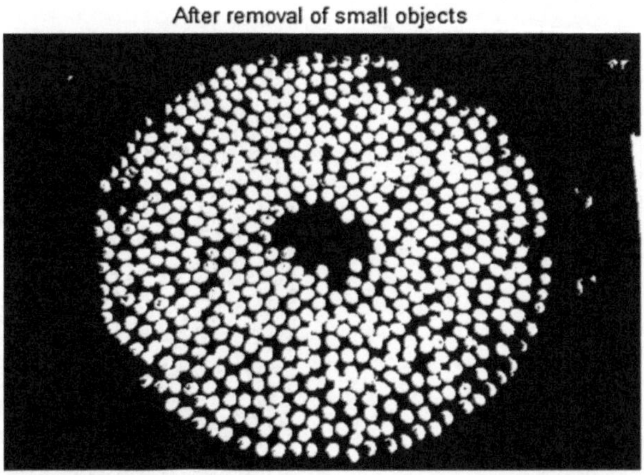

Fig. 6 After removal of small objects

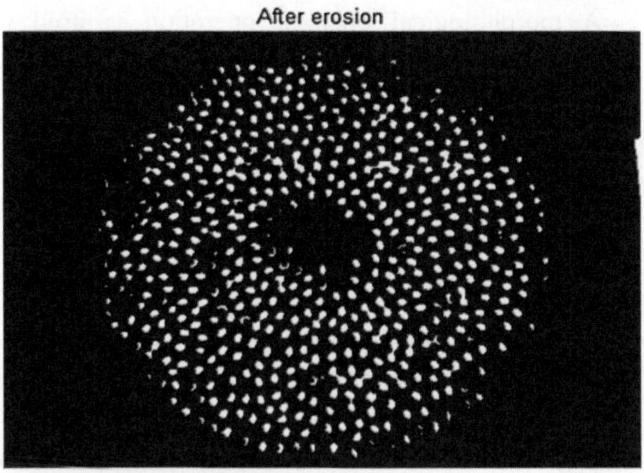

Fig. 7 After erosion

Table 1 Counting results using proposed method

Image	Correct amount of eggs	Estimated amount of eggs by proposed algorithm
1	585	582
2	622	627
3	541	540
4	695	684
5	525	532
6	518	536

some of the connected regions get eliminated. Figure 7 shows the result of erosion due to which some of connected eggs are removed.

With the binary image, a connected components algorithm is applied to label the connected regions of the image. This algorithm puts a different label at each connected white area of the image. With this labeling, it is possible to evaluate each connected area. This gives the number of connected objects found in image which represents the number of eggs in image.

Discussion and Result

Table 1 presents the results of the explored algorithm applied to another five sample images with different count. The image labeled as '1' in Table 1 is the image previously presented in Fig. 1 with 585 eggs.

From the results given in Table 1, the overall accuracy for method is 93.31 %. These algorithms were tested for 19 different images.

Conclusion and Future Scope

The algorithm evaluated is simple and efficient. It is sensitive to the nonuniform illumination. If illuminating pattern varies significantly, this method does not show the accurate count of eggs. Using image processing techniques, we can find the number of eggs present in an image. This reduces the time of counting significantly. Accuracy of results depends upon size of objects, whether or not any object are touching (in which case they might be labeled as one object), accuracy of approximated background and the connectivity selected.

This work can be extended to count the number of eggs from sheet containing many DFLs. A dedicated hardware can be implemented to count the eggs using programmable logic devices such as FPGA and CPLD which can give rise to higher speed of computation. Web-based counting mechanism can also be implemented by using proposed method.

Acknowledgments This research is supported by BAIF research foundation. The authors are grateful to Dr. Sinha, Dr. Hugar, and Mr. Murkute, sericulture Central Research station, Urulikanchan (India) for their contribution in making database of different images for evaluation.

References

1. http://www.tnsericulture.gov.in/prototype2/briefnoteonesriculture.htm
2. http://kssrdi.org/technology/technology_view.asp?id=303
3. Mello, C.A.B., dos Santos, W.P., Rodrigues, M.A.B., Candeias, A.L.B., GusmãoImage, C.M.G.: Segmentation of Ovitraps for automatic counting of Aedes Aegypti Eggs. Paper presented at 30th annual international IEEE EMBS Conference Vancouver, British Columbia, Canada, August 20–24, 2008
4. Gusmão, G., Machado, S.C.S., Rodrigues, M.A.B.: A new algorithm for segmenting and counting Aedes aegypti Eggs in Ovitraps. Paper presented at 31st annual international conference of the IEEE EMBS Minneapolis, Minnesota, USA, 10
5. Pornpanomchai, C., Stheitsthienchai, F., Rattanachuen, S.: Object detection and counting system. In Proceedings of 2008 congress on image and signal processing. (2008)
6. Toh, Y.H., Ng, T.M., Liew, B.K. Automated fish counting using image processing. 978-1-4244-4507-3/09/$25.00 ©2009 IEEE
7. Pearson, T.C., Edwards, R.H., Mossman, A.P., Wood, D.F., Yu, P.C., Miller, E.L.: Insect egg counting on mass rearing oviposition pads by image analysis. Appl. Eng. Agric. **18**, 129–135. (2002). American society of Agricultural Engineers ISSN 0883-8542
8. Haouari, A., Chassery, J.M.: A two pass labeling algorithm for automatic schistosome egg detection and counting. CH2614-6/88/0000/0827$01.00 0 1988 IEEE. (1998)
9. Liao, Q., Chehdi, K., Lin, X., Zhang, Y.: Identification of pelagic eggs by image analysis. In: Proceedings of ICSP ‘ 96
10. Gonzalez, R.C., Woods, R.: Digital Image Processing, 3rd ed. Prentice-Hall, (2007)

Proficient Energy Consumption Algorithm Using HMAC and ANT Colony-Based Algorithm

G. Saravanan, J. V. Anand and M. Roberts Masillamani

Abstract In this research we deal with Ant colony-based clustering algorithm (BACCA) for wireless sensor networks which uses low energy adaptive cluster hierarchy (LEACH) as its prototype a good approximation of a proactive network protocol in the case of adaptive routing. Based on this we estimate the amount of energy consumed by these in each case then we use hybrid medium access control (HMAC) used along with TDMA Multiple hop in a single medium access control, end-to-end quality of service, Latency sensitive traffic flows, and high priority channel access. It is shown by computer simulation that these mechanisms result in a significant improvement in energy consumption. Similarly in the final half we do consider the bandwidth utilization using (SFF) and dynamic first fit (DFF) by which we minimize the bandwidth which is a constrain for many user in the case of adaptive routing.

Keywords BACCA · LEACH · HMAC

G. Saravanan (✉)
Department of Electronics and Communication Engineering, Sri Sairam Institute of Technology, Chennai, India
e-mail: saran_keerthana@yahoo.com

J. V. Anand
Department of Electronics and Communication Engineering, A C T College of Engineering and Technology, Kancheepuram District, Chennai, India
e-mail: jvanand87@gmail.com

M. Roberts Masillamani
Department of Computer Science and Engineering, Hindustan Institute of Technology and Science, Chennai, India
e-mail: deancs@hindustanuniv.ac.in

S. Sathiakumar et al. (eds.), *Proceedings of International Conference on Internet Computing and Information Communications*, Advances in Intelligent Systems and Computing 216, DOI: 10.1007/978-81-322-1299-7_13, © Springer India 2014

Introduction

In this paper, we deal with energy consumption for that purpose we consider low-energy adaptive clustering hierarchy (LEACH). Ant colony-based clustering algorithm (BACCA) for wireless sensor network which precedes as if (in ant colony optimization (ACO)), a colony of artificial ants is used to construct solutions guided by the pheromone trails and heuristic information. ACO was inspired by the foraging behavior of real ants. This behavior enables ants to find the shortest paths between food sources and their nest. Initially, ants explore the area surrounding their nest in a random manner. As soon as an ant finds a source of food, it evaluates the quantity and quality of food and carries some of this food to the nest. During the return trip, the ant deposits a pheromone trail on the ground. The quantity of pheromone deposited, which may depend on the quantity and quality of the food, will guide other ants to the food source. The indirect communication between the ants via the pheromone trails allows them to find the shortest path between their nest and food sources. This functionality of real ant colonies is exploited in artificial ant colonies in order to solve optimization problem in wireless sensor networks in BACCA [1] where adaptive routing is being carried out to find with minimum consideration in bandwidth with the help of static first fit and dynamic first fit [2].

Existing System

Clustering protocols have been investigated as either stand-alone protocols for ad-hoc networks [3, 4] or in the context of routing protocols. Routing protocols can also employ clustering. Clustering was proposed as a useful tool for efficiently pinpointing object locations. Many protocols proposed in the literature minimize energy consumption on routing paths while these approaches increase energy efficiency (GAF, SPAN, and ASCENT) they do not necessarily prolong network lifetime. So we do consider some scenarios. Clustering can be a side effect of other protocol operations. For example, in topology management protocols, such as GAF SPAN and ASCENT, nodes are classified according to their geographic location into equivalence classes.

Proactive Network Protocol

At each time when there is a change in topology, once the wireless sensor network heads are decided, the wireless head broadcasts the following parameters:

Report time (T_R). This is the time episode between consecutive reports sent by a node.

Attributes (A). This is a set of physical parameters which the user is interested in obtaining data about.

At every *report time*, the wireless sensor members sense the parameters specified in the attributes and send the data to the cluster- head. The wireless sensor head aggregates this data and sends it to the base station or the higher level cluster-head, as the case may be. This ensures that the user has a whole picture of the complete area covered by the network. For every change time in the network, $\mathbf{T_R}$ and $\boldsymbol{A}$ are transmitted afresh and so, can be changed.

Proposed Work

Low Energy Adaptive Cluster Hierarchy

Low-energy adaptive clustering hierarchy (LEACH) is a good approximation of a proactive network protocol, with some minor differences.

Once the clusters are formed, the cluster heads broadcast a TDMA schedule giving the order in which the cluster members can transmit their data. The total time required to complete this schedule is called the *frame time* TF. Every node in the cluster has its own slot in the frame, during which it transmits data to the cluster head. When the last node in the schedule has transmitted its data, the schedule repeats. The *report time* discussed earlier is equivalent to the *frame time* in LEACH. The *frame time* is not broadcast by the cluster head, though it is derived from the TDMA schedule. However, it is not under user control. Also, the attributes are predetermined and are not changed midway [5].

BACCA Algorithm

According to the above analysis, we are ready to get the BACCA algorithm: []

1. first round, the wireless sensor heads and the number of wireless sensor heads are determined by LEACH.
2. Those new wireless sensor heads broadcast the wireless sensor head message to all wireless sensor members around them.
3. the wireless sensor members receives the message from the wireless sensor head then manipulates according to which it chooses to join the cluster head or not.
4. After all data transmission has been finished, wireless sensor heads in this round compute their wireless sensor members' distance and energy pheromones according to distance pheromone or energy pheromone and, then compute these nodes' value of wireless sensor head selection function according to and choose the wireless sensor head for next round.
5. go to step (2) for next round of the loop until all wireless sensor members are exhausted.

Data Link Layer Affects TCP Performance and Recovery Algorithms Used

TDMA is allocated based on slotted aloha basis here conflict arises for scheduling. Negative acknowledgement and retransmission occurs at the MAC layer it calculates bit error rate, packet loss ratio based on per hop basis and energy consumption is being calculated. Finally sleep wake scheduling algorithm is being carried on [1] (Fig. 1).

Hybrid Medium Access Control

To overcome the difficulties of data link layer, we use HMAC. Previously, schedule-based algorithm was used it had a central node to manage and broadcast the schedule to other nodes but it exposed failures to entire network itself. So we proposed contention based approach since it can be used in a distributed fashion. It

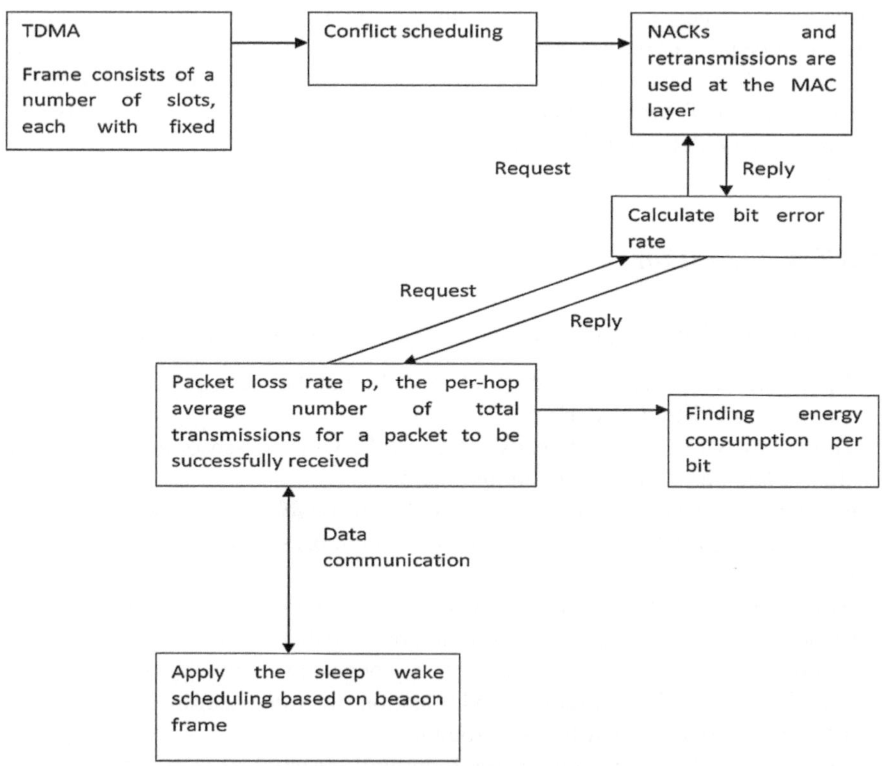

Fig. 1 Data link layer affects performance and recovery algorithm model

had drawbacks like idle listening period to overcome this we made sleeping mode (Fig. 2).

Contention based allows many users to access the same radio channel without precoordination

1. Each node turns on its radio during its own wake up slot and sleeps during all other wake up slots.
2. Each sender randomly picks up a data slot and announces the data slot number along with the receiver node identifier via a wake up message in the receiver wake up slot.
3. Upon reception of a WAKE UP message a node checks the embedded node identifier in the wakeup message. If its intended receiver, then the node turns on its radio for the incoming data packets for the specified data slot otherwise it just sleeps.

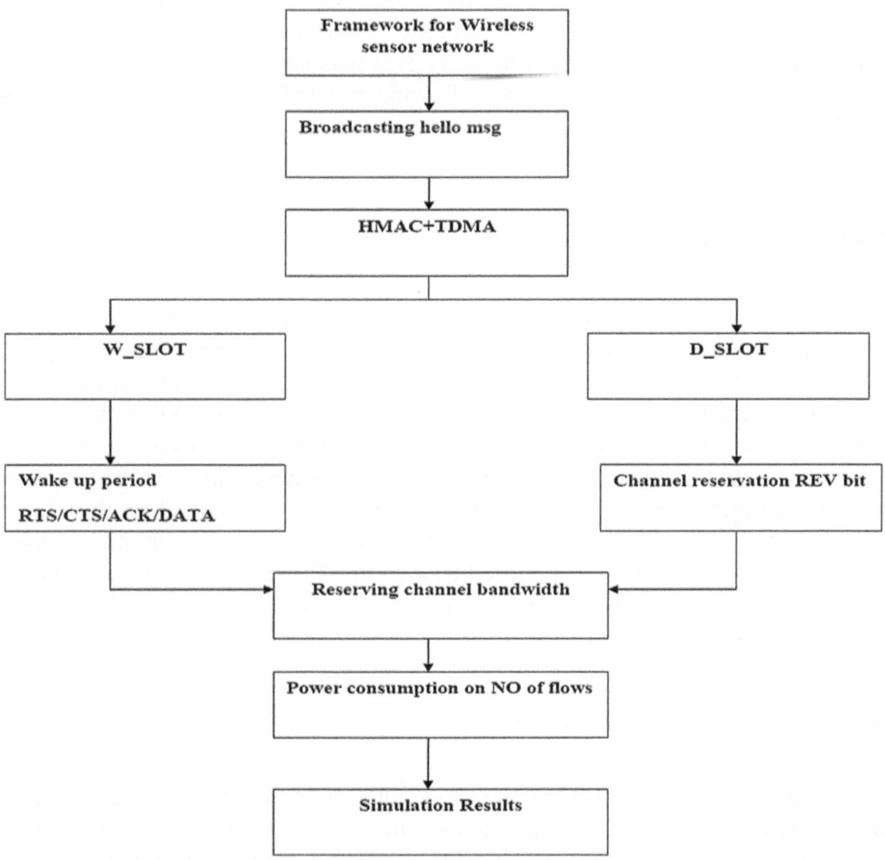

Fig. 2 Flowchart for HMAC

4. If any collision occurs in a node's wakeup slot, then the node turns on its radio for duration long enough to receive an RTS packet at the beginning of each data slot for a possible incoming data packet. If the node learns that is the intended receiver from the received RTS message, and then it keeps the radio on to receive the data packet; otherwise, it returns to sleep in the remaining period of the data slot. This way, a node can minimize the extra energy cost under such a situation.
5. In each data slot, unicast data transmission must follow the well-known RTS/CTS/DATA/ACK scheme in IEEE802.11 to avoid the "hidden terminal problem," since two senders may choose the same data slot to send data to their receivers at the same time, and the transmissions happen to be in a common interference range.

HMAC also provides support for one-hop broadcast operation. When a node has data to broadcast, it sends out a "WAKEUP" message containing a broadcast address and a data slot in each wakeup slot. After receiving such "WAKEUP" messages, all neighbors will wake up in the same data slot to receive the broadcast message. In addition, different from unicast operation, when the sender finds an idle channel during the CS period in the chosen data slot, it will immediately send out the broadcast message without following the before mentioned RTS/CTS/DATA/ACK scheme.

Static First-Fit Algorithm

Static first-fit (SFF) sorts proxies and objects in ascending order of their bandwidth-space ratios. Each object (in ascending order) is assigned to the first proxy (also in ascending order) that has sufficient space to store this object. If this proxy has sufficient bandwidth to service this object, the corresponding amount of bandwidth is reserved at the proxy, and the object is removed from the un-stored object pool. On the other hand, if the proxy does not have sufficient bandwidth to service the object, the available bandwidth at the proxy is reserved for this object. The object is returned back into the un-stored object pool with the reserved bandwidth subtracted from its required bandwidth. The proxy is removed from the proxy pool since all of its bandwidth has been consumed [2].

Dynamic First-Fit Algorithm

DFF is similar to SFF, except that the bandwidth-space ratio of a component proxy is recomputed after an object is placed onto that proxy and proxies are resorted by their new bandwidth-space ratios (in SFF, the ratio is computed only once, at the beginning). The intuition behind DFF is that the effective bandwidth-space ratio of a proxy changes after an object is cached, and re computing this ratio may result in a better overall placement.

Results and Discussion

Design Implemented

Design is implemented using

1. Topology: wireless
2. Scheduling algorithms: Drop tail
3. Transport protocols: TCP, UDP
4. Routing: static and dynamic routing.

Test Cases

1. Channel – wireless channel
2. Propagation- Two ray ground propagation model
3. Network interface type – Physical/wireless physical
4. MAC type- 802.11
5. Interface queue type –CMU Pri Queue
6. Link layer type – logical link
7. Antenna model – Omni directional antenna
8. Number of mobile nodes- as desired
9. Routing Protocol- DSR and AMODV
10. Topography-640 *800
11. Data traffic type – CBR.

Topology for the Networks

Figs. 3, 4

Comparison of Performance Metric Between LEACH HMAC and BACCA

See Figs. 5, 6, 7, 8, 9, and 10

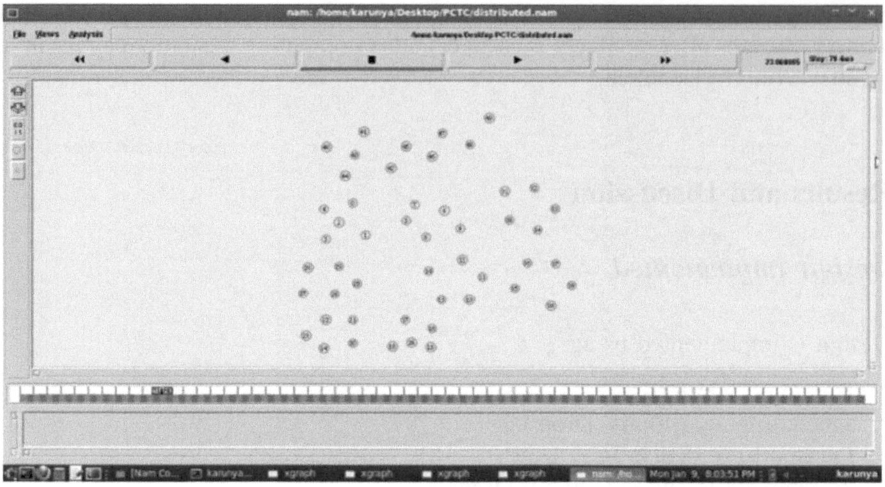

Fig. 3 We create 50 nodes initially

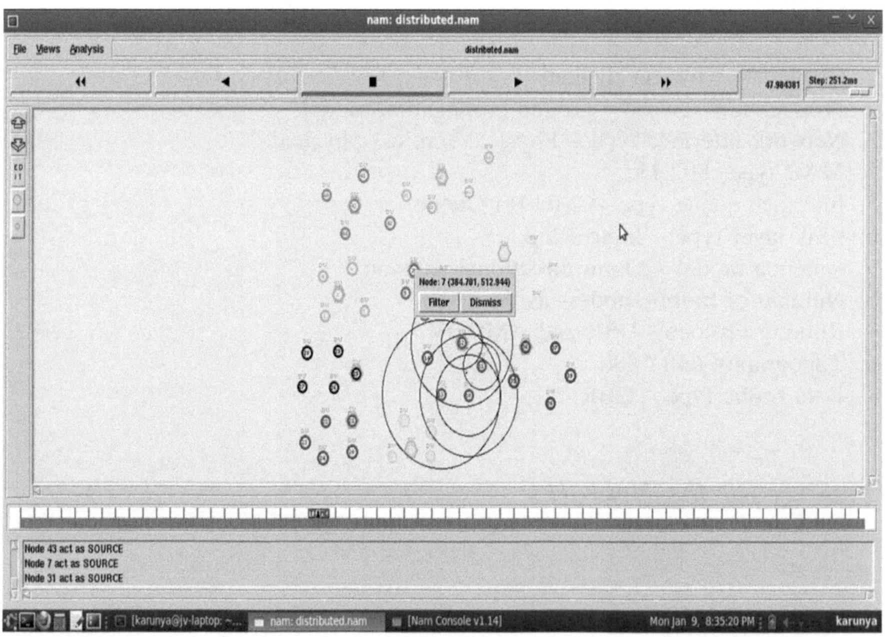

Fig. 4 There are 10 Source nodes and 40 destinations nodes in the source nodes are in the form of hexagon

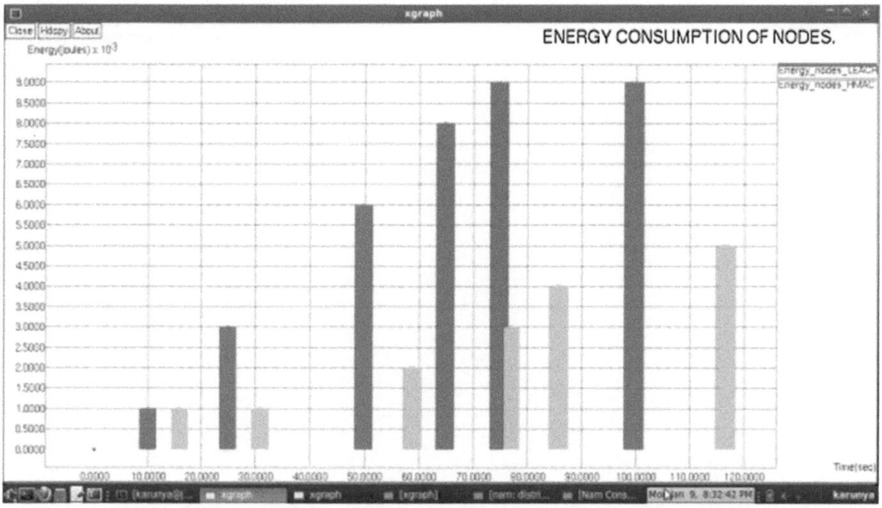

Fig. 5 It shows the energy consumption achieved using HMAC. This measures the energy expended per delivery data packet. It is expressed as

$$\frac{\Sigma \text{ ENERGY EXPENDED BY EACH NODE}}{\text{TOTAL NUMBER OF PACKETS DELIVERD}}$$

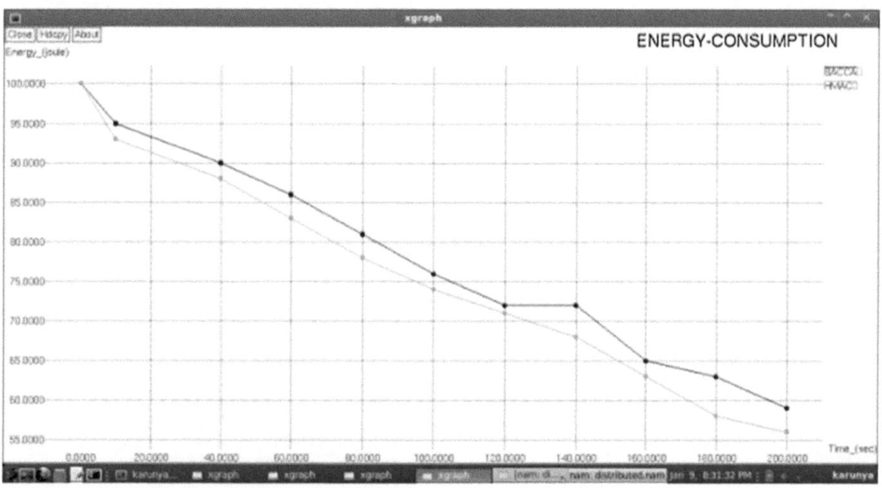

Fig. 6 It shows the overall energy consumption of the network. It is calculated by using the formula given from which we infer considerable amount of energy is consumed using HMAC

$$\frac{\Sigma \text{ ENERGY CONSUMED BY THE NETWORK}}{\text{TOTAL NUMBER OF PACKETS DELIVERED}}$$

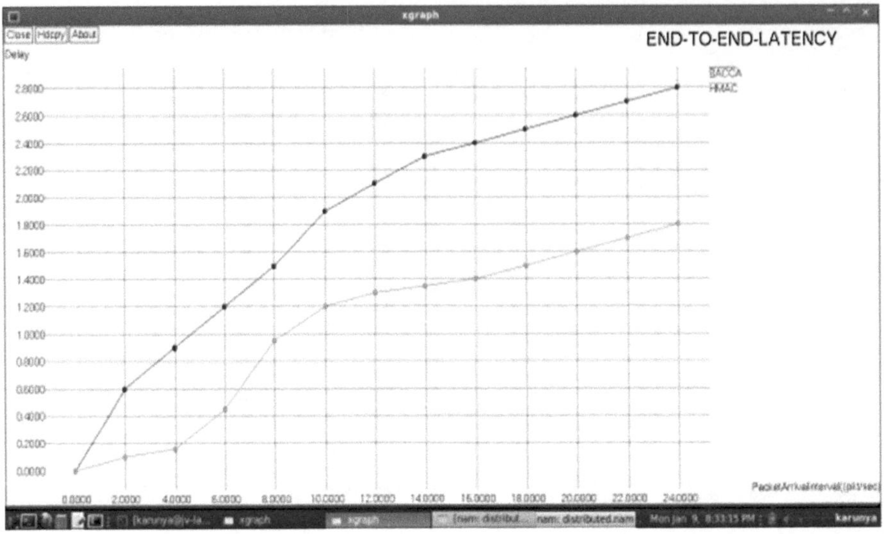

Fig. 7 It shows the overall end-to-end delay obtained by calculating Ending time – Starting time due to channel noise node mobility is significantly reduced in the overall network than HMAC

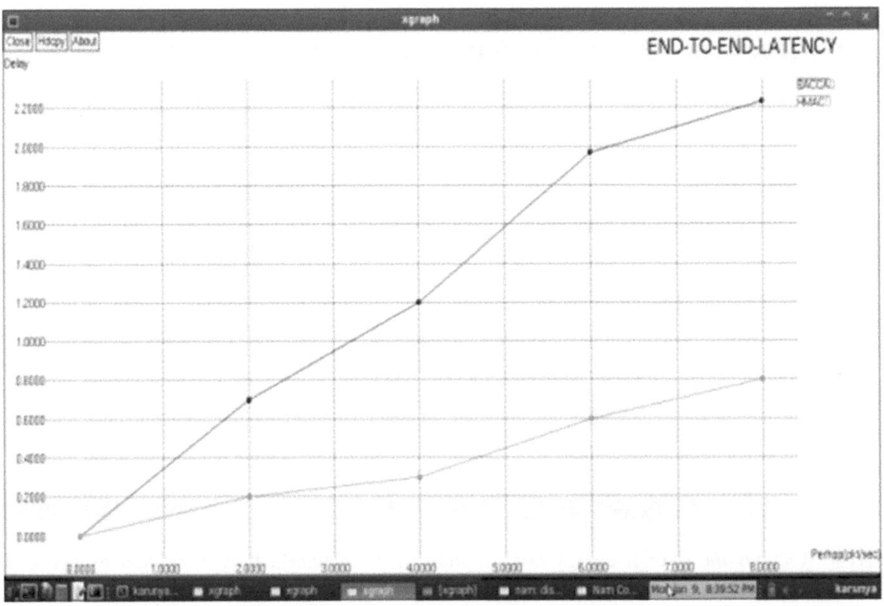

Fig. 8 The graph shows the one hop neighbor preceded by its following neighbors with delay taken in milliseconds and per hop taken in seconds

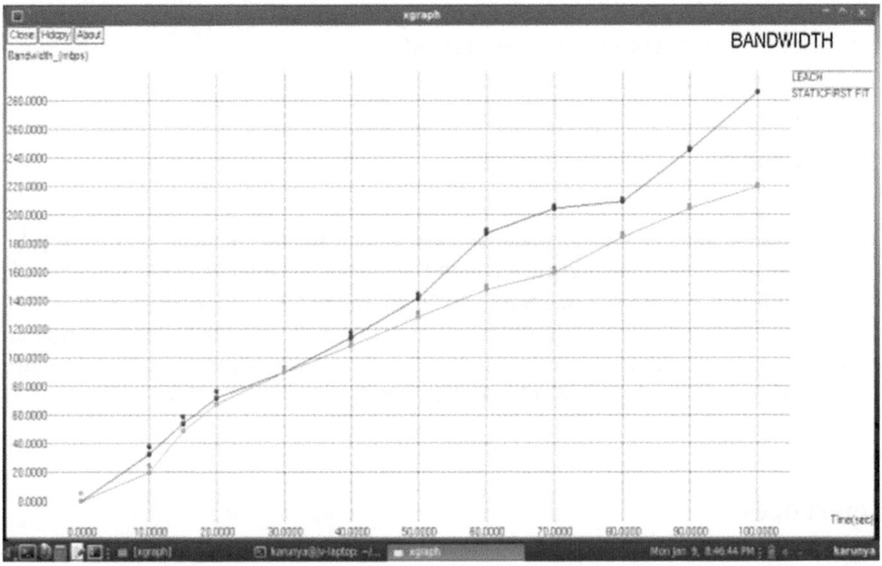

Fig. 9 It shows the overall bandwidth obtained by calculating the (data capacity of a link). Static first fit is reduced considerable than with the LEACH

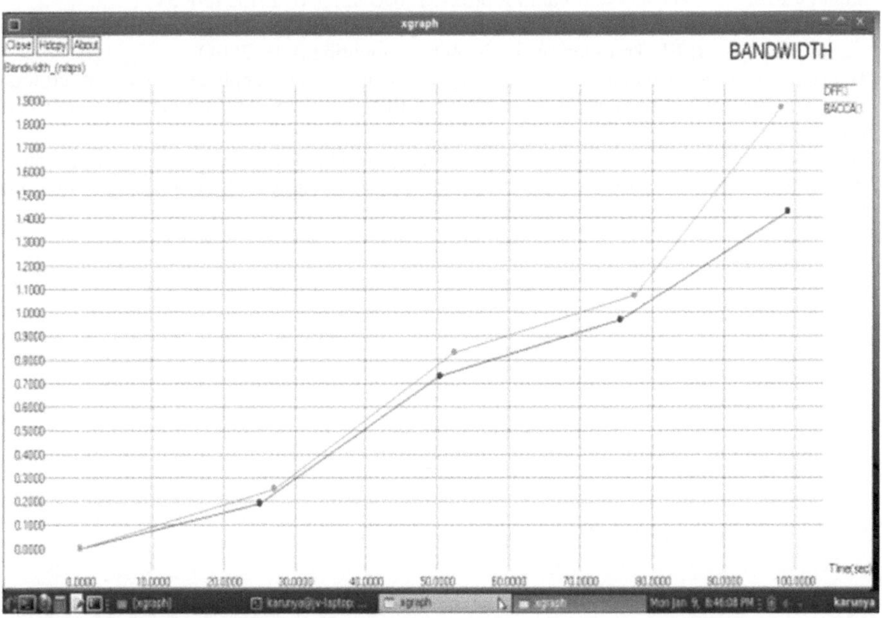

Fig. 10 shows the overall bandwidth obtained by calculating the (data capacity of a link). Static first fit is reduced considerable than with the LEACH

It is calculated by using the above formula from which we infer considerable amount of energy is consumed using HMAC.

Conclusion

Thus, in this research, we do optimize the performance of energy consumption with the help of various algorithm using network simulator 2.34, we do infer consider amount of energy is being reduced in HMAC, than in BACCA, and LEACH. In terms of bandwidth we investigate that SFF and DFF yields significant bandwidth reduction than the rest. Finally, we do consider delay in which we infer HMAC plays a vital role even with reduced delay.

References

1. Jiang, T., Zang, W., Zhao, C., and Shi, J. An energy consumption optimized clustering algorithm for radar sensor networks based on an ant colony algorithm. EURASIP J. Wirel. Commun. and Netw. (2000)
2. Guo, Y., Ge, Z., Urgaonkar, B., Shenoy, P., Towsley, D.: Dynamic cache reconfiguration strategies for a cluster-based streaming proxy. Proceedings of IEEE (2003)
3. Younis, O., Fahmy, S.: HEED: A hybrid, energy-efficient Distributed clustering in Ad-hoc sensor networks, IEEE Transactions on Mobile Computing **3**(4) (2004)
4. Heinzelman, W.R., Chandrakasan, A., Balakrishnan, H.: An application-specific protocol architecture for wireless microsensor networks. IEEE Trans. Wirel. Commun. **1**(4), 660–670 (October 2002)
5. Heinzelman, W., Chandrakasan, A., Balakrishnan, H.: Energy-efficient communication protocols for wireless microsensor networks. In: Proceedings of Hawaiian International Conference on Systems Science, (2000)

Secure Remote Access Fleet Entry Management System Using UHF Band RFID

N. Sathish and P. Ranjana

Abstract A Fleet management system based on UHF and RFID is proposed. This system is applied to a vehicle entering/leaving at the road gates. The system consists of a RFID tag present in the fleet, reader antenna, a reader controller, and the monitoring and the commanding software. The whole system sits on open source platform. The java code is used for controlling the platform. The band width of the UHF expands from 300 MHZ to 3 GHz which helps the system to detect the incoming fleet at distance of 30 m around the antenna even when the fleet travels at a speed of 45 km/h.

Keywords RFID · UHF

Introduction

Radio frequency identification (RFID) is an emerging technology with applications in several areas, from logistics to security. Usually, RFID cards have been used for people access control in office buildings, for public transportation billing, and even to store digital information in passports.

Active tags are normally used in cars, buses, and trucks because of their long reading range and reliability. However, the use of long range RFID for control access poses a serious problem: when the reader's detection range is difficult to control, it is very easy to make the reader presenting a wrong reading, and granting access without a valid request.

N. Sathish (✉) · P. Ranjana
School of Computing Sciences, Hindustan University, Chennai, Tamil Nadu, India
e-mail: nsathish.prs@gmail.com

P. Ranjana
e-mail: pranjana@hindustanuniv.ac.in

S. Sathiakumar et al. (eds.), *Proceedings of International Conference on Internet Computing and Information Communications*, Advances in Intelligent Systems and Computing 216,
DOI: 10.1007/978-81-322-1299-7_14, © Springer India 2014

This may happen when the tag enters an area next to the reading range. For instance, in a highway toll plaza, the toll reader can detect a car before it crosses the gate. So, if the car does not cross the gate for some reason, it will be anyway charged.

The same happens when a vehicle must get into premises, such as parking lots or garages. The vehicle can elude the system just by passing in front of the premises entrance and leaving. The reader has a high probability to detect the car tag in such a situation and make the system consider that the car is inside the garage. The objective of this work is to present and validate an approach to reduce the chance of a reader detecting a tag before it gets to a designated place.

Existing System

There is different type of vehicle entry management system available today. One of the existing vehicle entry systems uses security guards to manage the entry and exiting of the vehicles. This is an out dated system because it is literally inefficient for a large organization which could expect at least 1,000 vehicles per day. This system is does not provide conjunction control in the gate.

The other type of existing vehicle entry systems uses an IR sensor to detect the vehicles approaching the gate and open the door of. This is particularly effective if there is no security threat in the organization. The problem with this system is that it also allow attacker along with the other vehicles.

The latest existing vehicle entry system uses an RFID of lower frequency, which allow the system to sense a vehicle up to 9 meters from the base of the antenna. The system can detect vehicles that travel below the speed of 30 km/h. The existing system alerts the security personal to check with the unauthorized vehicle. The RFID tag can be detected by the antenna only when the RFID tag and antenna is placed in a precise position on the windshield.

Proposed Implementation

The proposed fleet entry management system uses ultra high frequency-based radio frequency identifier [1] that is more powerful, which will allow the system to sense the fleets arrival to a range of around 30 m, even with a speed of 45 km/h.

The proposed system also uses short range reader to find the fleets that are identified by the long range reader enter the premises. Only if the short range reader detects the fleets RFID, the gate will be opened.

The proposed system uses ethernet interface serial connection to communicate with the unauthorized person via vocal and video communication. This system can handle more number of vehicles that any existing vehicles management system can't. This system automates the work of checking and registering the vehicles which enters and exits the premises.

This system uses a video camera to take picture of the unauthorized persons for future reference. This system can be easily fitted in any type of vehicles which are allowed to enter the premises. With this system in place we can make the segregate the parking facility for each type of authorization level and type of vehicles.

Proposed Design

System Overview

The software product can be used to secure the fleet entry. The application is divided into five phases and each module has its own functions. The central feature of this project is the automation provided which is not available in the existing system. The application basically consists of the following phases:

1. Client to server and server to client communication.
2. Communication between hardware and clients.
3. Microcontroller program for reading RFID reader.
4. Microcontroller program to control and open gates.
5. Microcontroller program to control in and out sensor.

The working of these phases depends on the type of database and other such requirements. Automation is achieved by the proper functioning of these phases. Thus, the project will save a lot of time during the synchronization process and emphasizes focus on the main thing – design and development – by taking the tiring routine upon itself.

Phases of the System

Figure 1 shows the system framework. The personal computer and RFID controller are placed in guard room and they are connected by AT89C51.

Two gates are for vehicle entering and leaving control, two reader antennas are placed at high positions for broader view angle, two optionally monitoring cameras could placed near road side for capturing the vehicle's images. The gates and the monitoring cameras are controlled by I/O interface of personal computer. The reader antennas are connected with RFID controller by coaxial cables for less attenuation transmission. The length of coaxial cables depends on the coaxial cable's attenuation value. Normally, at least 10 m is required for the implementation.

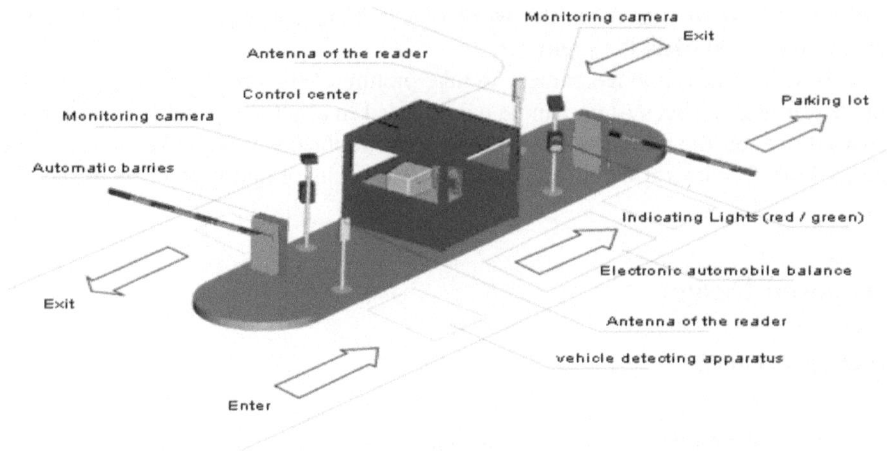

Fig. 1 System framework

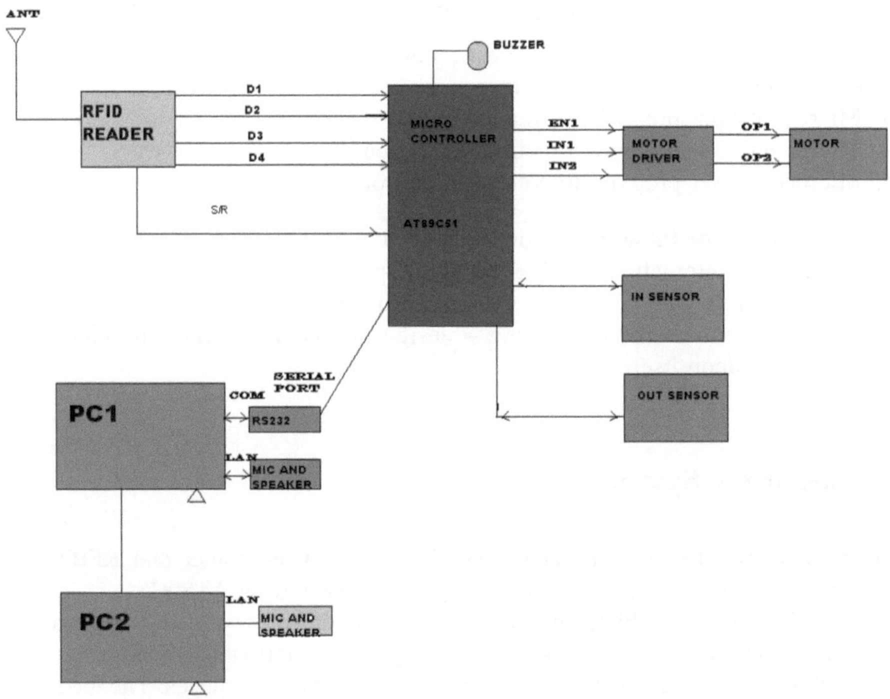

Fig. 2 Architecture of the system

Figure 2 shows the architecture of the system. The microcontroller "AT89C51" integrates the RFID reader with the computer. The microcontroller receives the signal from the RFID reader and sends the corresponding command to the software loaded in the computer.

Client-to-Server and Server-to-Client Communication

Java media framework is used to create a communication between client and server. The connectivity is done using ethernet interface serial communication [2]. To create a two-way communication media transmitter and simple audio player is used on both the client and the server.

Communication between Hardware and Clients

Java package helps to couple the hardware with the system. The microcontroller RS232 [3] converts the microcontroller data into voltage levels and vice versa. DB-9 or DB2 [4] pin connector is used to connect the hardware with system (Fig. 3).

Fig. 3 Flow diagram for a vehicle entering the premises

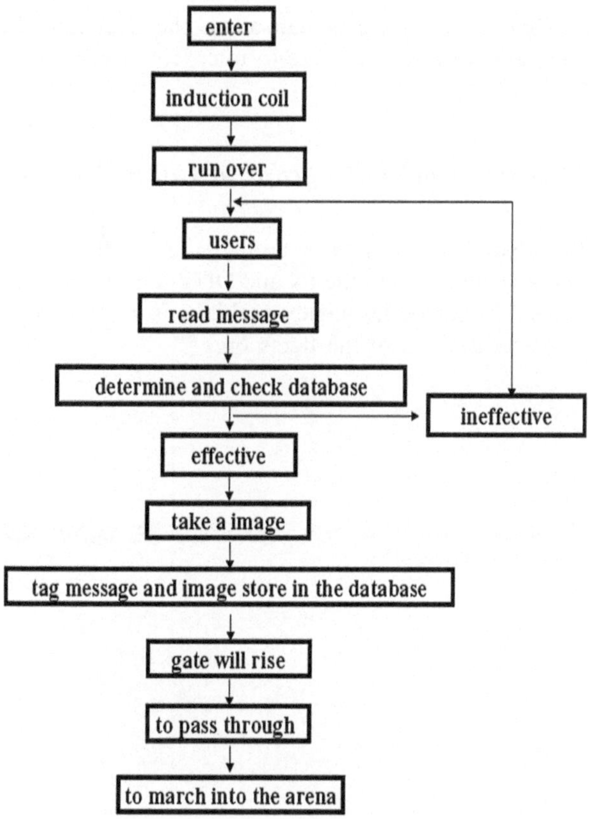

Microcontroller Program for Reading RFID Reader

RFID [5] is a technology which uses tags as a component in an integrated supply chain. A tag is a transponder which receives a radio signal and in response to it sends out a radio signal. Tag contains an antenna, and a small chip that stores a small amount of data tag is powered by the high power electromagnetic field generated by the antennas. The microcontroller is programmed in such a way that it should respond to the inputs of the RFID.

Microcontroller Program to Control and Open Gates

The microcontroller used here is "AT89C51" which is an 8051 controller [5].The microcontroller issues a command to the motor that in turn open the gate. Authorized vehicles enter the premises. The security officer issues a command to open the gates. We use a RS232 to interface the motor and the microcontroller.

Figure 4 shows the alert to the security personal when an unauthorized fleets entering into the premises. The security personal can add the RFID tag to the database if available or can allow the fleet into the premises by sending open command through software to microcontroller.

Microcontroller Program to Control In and Out Sensor

IR proximity sensor built around the TSOP 1738 module which is simple and effective and used in the IN and OUT sensor which are placed on either side of the gates. The carrier frequency of this sensor is between 32 kHz to 42 kHz. It is used to detect the flux of the fleets Fig. 5.

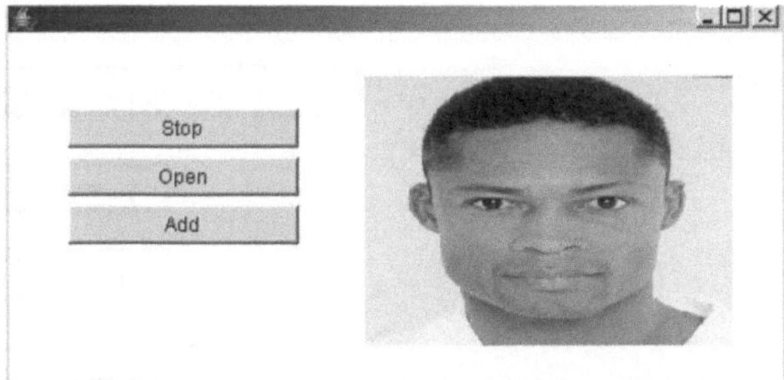

Fig. 4 The entering message displays window

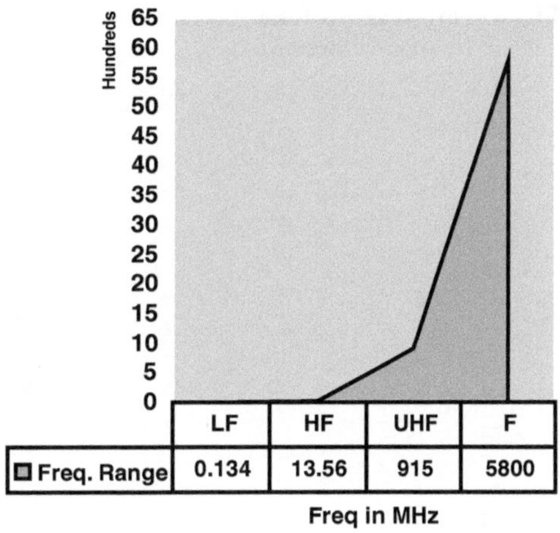

Fig. 5 Frequency graph

Dataflow Diagram

All the fleets entering the premises with the RFID tag will read by the antenna and the data will be compared with all the records from the database as shown in Fig. 3.

If the record is available in the database, then the information about the fleet and a image will be taken and stored in the entry table. Then the control will be sent to the microcontroller, which will control the open and close operations of the gate.

If the record is not available in the database, then the control will be sent to the security personal located anywhere in the premises.

Figure 6 describes the process for the fleets leaving the premises. When a RFID is received by the antenna, the tag will be read and checked in the database.

If the user is permanent, the fleet is allowed to move out from the premises. If not, then the parking pay will be deducted from the use, and then the fleet is allowed to leave the premises.

Experimental Analysis

The proposed system consists of both long range and short range readers. The long range reader detects the fleet about 30–40 m and sends to the system where the fleets unique RFID is matched with the data stored in the database.

If the RFID of the fleet is available in the database, then the system will wait for the short range RFID reader to detect the fleet. Only then the system will enter the details in the log file. Long range RFID has an accuracy rate of 80 %, whereas short range RFID has an accuracy rate of 100 %.

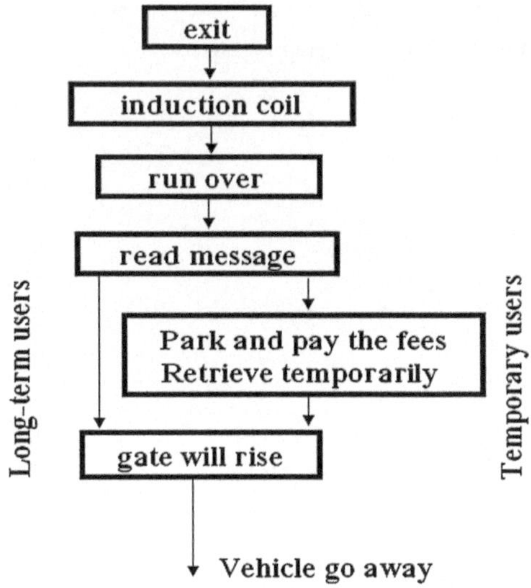

Fig. 6 Flow diagram for a vehicle leaving the premises

If the short range RFID reader did not detect the RFID tag that is detected by the long range reader, then the log will not be created for the fleet and the gate will not be opened.

Without using the short range RFID reader, the fleets that came near the entry gate may not be passed through will also be stored in the log file. Using the system, the log will be created for the vehicle that enters the gate.

The java platform has to be developed such a way that it can handle multiple requests at a time and also the request has to be in hold for some time till the short range reader detects the fleet.

Conclusion

Thus, we developed a web-based application which provides a common platform for several features. It mainly provides user to allow the vehicles IN or OUT without any security personal at the gate. It also provides the facility to chat with the unauthorized person who tries to enter the gate.

Security has been a bigger concern these days, in current scenario the security of the premises is handled by the security guards. The security guards can be easily corrupted which will make the premises vulnerable to attack like terrorist attack, data theft, and vehicle theft. Our system reduces the risk as the fleet entry is automated and the intervention of the security guards are kept minimal.

Future Work

Future enhancements are possible in this paper. Some of the possibilities are:

- Our project can be future improved by adding CCTV camera
- SMS alerts
- Video conferencing
- Crystal reports of the historical data

References

1. Tseng, J.-D., Wang, W.-D., Ko, R.-J.: An UHF band RFID vehicle management system, Anti-counterfeiting, security, identification, IEEE International Workshop at Xiamen, Fujian, pp. 16–18 April (2007)
2. Hall, M., Brown, L., Chaikin, Y.: Core Servlets and Java Server Pages published by Prentice Hall, vol. 2: Advance Technology, 2nd edn. pp. 519–555 (2008)
3. Peckol, J.K.: A Contemporary Design Tool, 1st edn. Published by John Wiley & Sons University of Washington (2008)
4. Chong, R., Hakes, I., Ahiya, R.: Foreword by Dr. Arvind Krishna, DB2 Express C, 3rd edn. Developed by IBM pp. 18–25 (2009)
5. Finkenzeller, K.: Fundamentals and Applications in Contactless Smart Cards and Identification RFID, 2nd edn. (2003)

Distributed Data Mining in the Grid Environment

C. B. SelvaLakshmi, S. Murali, P. Chanthiya and P. N. Karthikayan

Abstract Grid computing has emerged as an important new branch of distributed computing focused on large-scale resource sharing and high-performance orientation. In many applications, it is necessary to perform the analysis of very large data sets. The data are often large, geographically distributed and its complexity is increasing. In these areas, grid technologies provide effective computational support for applications such as knowledge discovery. This paper is an introduction to grid infrastructure and its potential for machine learning tasks.

Keywords Grid computing · Knowledge grid · Data mining · Distributed data mining

Introduction

Grid Computing

A parallel processing architecture in which CPU resources are shared across a network, and all machines function as one large supercomputer, it allows unused CPU capacity in all participating machines to be allocated to one application that is

C. B. SelvaLakshmi (✉) · S. Murali · P. Chanthiya · P. N. Karthikayan
Velammal College of Engineering and Technology, Madurai, India
e-mail: cbselak08@gmail.com

S. Murali
e-mail: muralicse2008@gmail.com

P. Chanthiya
e-mail: chanthiyapuhal@gmail.com

P. N. Karthikayan
e-mail: karthikayan.it@gmail.com

S. Sathiakumar et al. (eds.), *Proceedings of International Conference on Internet Computing and Information Communications*, Advances in Intelligent Systems and Computing 216,
DOI: 10.1007/978-81-322-1299-7_15, © Springer India 2014

extremely computation intensive and programmed for parallel processing. Grid computing is also called peer to peer computing and distributed computing. The grid computing gives us yet another way of sharing the computer resource and yields us the maximum benefit at the time and speed efficiency. Grid computing enables multiple applications to share computing infrastructure, resulting in much greater flexibility,cost, power efficiency, performance, scalability and availability at the same time.

Data Grid

A data grid is a grid computing system that deals with the data controlled sharing and management of large amount of distributed data. A data grid can include and provide transparent access to semantically related data resources that are different managed by different software systems and are accessible through different protocols and interfaces.

Distributed Data Mining

Distributed data mining deals with the problem of data analysis in environments of distributed computing nodes.

Distributed Data Mining and Grids

Today many organizations, companies, and scientific centers produce and manage large amounts of complex data and information. Climate data, astronomic data and company transaction data are just some examples of massive amounts of digital data repositories that today must be stored and analyzed to find useful knowledge in them. This is particularly true in grid-based knowledge discovery [1], although some research and development projects and activities in this area are going to be activated mainly in Europe and USA, such as the Knowledge Grid (K-Grid), the Discovery Net, and the AdAM project. In particular, the K-Grid [2] that we shortly discuss in the next section provides a middleware for knowledge discovery services for a wide range of high performance distributed applications. Examples of large and distributed data sets available today include gene and protein databases, network access and intrusion data, drug features and effects data repositories, astronomy data files, and data about web usage, content, and structure. Workflows are mapped on a grid, assigning its nodes to the grid hosts and using interconnections for communication among the workflow components (nodes). In the latest years, through the Open Grid Services Architecture (OGSA), the grid community defined grid services as an extension of Web services for providing a standard model for using the grid resources and composing distributed applications as composed of several grid services. OGSA provides an extensible set of services that virtual organizations can aggregate in various ways defines uniform exposed-service semantics, the so-called grid service, based on concepts and technologies from both the grid computing and Web services communities. Recently the Web Service Resource Framework (WSRF) was defined as a standard specification of grid services for providing interoperability with standard Web services so building a bridge between the grid and the Web.

Grid Services for Distributed Data Mining

The Service Oriented Architecture (*SOA*) is essentially a programming model for building flexible, modular, and interoperable software applications. SOA enables the assembly of applications through parts regardless of their implementation details, deployment location, and initial objective of their development. Another principle of SOAs is, in fact the reuse of software within different applications and processes. The grid community adopted the OGSA as an implementation of the SOA model within the grid context. OGSA provides a well-defined set of basic interfaces for the development of interoperable grid systems and applications [3]. OGSA adopts Web Services as basic technology. Web Services are an important paradigm focusing on simple, Internet-based standards, such as the Simple Object Access Protocol (SOAP) and the Web Services Description Language (WSDL), to address heterogeneous distributed computing. Web service defines techniques for describing software components to be accessed, methods for accessing these components, and discovery mechanisms that enable the identification of relevant service providers. OGSA defines standard mechanisms for creating, naming, and discovering transient grid service instances. The WS-Resource Framework (WSRF) was recently proposed as a refactoring and evolution of grid services aimed at exploiting new Web Services standards, and at evolving OGSI based on early implementation and application experiences [4]. WSRF provides the means to express state as stateful resources and codifies the relationship between Web Services and stateful resources in terms of the implied resource pattern, which is a set of conventions on Web Services technologies, in particular XML, WSDL, and WS-Addressing. A stateful resource that participates in the implied resource pattern is termed as WS-Resource. The framework describes the WS-Resource definition and association with the description of a Web Service interface, and describes how to make the properties of a WS-Resource accessible through a Web Service interface. Through WSRF is possible to define basic services for supporting distributed data mining tasks in grids. Those services can address all the aspects that must be considered in data mining and in knowledge discovery processes from data selection and transport to data analysis, knowledge models representation and visualization. To do this, it is necessary to define services corresponding to single steps that compose a KDD process such as preprocessing, filtering, and visualization;[2] single data mining tasks such as classification, clustering, and rule discovery;[2] distributed data mining patterns such as collective learning, parallel classification and meta-learning models. At the same time, those services should exploit other basic grid services for data transfer and management such as Reliable File Transfer (RFT), Replica Location Service (RLS), Data Access and Integration (OGSA-DAI) and Distributed Query processing (OGSA-DQP). Moreover, distributed data mining algorithms can optimize the exchange of data needed to develop global knowledge models based on concurrent mining of remote data sets. This approach also preserves privacy and prevents disclosure of data beyond the original sources. Finally, grid basic mechanisms for handling security, monitoring, and scheduling distributed tasks can be used to provide efficient implementation of high-performance distributed data analysis.

The K-Grid Framework

The K-Grid framework is a system implemented to support the development of distributed KDD processes in a grid [2]. It uses basic grid mechanisms to build specific knowledge discovery services. These services can be developed in different ways using the available grid environments. The K-Grid provides users with high-level abstractions and a set of services by which is possible to integrate grid resources to support all the phases of the knowledge discovery process, as well as basic, related tasks like data management, data mining, and knowledge representation. In this implementation, each K-Grid service (K-Grid service) is exposed as a Web Service that exports one or more operations (OPs), by using the WSRF conventions and mechanisms. The operations exported by high-level K-Grid services (data access services (DAS), tools and algorithms access services (TAAS), execution plan management services (EPMS), and result presentation services (RPS)) are designed to be invoked by user-level applications, whereas operations provided by core K-Grid services (knowledge directory services (KDS) and resource access and execution services (RAEMS)) are thought to be invoked by high-level and core K-Grid services. Fig. 1.

In the WSRF-based implementation of the K-Grid, each service is exposed as a Web Service that exports one or more operations (OPs), by using the WSRF conventions and mechanisms. The operations exported by high-level K-Grid services are designed to be invoked by user-level applications only, whereas the operations provided by Core K-Grid services are thought to be invoked by high-level as well as Core K-Grid services. Users can access the K-Grid functionalities by using a client interface located on their machine. The client interface can be an integrated visual environment that allows for performing basic tasks (e.g., searching of data and software, data transfers, simple job executions), as well as for composing distributed data mining applications described by arbitrarily complex execution plans. The client interface performs its tasks by invoking the appropriate

Fig. 1 Interactions between a client and the knowledge grid environment

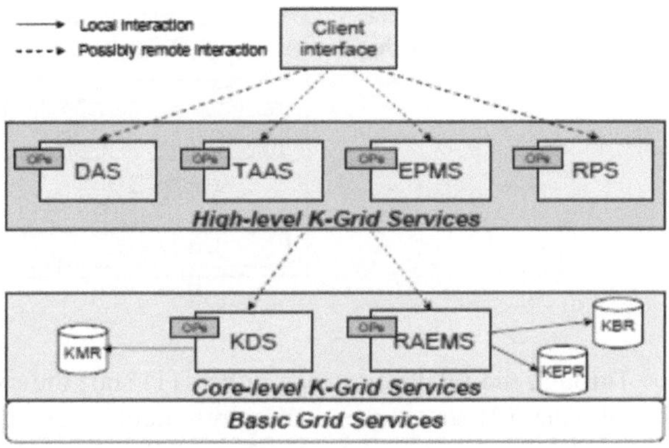

operations provided by the different high-level K-Grid services. Those services may be in general executed on a different grid node; therefore the interactions between the client interface and high-level K-Grid services are possibly remote.

Weka4WS

Weka4WS is a framework that extends the widely used open source Weka toolkit [5] for supporting distributed data mining on WSRF-enabled grids. Weka4WS adopts the WSRF technology for running remote data mining algorithms and managing distributed computations. The Weka4WS user interface supports the execution of both local and remote data mining tasks. On every computing node, a WSRF compliant Web Service is used to expose all the data mining algorithms provided by the Weka library. The Weka4WS software prototype has been developed by using the Java WSRF library provided by Globus Toolkit (GT4). All involved grid nodes in Weka4WS applications use the GT4 services for standard grid fu and so on. We distinguish those nodes in two categories on the basis of the available Weka4WS components: user nodes that are the local machines providing the Weka4WS client software; and computing nodes that provide the Weka4WS Web Services allowing for the execution of remote data mining tasks. Data can be located on computing nodes, user nodes, or third-party nodes (e.g., shared data repositories). If the dataset to be mined is not available on a computing node, it can be uploaded by means of the GT4 data management services.

Figure 2 shows the software components of user nodes and computing nodes in the Weka4WS framework. User nodes include three components: Graphical User Interface (GUI), Client Module (CM), and Weka Library (WL). The GUI is an extended Weka Explorer environment that supports the execution of both local and remote data mining tasks. Local tasks are executed by directly invoking the local WL, whereas remote tasks are executed through the CM, which operates as an intermediary between the GUI and Web Services on remote computing nodes.

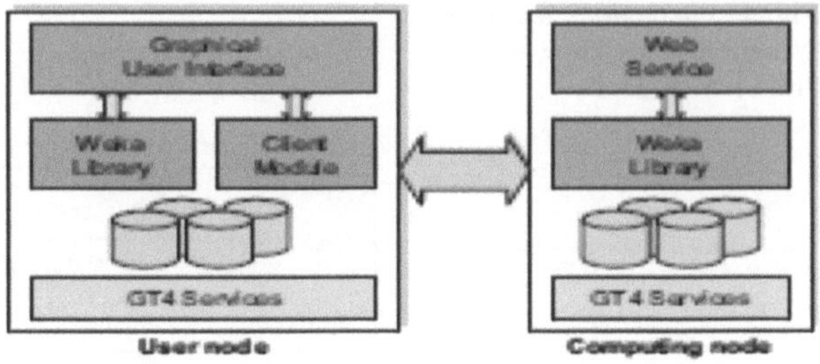

Fig. 2 Software components of user nodes and computing nodes

Through the GUI a user can either: (1) start the execution locally by using the Local pane; (2) start the execution remotely by using the Remote pane. Each task in the GUI is managed by an independent thread. Therefore, a user can start multiple distributed data mining tasks in parallel on different Web Services, this way taking full advantage of the distributed grid environment. Whenever the output of a data mining task has been received from a remote computing node, it is visualized in the standard Output pane. A recent paper [6] presents the architecture, details of user interface, and performance analysis of Weka4WS in executing a distributed data mining task in different network scenarios. The experimental results demonstrate the low overhead of the WSRF Web service invocation mechanisms with respect to the execution time of data mining algorithms on large data sets and the efficiency of the WSRF framework as a means for executing data mining tasks on remote resources. By exploiting such mechanisms, Weka4WS provides an effective way to perform compute-intensive distributed data analysis on large-scale grid environments. Weka4WS can be downloaded from http://grid.deis.unical.it/weka4ws.

Conclusion

The development of practical grid computing techniques will have a profound impact on the way data is analyzed. In particular, the possibility of utilizing grid-based data mining applications is very appealing to organizations wanting to analyze data distributed across geographically dispersed heterogeneous platforms. Grid-based data mining would allow companies to distribute compute intensive analytic processing among different resources. Moreover, it might eventually lead to new integration and automated analysis techniques that would allow companies to mine data where it resides. This is in contrast to the current practice of having to extract and move data into a centralized location for mining processes that are becoming more difficult to conduct due to the fact that data is becoming increasingly geographically dispersed, and because of security and privacy considerations.

References

1. Berman, F.: From Teragrid to Knowledge Grid. Commun. ACM. **44**(11), 27–28 (2001)
2. Cannataro, M., Talia, D.: The knowledge grid. Commun. ACM. **46**(1), 89–93 (2003)
3. Foster, I., Kesselman, C., Nick, J., Tuecke, S. The physiology of the grid. In: Berman, F., Fox, G., Hey, A. (eds.) Grid Computing: Making the Global Infrastructure a Reality, pp. 217–249. Wiley (2003)
4. Czajkowski, K. et al. The WS-Resource Framework Version 1.0. http://www-106.ibm.com/developerworks/library/ws-resource/wswsrf.pdf
5. Witten, H., Frank, E. Data Mining: Practical Machine Learning Tools with Java Implementations. Morgan Kaufmann
6. Talia, D., Trunfio, P., Verta, O. Weka4WS: a WSRFenabled Weka toolkit for distributed data mining on grids. In Proceedings of PKDD 2005, LNAI vol. 3721, pp. 309–320. Springer-Verlag, Porto, Portugal, October (2005)
7. Cannataro, M., Congiusta, A., Mastroianni, C., Pugliese, A., Talia, D., Trunfio, P.: Grid-based data mining and knowledge discovery. In: Zhong, N., Liu, J. (eds.) Intelligent Technologies for Information Analysis, pp. 19–45. Springer-Verlag, (2004)
8. Cannataro, M., Talia, D.: Semantics and knowledge grids: building the next generation grid. IEEE Intell. Syst. **19**(1), 56–63 (2004)
9. Kargupta, H., Kamath, C., Chan, P. Distributed and parallel data mining: emergence, growth, and future directions, In: Advances in Distributed and Parallel Knowledge Discovery, pp. 409–416. AAAI/MIT Press (2000)

An Efficient Method for Improving Hiding Capacity for JPEG2000 Images

T. Shahida and C. C. Sobin

Abstract Information hiding techniques have recently become important in a number of application areas. The redundancy of digital media, as well as the characteristic of human visual system, makes it possible to hide messages. In this paper, a detailed review on various information hiding techniques discussed both in spatial and wavelet domain and focused on the problem of how to enhance the hiding capacity for JPEG2000 images as it is really a challenging problem because of limited redundancy and bit stream truncation. Spatial domain techniques spanning from Least Significant Bit (LSB) method, Least Pixel Adjustment Process (LPAP), and wavelet-based techniques include, Discrete Cosine Transform (DCT) method, Discrete Wavelet Transform (DWT) based techniques and spread spectrum techniques. Finally, JPEG2000 Architecture and high capacity steganography scheme for JPEG2000 Baseline system are described. Hiding capacity is very important for efficient covert communications. Available redundancy is very limited in JPEG2000 compressed images. So it is necessary to enlarge the hiding capacity and also it is very difficult to hide the information because of the bit stream truncation. Here a high-capacity steganography scheme is proposed for JPEG2000 baseline system, which uses bit-plane encoding procedure multiple times to solve the problem due to bit stream truncation and redundancy evaluation method, is used to increase hiding capacity.

Keywords Steganography · JPEG2000 · Bit-plane encoding

T. Shahida (✉) · C. C. Sobin
Department of Computer Science and Engineering, MES College of Engineering, Malappuram, Kerala, India
e-mail: shahida.faiha@gmail.com

C. C. Sobin
e-mail: sobincc@gmail.com

S. Sathiakumar et al. (eds.), *Proceedings of International Conference on Internet Computing and Information Communications*, Advances in Intelligent Systems and Computing 216, DOI: 10.1007/978-81-322-1299-7_16,

Introduction

Steganography is the art of science to hide data in a cover image, if image is used as the communication media. Steganography is coming as an application under information security. It should have high imperceptibility, security level, and payload. The goal of Steganography is to mask the presence of thereby hiding the secret information invisible to others. Today, computer and network technologies provide easy to use communication channels for Steganography. Because of the large development of multimedia applications, the digital media files like images, audio and video files are used as the carrier contents of the Steganography.

Literally Steganography means covered writing. From the Greek words stegano means covered and graphos means to write. In encryption the encrypted messages have been intercepted. The interception of the message is like a damaging because it tells an enemy that both parties are communicating in secrets. The opposite approach is taken by Steganography in which nobody can see that both parties are communicating in secrets. Steg analysis is the art of discovering and rendering such covert messages. Information hiding techniques are generally divided into two groups. One is called spatial domain methods which operate in the pixel value of the image and another is in the frequency domain which operates in the frequency coefficients of the message.

Three factors are considered while the designing of information hiding scheme. They are capacity, security, and robustness. Security means invisibility and keeping undetectable. Capacity refers to the maximal secure payload. Robustness relates to the amount of modification the stego-object can withstand before an adversary can destroy the hidden information. Generally robustness is mainly considered in watermarking rather than Steganography. Steganography is employed in various useful applications, e.g, copyright control of materials, enhancing robustness of image search engines and smart IDs (identity cards) where individual's details are embedded in their photographs. Other applications are video-audio synchronization, companies safe circulation of secret data, TV broadcasting, TCP/IP packets (for instance, a unique ID can be embedded into an image to analyze the network traffic of particular users), and also checksum embedding.

Requirements of Hiding Information

For hiding data in a given object there exist many different protocols and embedding techniques. However, all of the protocols and techniques must satisfy a number of requirements so that Steganography can be applied correctly. List of main requirements that Steganography techniques must satisfy are as follows:

1. The integrity of the hidden information after it has been embedded inside the stego object must be correct. The secret message must not change in any

way, such as additional information being added, loss of information or changes to the secret information after it has been hidden. If secret information is changed during Steganography, it would defeat the whole point of the process.
2. The stego object must remain unchanged or almost unchanged to the naked eye. If third party notices the significant changes in stego object, he will try to extract or to destroy it.
3. In watermarking, changes in the stego object will not have an effect on the watermark. Imagine if you had an illegal copy of an image and manipulate it in various ways. These manipulations can be simple processes such as resizing, trimming or rotating the image. The watermark inside the image must survive these manipulations, otherwise the attackers can very easily remove the watermark and the point of Steganography will be broken.
4. Finally, we always assume that the attacker knows that there is hidden information inside the stego object.

Information hiding techniques are generally divided into two groups. One is called spatial domain methods which operate in the pixel value of the image and another is in the frequency domain which operates in the frequency coefficients of the message. Techniques in spatial domain are listed below.

Least Significant Bit Method

The simplest method for information hiding in spatial domain is Least Significant Bit (LSB) method. The basic principle in such a method is to manipulate the LSBs of the cover image with the secret image. Apart from the ease of implementing this method yields high capacity. There are various examples of LSB schemes can be found in the literature [1, 2]. But the problem with such an approach is that if one knows earlier that LSB method is used for information hiding it is very easy to extract the information.

Local Pixel Adjustment Process

A variation to LSB method was proposed by Wang [3]. In this method, the image quality is improved compared to the LSB method. But it is not an optimal one because it considers only the last three significant bits and ignores the fourth one. So a new method was proposed by Chan and Cheng [1] combining simple LSB substitution and LPAP. Experimental results show that this method enhances image quality with low extra computational complexity.

The other method is in the frequency domain in which messages are embedded in to the coefficients of the messages. These methods overcome the issues related to the robustness and imperceptibility found in spatial domain.

Discrete Cosine Transform

The DCT is one of the commonly used data hiding technique in frequency domain. It is used in JPEG compression. The basic principle is given below.

1. The input image is split into 8 × 8 squares.
2. Each of the squares is transformed via DCT which outputs an array of 63 coefficients.
3. Quantizer rounds each of the 63 coefficients, means some of the data is lost.
4. Unimportant coefficients are rounded to zero.
5. The coefficients are further compressed via Huffman coding.
6. Decompression is done via inverse DCT.

DCT is having extensive use in the image and video compression because for any person who just looks at the pixel value of the image would be unaware that anything missing. Also the hidden data is evenly distributed among the source image makes it so robust. Although this method is much simple one it is vulnerable to noise.

Spread Spectrum Hiding Technique

Spread spectrum systems encodes data as binary sequences which sound like noise but can be recognized at the receiver end with a correct key. The technique has been used in military from 1940, because the fact that signals is hard to jam or intercept. Today advancement of spread spectrum systems allow us to apply to JPEG2000 images directly but without considering bit-stream truncation [4]. Spread spectrum hiding techniques can be applied in JPEG2000 directly, without consideration on bit stream truncation [5]. The receiver extracts the hidden information by correlation detection. It is not necessary to keep all the embedded messages available in correlation detection. However, spread spectrum preprocessing will decrease hiding capacity significantly. Therefore, spread spectrum technology is often used in digital watermarking applications rather than covert communications.

Discrete Wavelet Transform

Wavelet domain-based techniques have used extensively in various fields including approximation theory, signal processing, physics astronomy, and image processing applications. The main advantage of wavelet-based techniques is it will increase the hiding capacity as well as robustness. The hierarchical nature of Wavelet representation allows multi-resolution detection of hidden image, which is a Gaussian distributed random vector added to all the high bands in the wavelet domain. It is shown that even if distortion happened due to compression, still the hidden image can be correctly identified at each resolution in the Discrete Wavelet Transform (DWT).

A wavelet domain Steganography algorithm based on redundancy evaluation, named as RES, is proposed for the JPEG2000 baseline system. The compatibility and reversibility can be taken into account by the process of uniform quantization and rate-distortion optimized truncation. Every embedding point and its intensity are adjusted for enhancing the information hiding volume. The experiment shows that the proposed method is feasible and effective.

Redundancy of digital media suggests the feasibility of carrying secret message bits. Secret message can be transmitted within various digital media such as imagery and audio by information hiding technology. But most digital media has been compressed to communicate effectively. The information hiding volume is badly limited by the poor available redundancy of compressed images.

JPEG2000, the latest image compression standard, offers higher compression performance to JPEG and puts emphasis on scalable compressed representations. JPEG2000 is based on DWT rather than DCT. DWT tends to be more flexible than DCT. Unlike JPEG, the introduced image coding system JPEG2000 allows wavelets to be employed for compression instead of the DCT. This makes DWT-based Steganography the future leading method.

JPEG2000 Architecture

International Telecommunication Union (ITU) and the International Organization for Standardization (ISO) have been working together to establish a joint international standard for the compression of gray scale and color still images. This effort has been known as JPEG, the Joint Photographic Experts Group. The "joint" in JPEG refers to the collaboration [6]. The JPEG 2000 standard provides a set of features as listed below.

1. Superior low bit-rate performance. This standard must offer high performance when compared with current standards at low bit-rates (e.g., below 0.25 bpp for highly detailed gray-scale images). This standard significantly improved low bit-rate performance without sacrificing performance.
2. Lossless and lossy compression. It is desired to provide lossless compression at the decoding side. In medical image, it is not desirable to loss the part of image at the decoding side. The standard should have the property of creating embedded bit stream. It should allow progressive lossy to lossless build-up.
3. Progressive transmission by pixel accuracy and resolution. Image can be reconstructed with increasing pixel accuracy by using progressive transmission. Spatial resolution is essential for many applications. This feature allows the reconstruction of images with different resolutions and pixel accuracy, as

needed or desired, for different target devices. World Wide Web, image archival, and printers are some application examples.

4. Region-of-Interest Coding. In an image, some parts are more important than others. This feature allows users to define certain region of interests in the image to be coded and transmitted with better quality and less distortion than the rest of the image.
5. Random code stream access and processing. User-defined region of interests in the image can be randomly accessed and/or decompressed with less distortion than the rest of the image by using this features. Also, random code stream processing could allow operations such as rotation, translation, filtering, feature extraction, and scaling. The standard supports for compound documents and computer graphics.

Architecture of JPEG2000 is described in the figure below.

There are two primary paths and several options for encoding. If reversible or lossless coding is desired, then the Reversible Component Transform (RCT) is used with the 5-3 wavelet, and quantization is by truncation. If purely lossy coding is desired at the highest quality for a given bit rate, then the YCbCr transform is used, with the 9-7 wavelet, and arbitrary quantization by division is possible in addition to truncation. With either of these main paths there are several options for Region of Interest (ROI), different coding options to trade complexity and performance, and choices about the amount of scalability in the bit stream. First, apply discrete transform on the source image data. Then quantize the transform coefficients and entropy coded, before forming the output code stream, which is also called bit stream.

The decoder is the reverse of the encoder Fig. 1. The output code stream or bit stream is first entropy decoded, dequantized and inverse discrete transformed, which yields the formation of reconstructed image data.

Before describing the details of each block of encoder in Fig. 1 it should be mentioned that the standard works on image tiles. The term tiling means the partition of the original (source) image into rectangular non-overlapping blocks (tiles).

Each block is then compressed independently, as though they were entirely distinct images. Before doing the computation of the forward DWT on each image tile, we have to shift (DC level shift) all samples of the image tile component by subtracting the same quantity (i.e., the component depth). DC level shifting is

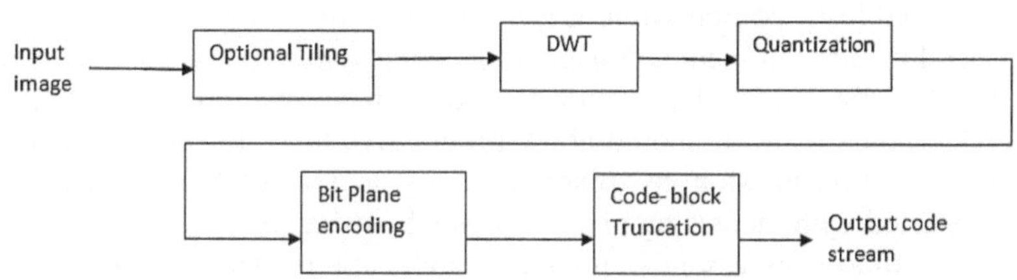

Fig. 1 JPEG 2000 Architecture

performed on samples of components that are unsigned only. If color transformation is used, it is performed prior to computation of the forward component transform. Otherwise it is performed prior to the wavelet transform.

At the decoder side, inverse DC level shifting is performed on reconstructed samples of components that are unsigned only. If used, it is performed after the computation of the inverse component transform. Arithmetic coding is used in the last part of the encoding process. The MQ coder is adopted in JPEG2000. This coder is basically similar to the QM coder adopted in the original JPEG standard. The encoding procedure is as follows:

- The source image is decomposed into components.
- The image and its components are decomposed into rectangular tiles. The tile-component is the basic unit of the original or reconstructed image.
- The wavelet transform is applied on each tile. The tile is decomposed in different resolution levels.
- These decomposition levels are made up of sub bands of coefficients that describe the frequency characteristics of local areas (rather than across the entire tile-component) of the tile component.
- The sub bands of coefficients are quantized and collected into rectangular arrays of code-blocks.
- The bit-planes of the coefficients in a code-block are entropy coded.
- The encoding can be done in such a way, so that certain ROI's can be coded in a higher quality than the background.
- Markers are added in the bit stream to allow error resilience.

The code stream has a main header at the beginning that describes the original image and the various decomposition and coding styles that are used to locate, extract, decode, and reconstruct the image with the desired resolution, fidelity, region of interest, and other characteristics.

The optional file format describes the meaning of the image and its components in the context of the application.

High Capacity Steganography Scheme

For JPEG2000 compressed images, limited redundancy and bit stream truncation makes it difficult to hide information [7]. After analyzing the challenge of covert communication in JPEG2000 image codec, Su and Kuo presented a Steganography scheme to hide high volumetric data into JPEG2000 bit stream. In order to avoid affection of bit stream truncation, their method was not designed for the standard baseline system of JPEG2000. It was limited to the simplified version of JPEG2000, named as lazy mode, in which the entropy coding procedure was completely bypassed. Later another, method was proposed by Liang et.al, but the improvements in hiding capacity were limited [8, 9]. In. In this paper, we are proposing a high capacity stenographic scheme by applying bit plane encoding thrice to solve the problem due to truncation and embedding points and intensity are adjusted by using redundancy evaluation method.

The algorithm for the proposed method is given below.

Step 1: for determining the embedding points and embedding intensity for a code block

- The wavelet coefficients greater than a given threshold are chosen as candidate embedding points.
- According to the rate distortion optimization, the lowest bit-plane which keeps unabridged after bit stream truncation is determined as the lowest embed allowed bit plane of the code block.
- The embedding points and embedding intensity are adjusted adaptively on the basis of redundancy evaluation.

Step 2: Scrambled synchronization information and secret messages are embedded into the selected embedding points from the lowest embed-allowed bit-plane to higher ones

Step 3: Secondary bit-plane encoding is operated after information embedding

Step 4: After embedding, we organize the bit stream according to the previous result of rate-distortion optimization

Step 5: Third level of bit plane encoding is done. By doing this, messages are embedded into bit-planes that would not be truncated by rate-distortion optimization

The integrity of the embedded message is ensured at the cost of increased computational complexity and slightly changed compression ratio. The thrice bit-plane encoding procedure is explained to execute the bit-plane encoding thrice, whereas the rest parts, such as wavelet transform, quantization, rate-distortion optimization, bit stream organization are executed only once [10].

Results

Information embedding is illustrated in Fig. 2. The image used as cover media is a gray image called Lenna and the secret image is the logo of the Civil Aviation University, China. First, the lowest bit-plane with complete information of all its three coding passes can be determined easily in the procedure of entropy decoding. Then the embedding points and their intensity are determined by the method similar to the encoder. Finally, both synchronization information and secret messages are extracted which is shown in Fig. 3, which is not able to distinguished by the previous one [11, 12].

Figure 4 shows the plot of the existing methods versus proposed method which uses multiple level of bit plane encoding. From the figure, it is clear that the hiding capacity has been increased [13].

Fig. 2 Original image used as cover and secret image

Fig. 3 Retrieved image and secret image

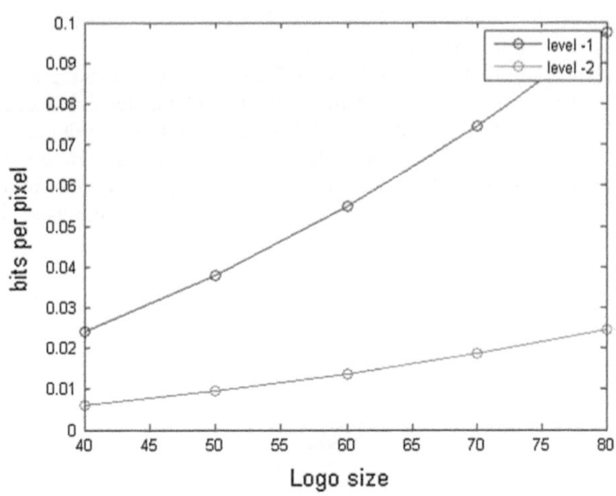

Fig. 4 Plot of hiding capacity

Conclusion

There is various information hiding schemes are in literature, but none fails to focus on the latest JPEG2000 images. Increasing hiding capacity is a challenging problem for JPEG2000 images because of limited redundancy and bit stream truncation. In this paper a high capacity stenographic scheme proposed for JPEG2000 images, which uses multiple level of bit plane encoding to solve the problem of bit stream truncation and redundancy evaluation technique for improving the hiding capacity.

References

1. Chan, C.K., Cheng, L.M.: Hiding data in images by simple LSB substitution. Pattern Recognit. **37**(3), 469–474 (2004)
2. Wu, H.C., Wu, N.I., Tsai, C.S., Hwang, M.S.: Image steganographic scheme based on pixel-value differencing and LSB replacement methods. IEE Proc. Vis., Image. Signal Process. **152**(5), 611–615 (2005)
3. Zhang, L., Wang, H., Wu, R.: A high-capacity Steganography scheme for JPEG2000 baseline system. IEEE Trans. Image Process. **18**(8), 1797–1803 (2009)
4. Grosbois, R., Ebrahimi, T.: Watermarking in the JPEG2000 domain. In Proceedings of IEEE 4th Workshop on Multimedia Signal, France, pp. s339–s344, 2001
5. Ingemar, J.C., Kilian, J., Leighton, F.T., Shamoon, T.: Secure spread spectrum watermarking for multimedia. IEEE Trans. Image Process. **6**(12), 1673–1687 (1997)
6. Christopoulos, C., Skodras, A., Ebrahimi, T.: The JPEG2000 still image coding system: an overview. IEEE Trans. Consum. Electron. **46**(4), 1103–1127 (2000)
7. Tolbal, F., Ghonemy, A., Taha, A., Khalifa, S.: Using integer wavelet transforms in colored image Steganography. Int. J. Intell. Cooper. Inform. Syst. **4**(2), 230–235 (2004)
8. JPEG2000Part 1: Final Committee Draft Version 1.0, ISO/IEC. FCD information hiding.15444-1, 2000.
9. Su, P.C., Kuo, C.C.J.: Steganography in JPEG2000 compressed images. IEEE Trans. Consum. Electron. **49**(4), 824–832 (2003)
10. Andersonand, R.J., Petitcolas, F.A.P.: On the limits of Steganography. IEEE J. Sel. Areas Commun. **16**, 474–481 (1998)
11. Cummins, J., Diskin, P., Lau, S., Parlett, R.: Steganography and Digital Watermarking. The University of Birmingham, School of Computer Science, London, UK (2004)
12. Cheddad, A., Condell, J., Curran, K., Kevitt, P.M.: Digital image Steganography: survey and analysis of current methods. IEEE Trans. Image Process. **6**(12), 1673–1687 (1997)
13. Abdelwahab, A.A., Hassaan, L.A.: A discrete wavelet transform based technique for image data hiding. Paper presented at radio science conference, NRSC 2008, Egypt, pp. 1–9, 2008

Electronic Customer Relationship Management (e-CRM): Data Integration for Technical Institutions

Kini K. Shashidhar and D. H. Manjaiah

Abstract Educational institutions worldwide are undergoing funda- mental shifts in how they operate and interact with their "customers": students, alumni, donors, faculty members, and staff members. Kotler and Fox [32] state that "the best organization in the world will be ineffective if the focus on 'customers' is lost. First and foremost is the treatment of individual students, alumni, parents, friends, and each other internal customers). Every contact counts!". Many organizations are familiar with using CRM (Customer relationship management) to manage and enhance the customer relationship. Good customer relationship can bring great benefits and a competitive advantage to organization. And in this era of technology, CRM that consists of e-CRM (Electronic customer relationship management) is acknowledged as another potential solution for business. The focus is currently shifting from improving internal operations to concentrating more on customers. Technical education customers are demanding more attention and immediate service-that is, "Internet time". Proactive institutions are now adjusting their practices by refocusing their efforts externally. Because of the need to concentrate more on customers, many institutions are once again turning to technology-this time to customer relationship management (CRM) software. CRM goes several steps further than ERP by helping institutions maximize their customer-centric resources. The purpose of this study is how Electronic Customer Relationship (e-CRM) will help the technical institutions to integrate the data from customer touch points.

Keywords Customer relationship management (CRM) · Electronic customer relationship (e-CRM) · Information technology (IT) · Information system (IS) · Enterprise resource planning (ERP)

K. K. Shashidhar (✉) · D. H. Manjaiah
Department of Computer Science, Mangalore University, Mangalagangotri, Karnataka 574 199, India
e-mail: skinipa@gmail.com

D. H. Manjaiah
e-mail: ylm321@yahoo.co.in

S. Sathiakumar et al. (eds.), *Proceedings of International Conference on Internet Computing and Information Communications*, Advances in Intelligent Systems and Computing 216,
DOI: 10.1007/978-81-322-1299-7_17, © Springer India 2014

Introduction

Good relationship among a customer and an organization creates higher customer satisfaction. Almost all businesses focus on enhancing this relationship because customers are [26]. customer relationship management (CRM) systems are capable of increasing the satisfaction of customers and creating the competitive advantage that companies need to attract customers. As businesses are looking for the better way to communicate and interact with customers nowadays, many of them choose electronic Customer Relationship Management (e-CRM) . Since the competition among businesses is getting higher, many businesses concern to develop products and services to match customer needs. e-CRM allows a business to understand customer behavior and forecast customer needs easier through online activities and able to improve long-run profitability, thus it becomes more popular (Christopher et al. 1991 cited in [2]).

According to Dotan [16] e-CRM can improve the levels of interaction between customer and service. The common goal of using e-CRM is to improve customer relationship via improving customer service and retaining profitable customers. In addition, e-CRM is able to create loyalty and extent customer life cycle by increase customer value and satisfaction.

e-CRM Background

Electronic Customer Relationship Management is referred to the marketing activities, tools and techniques via the Internet network which are able to build and enhance relationship between organization and customers ([33], p. 241). It is sometime referred to web-enabled or web-based CRM [2]. Since the use of IT plays the important role to deliver products and services to customer (customer does not use CRM), business decides to use e-CRM to support the multiple electronic channels to contact and communicate with its customers [9]. Same as CRM, e-CRM objective is to gather information from customers and adjust service level to match with specific needs which will able to enhance customer relationship [27]. e-CRM helps organizations to enable specific products and services to reach customer needs through Internet access [23, 45]. This is one of the opportunities that organizations received because it can retain profitable and valuable customers by fulfill their requirements [24]. Customer satisfaction, customer retention and customer loyalty are three sig- nificant components of customer relationship that organization want to achieve which is the significant goals of CRM [39]. The customer relationship that e-CRM created can help organization maintain their profitable customers and also create loyalty among customers.

Yesterday's trends are reoccurring today as organizations continue to leverage their data resources by developing and deploying data mining technologies to enhance their decision-making capabilities [16]. To address this need, organizations are implementing organizational data mining (ODM) technologies, which are defined as technologies that leverage data mining tools to enhance the decision-making process by transforming data into valuable and actionable knowledge to gain a competitive advantage [41]. ODM spans a wide array of technologies, including but not limited to e-business intelligence, data analysis, CRM, predictive analytics, dashboards, web portals, etc.

As a result of these marketplace trends, organizations must begin implementing customer-centric metrics as opposed to solely adopting product-centric metrics [11]. This scenario has triggered increased interest in the implementation and use of customer-oriented ODM technologies such as CRM systems. CRM can be defined as the adoption, through the use of enabling technology, of customer-focused sales, marketing, and service processes [18]. Customer relationship management is the process that manages the interaction between a company and its customers. The goal of customer relationship management is to create a long-term, profitable relationship with all of an orga- nization's customers. It is more than just a software package- it is a technology- enabled business process. Customer relationship management has become a key process in the strengthening of customer loyalty and in helping businesses obtain greater profit from low-value customers. The manner in which companies interact with their customers has changed greatly over the past decade. Customers no longer guarantee their loyal patronage, and this has resulted in organizations attempting to better understand them, predict their future needs, and decrease response times in fulfilling their demands. Customer retention is now widely viewed by organizations as a significant marketing strategy in creating a competitive advantage, and rightly so. Research suggests that as little as a 5 % increase in retention can provide a 95 % boost in profits, and repeat customers generate over twice as much gross income as new customers [48].

Methodology

Many studies [1, 3, 7, 8, 13, 35, 40] have shown that data has been ranked as one of the top priorities for information services (IS) executives. With the growth of web-based technologies, the collection and storage of data- both internal and external–has increased dramatically. Internal data refers to data generated from systems within an organization, such as legacy and online transactional processing (OLTP) systems. External data refers to data that is not generated by systems within an organization, such as government census data, industry benchmark data, consumer psychographic data and economic data. For instance, consumer demographic and psychographic data is available for each of the 200+ million adults in the United States, and product-based data is available for the millions of businesses in the United States. If this data is collected, integrated and formatted properly, it can prove to be immensely beneficial to a firm in better understanding its customers [43]. External data should be leveraged in a CRM system to the extent that it adds additional value to the existing internal organizational data.

More recent studies have shown favorable CRM outcomes with data integration, and from the opposite view, significant failure rates of CRM projects that ignore it. Technical issues such as capturing the wrong customer information, using misleading metrics and underestimating the difficulties involved in data mining, data cleansing and data integration are major barriers in implementing and managing successful CRM projects [25, 30, 38].

Companies approach consumers through various marketing channels. Traditionally, each channel or functional area has been managed separately, and all data pertaining to a channel is housed in its own system in a proprietary format [16, 44]. Technically, data integration can be defined as the standardization of data definitions and structures through the use of a common conceptual schema across a collection of data sources [21, 34]. This implies that data is accessible across functional areas, making data in different corporate databases accessible and consistent [36]. While collecting the data from the website organization has to consider various technological and design issues such as data requirements, data quality, data inconsistencies, synchronization, security, etc. Once these issues are addressed, an organization must present the data in a way that is consistent and conducive to viewing across heterogeneous enterprise departments [28].

This finding raises a number of allocation questions. How do organizations determine which marketing media to use, where their customers spend most of their time, and what their customers' lifestyles are? To better answer these questions, online marketers must build a 360° (holistic) view of their customers in order to track purchasing behaviors, preferences, likes and dislikes. This holistic view requires organizations to integrate their data to track every customer transaction (customer purchases, returns and complaints) in all customer touch-points (stores, email, mobile, search marketing, social media and direct mail).

Forrester Research predicts online retail sales will account for 8 % of all U.S. retail sales in 2014, up from 6 % last year. More impressive is that by 2014, more than half of total retail sales (53 %) will be affected by the web-for example, consumers going online to do product research or contact customer service [17]. In another survey, e-business executives report rising costs of acquiring customers online-current online acquisition costs total half of store acquisition costs, an increase from one-third of the cost reported a year ago. To minimize these marketing costs, organizations should concentrate on satisfying and serving existing customers and understanding the engagement of those customers with their companies [29]. These findings suggest that if you want to compete in today's marketplace and increase profitability in the coming years, you need to go beyond web cookies and meta-tags-you need to build an integrated offline and online customer profile.

Most companies/Institutions now realize and understand the value of collecting customer data but are faced with the challenges of using this knowledge to create intelligent pathways back to the customer. Most data mining technologies and techniques for recognizing patterns within data help businesses sift through the meaningless data and allow them to anticipate customers' requirements and expectations while more profitably managing channel partnerships and similar relationships. These technologies also enable companies to maintain customer privacy and confidentiality while gaining the benefits of profiling, calculating the economic value of the CRM system, and discovering the key factors that make or break the CRM project. By integrating these data mining tools with CRM software, organizations are able to analyze very large databases to extract new customer insights for stronger and more profitable relationships.

Similarly, electronic customer relationship management can be defined as the process of acquiring a thorough understanding of an organization's online visitors and/or customers in order to offer them the right product at the right price. e-CRM analytics is the process of analyzing and reporting online customer/visitor behavior patterns with the objective of acquiring and retaining customers through stronger customer relationships. Prior research has found that in order to understand online customers, a company must integrate its data from both online and offline sources [37]. More recent research [10, 31] has also concluded that system and data integration are critical success factors in e-CRM and CRM initiatives.

In a similar light, our paper demonstrates that a technical institutions cannot thoroughly understand its customers if it neglects integrating its customers' behavioral data from both online and offline channels. In order to have this complete customer viewpoint, it is imperative that organizations/Institutions integrate data from each customer touch-point. Our paper elaborates on this key issue of integrating data from multiple sources and its enabling role in facilitating successful and value-creating e-CRM analytics. A new frame of e-CRM value will be helpful to identify the data integration for better decision making in the technical institutions.

A website is developed for this purpose and the user touch points are recorded. This will help the customer to get all the details of technical institutions without physically coming to the institutions and they will get all the information about their requirements. Their touch points in all the areas are recorded.

Results and Discussion

Extensive research and case studies have shown that data integration is one of several critical factors in successful CRM implementations. To realize measurable business value, firms must combine physical resources (such as computers and networks) and informational resources (online and offline customer databases, call records, email correspondence and other customer service interactions) in their CRM systems [19]. With today's demanding customers communicating through multiple marketing channels, organizations must be cognizant of customer preferences to optimally manage their delicate yet vital relationship with them.

The results conveys that

- The more data sources a company integrates, the better the customer insight, thus creating more value for the company.
- Integrating online data with data from the firm's offline operations will lead to better customer insight, thus creating more value for the company.

Timeliness of data is an important component of user satisfaction [1, 4, 14]. Users need to have up-to-date information about customers' needs and preferences [46] to thoroughly understand and satisfy those needs. Traditional customer-centric measures such as recency, frequency and monetary statistics should be captured and incorporated into CRM analytics. Without integrated data (from online and offline sources), these statistics will not accurately represent the customer.

A recent survey of 231 online marketers by an innovative Internet market- ing company found that businesses that blog multiple times a day acquire more customers than those who blog less frequently. In fact, 100 % of companies who blog multiple times a day have generated customers from their blog com- pared to 90 % of respondents who blog daily and 69 % of respondents who blog two or three times a week [22]. This finding shows the additional value obtained by frequently updating and refreshing marketing and e-CRM data.

Traditionally, it was acceptable for organizations to update their customer database on a monthly or quarterly basis. But in today's fast-paced electronic economy where critical decisions are made daily, companies strive for more current information, requiring systems to update their databases much more frequently (daily, hourly, or in real time). This leads us to our next proposition:

- Data that is more frequently refreshed will lead to better customer insight, thus creating more value for the company.

Past experiences or product quality are not the only reasons why customers make purchases. There are factors external to an organization such as new marketplace competitors, economic factors, competitor promotions, online social media and other similar factors that alter our buying preferences. The explosive growth of social media and its user-generated content are now becoming more effective at driving sales than traditional marketing channels. Consider the following statistics that support the growing importance of leveraging online and external data sources:

- Over 40 % of marketers using social media sites Twitter, LinkedIn, Facebook and company blogs have generated a customer from that channel [22].
- Over half (51 %) of consumers are using the Internet before making a purchase in shops, educating themselves on the products and best deals available [5].
- Brands with the highest "social media activity" (including reviews) increased revenues by as much as 18 % [5].

In his book Web Farming [20], Richard Hackathorn advocates that organizations must integrate external data into their data warehouse to gain a complete picture of its business. Sources of external data may include government databases, customer demographic and lifestyle data, online customer preferences, census data, geographic data and weather data.

- Integrating external data with internal data will lead to better customer insight, thus creating more value for the company.

In many instances, companies focus their limited resources on their core competencies and outsource many remaining business functions, sometimes retaining the services of application service providers (ASP) and specialized hosting partners to manage online and ecommerce functions [16]. Whether an organization's business processes are performed in-house or out- sourced, the collaboration and integration of systems and data from multiple functional areas is complex and difficult. A prior Data Warehousing Institute Industry Report [16] found that organizations are challenged when integrating web technologies into their existing legacy and IT systems. Some of the reasons behind this challenge are scalability issues, managing large clickstream databases, immaturity of technology, lack of experience, and the complexity of modeling web data for analysis. But despite the integration challenges, the benefits realized are significant.

For successful CRM analytics, an enterprise-wide, customer-centric data repository should be utilized rather than a channel specific data repository [1, 6, 28, 46, 47] suggests an enterprise-wide, customer-centric data warehouse should be the foundation of any CRM initiative.

- Deploying an enterprise-wide data warehouse as the CRM backbone will lead to better customer insight, thus creating more value for the Organization/ Institution.

Research in customer relationship management is growing as it is gaining greater acceptance within organizations. Customer relationship management has received considerable attention from researchers in many diverse disciplines. Although there is a growing pool of literature that addresses many aspects of the application of customer relationship management for business solutions, there are few scholarly publications that focus on the study of customer relationship management from an e-commerce perspective. Given the complexity of the issues involved in data integration, the enormous benefits that electronic customer relationship management can offer, and the role data integration plays in achieving e-CRM's goals. Using e-CRM value frame work (Fig. 1) to study data integration issues and their impact on the overall value attained from e-CRM projects. The results of our analysis reveal that four of the five factors support this new framework and have a significant influence on creating value for an organization.

It is concluded that e-CRM supports the collections of customer information and the process with customers through Internet so the core technologies of e-CRM will be network and IT facilities. Organization should provide an efficient network and IT facilities sufficient for the usage of employees and able to support

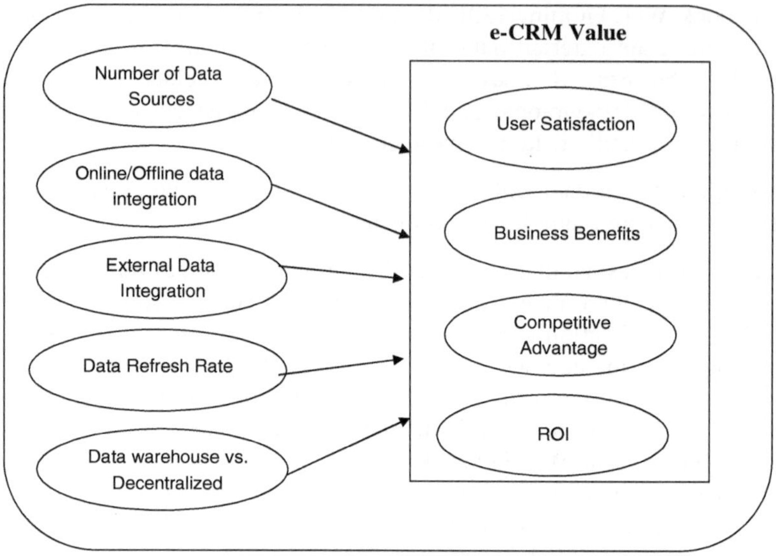

Fig. 1 e-CRM Value Framework

the use of e-CRM. In this fast growing technology society, e-CRM technologies must be one of other priorities that organization should consider. This is because e-CRM is the system that runs 24 × 7 operations. To maintain e-CRM and improve customer relationship continuously, organizations must run and operate the network and customer databases smoothly.

Acknowledgments The authors would like to thank UGC for providing fund for this project with an entitled“Design Tool of IPv6 Mobility for 4G-Networks”, under eleventh plan of Major Research Project scheme(Ref.No F.No 36- 167/2008(SR) dated 26.03.2009). Thanks also goes to the dedicated research group in the area of Advanced Computer Networking, Cloud and Grid computing and e-CRM at the Dept of Computer Science, Mangalore University, Mangalore, India, for many stimulating discussions for further improvement and future aspects of the project. Lastly but not least the author would like to thank everyone, including the anonymous reviewers.

References

1. Adams, C.R., Song, J.H.: Integrating decision technologies. MIS Q**13**. 199–209 (1989)
2. Aileen, K.: Electronic customer relationship management (eCRM): opportunities and challenges in a digital world. Ir. Mark. Rev. **18**(1–2), 58–68 (2006)
3. Ball, L., Harris, R.: SMIS member: a membership analysis. MIS Q. **6**(1), 19–38 (1982)
4. Ballou, D., Wang, R., Pazer, H., Tayi, G.K.: Modeling information manufacturing systems to determine information product quality. Manage. Sci. **44**(4), 462–484 (1998)
5. Bazaarvoice: Social commerce industry statistics. http://www.bazaarvoice.com/resources-sn/statistics-sn (2010)

6. Beck, B., Summer, J.: Data warehousing horizons: CRM: not just operational and collaborative. DMReview, September (2001)
7. Brancheau, J.C., Wetherbe, J.C.: Key issues in information systems management. MIS Q. **11**(1), 23–46 (1987)
8. Brancheau, J.C., Janz, B.D., Wetherbe, J.C.: Key issues in information systems management: 1994–95 sim delphi results. MIS Q. **20**(2), 225–242 (1996)
9. Bernett, H.G., Kuhn, M.D.: The emergence of electronic customer relationship management. Telecommun. Rev. **13**, 91–96 (2002)
10. Chen, Q., Chen, H.: Exploring the SFactors of eCRM strategies in practice. Database Mark. Customer Strat. Manage. **11**(4), 333–343 (2004)
11. Cutler, M., Sterne, J.: E-metrics: business metrics for the new economy. NetGenesis Corp. www.netgen.com/emetrics (2001)
12. Dean, E., Heather, J., Dane, G.: Customer relationship management strategic application and organizational effectiveness: an empirical investigation. J. Strateg. Mark. **17**(1), 75–96 (2008)
13. Dickinson, G.W., Leithesier, R.L., Wetherbe, J.C., Nechis, M.: Key information systems issues for the 1980s. MIS Q. **8**, 135–159 (1984)
14. Doll, W.J.,Torkzadeh, G.: The measurement of end-user computing satisfaction. MIS Q. 259–274 (1988)
15. Dotan A.: E-crm implementation: a comparison of three approaches. management of innovation and technology. In: 4th IEEE International Conference, pp. 457–462.50 21–24 Sept. (2008)
16. Eckerson, W., Watson, H.: Harnessing customer information for strategic advantage: technical challenges and business solutions. Industry study 2000, executive summary. The Data Warehousing Institute (2001)
17. Engleman, E.: Forecast points to steady climb for U.S. online retail sales. http://www.techflash.com/seattle/2010/03/forecastforonlineretailsalessteadyupwardclimb.html (2010)
18. Forsyth, R.: Successful CRM: global trends. http://www.crm-forum.com/library/conf/con-031/ Accessed 22 June 2001
19. Foss, Bryan, Stone, Merlin, Ekinci, Yuksel: What makes for CRM system success - or failure? J. Database Mark. Customer Strat. Manage. **15**(2), 68–78 (2008)
20. Hackathorn, Richard: Web farming for the data warehouse. Morgan Kaufmann Publishers, San Francisco (1998)
21. Heimbigner, D., McLeod, D.: A federated architecture for information management. ACM Trans. Off. Inf. Syst **3**(3), 253–278 (1985)
22. HubSpot : 2010 State of inbound marketing report. http://blog.hubspot.com/blog/tabid/6307/bid/5620/Companies-That-Blog-More-Have-More-Consistent-Sales.aspx (2010)
23. Hugh, W., Elizabeth, D., Malcolm, M.: Factors for success in customer relationship management (CRM) systems. J. Mark. Manage. **1**(18), 193–219 (2002)
24. Ing-long, W., Ching-Yi, H.: A strategy-based process for effectively determining system requirements in eCRM development. Inf. Softw. Technol. **51**(9), 1308–1318 (2009)
25. Jain, R., Jain, S., Dhar, U.: CUREL: a scale for measuring customer relationship management effectiveness in service sector. J. Serv. Res. **7**(1), 37–58 (2007)
26. Jayanthi, R., Vishal, B.: Principles for successful aCRM in organization. Direct Mark. Int. J. **2**(4), 239–247 (2009)
27. Jerry, F., Nicholas, C.R.: Electronic customer relationship management P. 51. M.E. Sharpe Inc., New York (2006)
28. Johnson, B.: Fault lines in CRM: new E-commerce business models and channel integration challenges, vol. 1, CRM Project, 1 Jan 2000
29. Johnson, C., Davis, E.: 2009 Cost of eBusiness operations and customer acquisition. Forrester research. http://www.forrester.com/rb/Research/2009 Cost of eBusiness operationsand customer/q/id/54509/t/2. Accessed 20 May 2009
30. Kale, S.H.: CRM failure and the seven deadly sins. Mark. Manage. **13**, 42–46 (2004)

31. King, S.F., Burgess, T.F.: Understanding success and failure in customer relationship management. Ind. Mark. Manage. **37**, 421–431 (2008)
32. Kotler, P., Fox, K.: Strategic marketing for educational institutions. Prentice Hall, Englewood Cliffs (1995)
33. Lee-Kelly, L., David, G., Robin, M.: e-CRM analytics: the role of data integration. J. Electron. Commer. Organ. **1**(3), 73–90 (2003)
34. Litwin, W., Mark, L., Roussopoulos, N.: Interoperability of multiple autonomous databases. ACM Comput. Surv. **22**(3), 267–293 (1990)
35. Martin, E.W.: Critical success factors of chief MIS/DP executives. MIS Q. **6**, 1–9 (1982)
36. Martin, J.: Information Engineering, Savant Research Studies, Carnforth, Lancashire. In: Brown, A. (ed.) Proceedings of the 9th European Conference on Information Technology (1986)
37. Mena, J.: Beyond the shopping cart. Intelligent Enterprise, 8 March (2001)
38. Missi, F., Alshawi, S., Fitzgerald, G.: Why CRM efforts fail: a study of the impact of data quality and data integration. In: Proceedings of the 38th Annual Hawaii international conference on system sciences (HICSS'05)—track 8, vol. 8, p. 216c.(2005)
39. Mohamed, K., Ning, S.: Effects of electronic customer relationship management on customer satisfaction: a temporal model. In: 38th Hawaii international conference on system sciences (2005)
40. Neiderman, F., Brancheau, J.C., Wetherbe, J.C.: Information systems management issues for the 1990s. MIS Q. **15**, 474–500 (1991)
41. Nemati, H.R., Barko, C.D.: Issues in organizational data mining: a survey of current practices. J. Data Wareh. Winter. **6**(1), 25–36 (2001)
42. Ravi, K., Macia, R.: E-Business 2.0 road map for success. Addison-Wesley, Boston. Cited in Aileen, K., 2006. Electronic customer relationship management (eCRM): opportunities and challenges in a digital world. Ir. Mark. Rev., 18, No.1–2, 58–68. 52 (2001)
43. Rendleman, J.: Customer data means money. Information week, 20 August 2001
44. SAS Institute: Staples: loyal customers and killer marketing. SAS COM Magazine, September/October. www.sas.com (2001)
45. Schneider, G.: Electronic commerce: the second wave, 5th edn. Thompson Learning, Boston Massachusetts (2004)
46. Swift, R.: Analytical CRM powers profitable relationships: creating success by letting customers guide you. DMReview, February (2002)
47. Vasset, D.: Using the data warehouse to drive your CRM effort. CIO.com. Accessed 23 Feb 2001
48. Winer, R.S.: A framework for customer relationship management. Calif. Manage. Rev. **43**(4), 89–106 (2001)

An Efficient Image Fusion Technique Using Wavelet with PCA

C. M. Sheela Rani

Abstract Image fusion is a process in which high-resolution Panchromatic Image (PAN) is combined with a low resolution Multispectral Image (MS) to form a new single image which contains both the spatial information of the PAN image and the spectral information of the MS image. By applying wavelet transform alone, the fusion result is often not good. Hence, when a wavelet transform is integrated with any traditional fusion method the fusion results are better. The decimated and undecimated wavelets used in image fusion can be categorized into three classes: Orthogonal, Biorthogonal, and Nonorthogonal. In this study, a fusion technique is proposed which uses both wavelet and PCA method for fusing the IRS-1D images using LISS III scanner for the locations Vishakhapatnam and Hyderabad, India. The proposed fusion results are compared using statistical performance measures and analyzed. It was ascertained that the wavelet with PCA is superior to the other wavelet transform methods.

Keywords Image fusion · Wavelet transforms · Principal component analysis (PCA) · Intensity-hue-saturation (IHS) · Performance measures

Introduction

Image fusion is a tool where gray-level high-resolution PAN image is integrated with a colored low resolution MS image to produce a fused image which retains the most desirable characteristics of PAN and MS images. The new image contains both the high-resolution spatial information about the PAN image and the spectral information about the MS image. The objective is to keep maximum spectral

---

C. M. Sheela Rani (✉)
Department of Computer Applications, Nalla Malla Reddy Engineering College, Hyderabad, India
e-mail: sheelarani11@yahoo.com

S. Sathiakumar et al. (eds.), *Proceedings of International Conference on Internet Computing and Information Communications*, Advances in Intelligent Systems and Computing 216,
DOI: 10.1007/978-81-322-1299-7_18,

information from the original MS image while increasing the spatial resolution. The fusion technique ensures that without introducing any artifacts or inconsistencies, all the important spatial and spectral information about the input images is transferred into the fused image. During recent years, many papers on Image Fusion based on Wavelet Transform have been published. King and Wang (2001) introduced a wavelet-based sharpening method that uses IHS transformation and biorthogonal wavelet decomposition. Hong and Zhang (2003) integrated IHS and wavelet to fuse Quickbird images and IKONOS images, and obtained promising results. Oguz Gungor and Jie Shan presented Evaluation of Satellite Image Fusion using Wavelet Transform method by testing a Quickbird and IKONOS images. Richard B.Gomez, Amin Jazeri, and Menas Kafatos proposed a Wavelet-based hyperspectral and multispectral image fusion between two spectral levels of a hyperspectral image and one band of multispectral image. Din-Chang Tseng, Yi-Ling Chen, and Michael S.C. Liu proposed Integer Wavelet Transform and Principal Component Analysis (PCA) to fuse low-resolution Landsat TM Multispectral images and SPOT Panchromatic (PAN) Image to generate spectrum-preserving high-resolution Multispectral images. Qu Jishuang, Wang Chao used a Wavelet Package-based Data Fusion method for Multitemporal Remote Sensing Image Processing. G. Hong and Y. Zhang (2008) presented Comparison and Improvement of Wavelet-based Image fusion to fuse Quickbird images and IKONOS images.

Generally, Image fusion is performed at three different levels at which the fusion takes place: pixel, feature, and decision level [1]. Nowadays, the technique of Image fusion is used in many areas such as remote sensing and medical imaging which leads to enhance interpretation by a human observer. This paper is organized into six sections where Sect. " Image Fusion Based on Wavelet Transform" describes the Image fusion based on wavelet transform; Sect. "Integration of Wavelet Transformation" describes the Integration of wavelet transform with a traditional method IHS and PCA; Sect. "Quality Assessment Techniques" deals with quality assessment techniques; Sect. "Results and Discussions" gives the results and discussions; and Sect. "Conclusions" draws the conclusions.

Image Fusion Based on Wavelet Transform

Image fusion using wavelet transforms provide multiscale and multiresolution analysis functions. Wavelet technique outperforms the standard fusion technique in spatial and spectral quality. The decimated wavelet transforms are discrete wavelet transforms (DWT) [2] and the undecimated wavelet transforms are stationary wavelet transforms (SWT). There are three general classes of wavelets—orthogonal, biorthogonal, and nonorthogonal in both DWT and SWT. The wavelet filters used for orthogonal wavelets are: Haar and Daubachies, for biorthogonal wavelets: Bior1.1, Bior1.3, Bior1.7 and for nonorthogonal wavelets: Meyers, Coiflets and

Symlets. The 2-D discrete wavelet transformation is used for fusing images. In this paper, the six wavelet fusion methods are discussed, i.e., orthogonal wavelet fusion with decimation (ORTH), biorthogonal wavelet fusion with decimation (BIOR), orthogonal wavelet fusion with undecimation (UORTH), biorthogonal wavelet fusion with undecimation (UBIOR), wavelets integrated with IHS transform (WIHS), and wavelets integrated with PCA transform (WPCA).

Additive-Based Image Fusion Method

Perform histogram match process between PAN and MS images to obtain three new PAN images. Use the wavelet Transform to decompose new PAN images and different bands of MS image. Add the detail images of the decomposed PAN image at different levels to the corresponding details of different bands in the MS image and obtain the new details component in the different bands of the MS image. Perform inverse wavelet transform on the bands of MS images, respectively, and obtain the fused image [3].

Integration of Wavelet Transformation

Wavelet transform is integrated with both IHS and PCA based on the same principle: to separate most of the spatial information of an MS image from its spectral information by means of linear transforms. The IHS transform separates the spatial information of the MS image as the intensity (I) component. In the same way, PCA separates the spatial information about the MSS image into the first principal component (PC1) [4].

Wavelet Transformation and IHS

The Intensity-Hue-Saturation (IHS) transform merges the images based on its ability to separate the spectral information of an RGB composition in its two components—Hue and Saturation, while isolating most of the spatial information in the Intensity component [5]. There are many algorithms to convert the color values (RGB) into values of Intensity, Hue, and Saturation [6–9]. This implies that the fusion will be applied to groups of three bands of the MS image. As a result of this transformation, we obtain the new Intensity, Hue, and Saturation components. The PAN image replaces the Intensity image. By applying the inverse transformation, the fused RGB image is obtained where the spatial details of the PAN image are incorporated.

Wavelet Transformation and PCA

Most sensors collect information in adjacent bands of the electromagnetic spectrum. In this context, PCA allows to synthesize the original bands creating new bands, the principal components, which pick up and reorganize most of the original information. In general, the first principal component (PC1) collects the information that is common to all the bands used as input data in the PCA, i.e., the spatial information, while the spectral information that is specific to each band is picked up in the other principal components [7]. This makes PCA an adequate technique when merging MS and PAN images. In this case, all the bands of the original MS image constitute the input data. As a result of this transform, we obtain non-correlated new bands, the principal components [10]. The PC1 is substituted by the PAN image, whose histogram has previously been matched with that of PC1. Finally, the inverse transform is applied to the whole dataset formed by the modified PAN image and the PC2 … PCn, obtaining in this manner the new fused bands with the spatial detail of PAN image incorporated into them.

Integration of Substitution Method with Wavelet Transform

Transform the MS image into IHS and PCA components. Perform histogram match between PAN image and Intensity component or PC1 to obtain a new PAN image. Decompose the histogram-matched PAN image and Intensity component or PC1 to wavelet planes, respectively. Replace the LLp of the New Pan with the LLI of the Intensity or PC1 decomposition and add the detail images of the New Pan to the corresponding detail images of the Intensity or PC1 decomposition and obtain LLI, LHPI, HHPI, and HLPI. Perform an inverse wavelet transform and generate a new Intensity or new PC1 component. Transform the new Intensity together with the Hue and Saturation components or new PC1 with PC2 and PC3 to get back into RGB space.

Quality Assessment Techniques

To assess the quality of fused image, some quality measures are required. The goal of image quality assessment is to supply quality metrics that can predict perceived image quality automatically. While visual inspection has limitations due to human judgment, qualitative approach based on the evaluation of “distortion” in the resulting fused image is more desirable for mathematical modeling [11].

Qualitative Measures

In mathematical modeling, qualitative measure is desirable. Qualitative measures are used to predict the quality of the perceived image. In this study, quality assessment using noise-based measures are used to evaluate the noise of the fused image by comparing to its original MS image. The following optimal noise-based measures are implemented to judge the performance of the above discussed fusion methods [12, 13].

1. Entropy: It is used to quantify the quantity of information contained in the fused image. A bigger value shows good fusion results.

$$H = -\sum_{i=0}^{L-1} h_F(i)\log_2 h_F(i) \tag{1}$$

   where h_F is the normalized histogram of the fused image and L is the number of gray levels.

2. Correlation Coefficient (CC): It is used to measure the similarities of the fused image to the corresponding original images.

$$\mathrm{CC} = \frac{E(XY) - E(X)E(Y)}{\sqrt{E(X^2) - E^2(X)}\sqrt{E(Y^2) - E^2(Y)}} \tag{2}$$

3. Mean Squared Error (MSE): It is used to measure the spectral distortion.

$$\mathrm{MSE} = \frac{\sum_{i=1}^{M}\sum_{j=1}^{N}\left(I_R(i,j) - I_F(i,j)^2\right)}{M * N} \tag{3}$$

   where I_R (i,j) denotes pixel (i,j) of the image reference and I_F (i,j) denotes pixel (i,j) of the fuse image, $M*N$ is the image size.

4. Peak Signal-to-Noise Ratio (PSNR): It is used to reveal the radiometric distortion of the final image compared to the original image.

$$\mathrm{PSNR}(dB) = 10\log_{10}(\frac{\mathrm{Peak}}{\mathrm{MSE2}}) \tag{4}$$

where peak is the maximum possible pixel value. Peak equals 255 for 8 bit images, 2047 for 11 bit images, and 65535 for 16 bit images.

5. Root Mean Squared Error (RMSE): It is used to measure the standard error of the fused image.

$$\mathrm{RMSE} = \sqrt{\mathrm{MSE}} \tag{5}$$

6. Mean absolute error (MAE): It is used to measure the average magnitude of the errors in a set of forecasts, without considering their direction. It measures accuracy for continuous variables.

$$\mathrm{MAE} = \frac{1}{M*N}\sum_{x=0}^{M-1}\sum_{y=0}^{N-1}(I_R(x,y) - I_F(x,y)) \tag{6}$$

where $I_{\mathrm{R}}(x, y)$ denotes pixel (x, y) of the image reference and $I_F(x, y)$ denotes pixel (x, y) of the fuse image, $M * N$ is the image size.

7. Mutual Information Measure (MIM): It is used to furnish the amount of information about one image in another. Given two images $M\ (i, j)$ and $N\ (i, j)$, MIM is defined as:

$$I_{MN} = \sum_{x,y} P_{MN}(x,y)\log\frac{P_{MN}(x,y)}{P_M(x)P_N(y)} \tag{7}$$

where $P_M\ (x)$ and $P_N\ (y)$ are the probability density functions in the individual images and $P_{MN}\ (x, y)$ is the joint probability density function.

8. Fusion Factor (FF): It is defined as

$$\mathrm{FF} = I_{\mathrm{AF}} + I_{\mathrm{BF}} \tag{8}$$

where A and B are given two images and F is the fused image. A higher value of FF indicates that fused image contains moderately good amount of information present in both the images.

9. Fusion Symmetry (FS): It is used to indicate the degree of symmetry in the information content from both the images.

$$\mathrm{FS} = \mathrm{abs}\left[\frac{I_{\mathrm{AF}}}{I_{\mathrm{AF}} + I_{\mathrm{BF}}} - 0.5\right] \tag{9}$$

The quality of the fusion technique depends on FS. When sensors are of good quality, FS should be as low as possible so that the fused image derives features from both the input images. If sensors are of low quality then it is better to maximize.

10. Fusion Index (FI): It is defined as

$$\mathrm{FI} = I_{\mathrm{AF}}/I_{\mathrm{BF}} \tag{10}$$

where I_{AF} is the mutual information index between MS image and fused image and I_{BF} is the mutual information index between PAN image and fused image. The quality of fusion technique depends on the degree of fusion index.

Results and Discussions

The above discussed six fusion methods were implemented and executed using Matlab 7.6.0 to compare their fusion results. The experiment was conducted and tested on IRS-1D PAN and MS images for the locations Vishakhapatnam and Hyderabad and the following results were evaluated. The undecimation orthogonal wavelet is used for WIHS and WPCA method (Table 1).

The following are the fused images for the location Vishakhapatnam using the methods ORTH, BIOR, UORTH, UBIOR, WIHS, and WPCA (Figs. 1, 2, 3, 4, 5, 6, 7 and 8).

The following are the fused images about the location Hyderabad using the methods ORTH, BIOR, UORTH, UBIOR, WIHS, and WPCA (Figs. 9, 10, 11, 12, 13, 14, 15, 16).

Table 1 Test case 1 for comparison of different metrics using different methods for Vishakhapatnam image

Metrics	ORTH	BIOR	UORTH	UBIOR	WIHS	WPCA
Entropy	7.7677	7.7728	7.7840	7.8048	7.8325	7.9008
CC	0.9671	0.9651	0.9806	0.9811	0.9731	0.9890
MSE	0.0117	0.0120	0.0094	0.0092	0.0099	0.0062
PSNR	67.459	67.353	68.400	68.475	68.179	70.240
RMSE	0.1080	0.1094	0.0969	0.0961	0.0994	0.0784
MAE	0.0820	0.0830	0.0741	0.0735	0.0771	0.0601
MIM	1.0995	1.0690	1.2081	1.1931	1.2167	2.2209
FF	2.1991	2.1379	2.4162	2.3862	2.4335	4.4418
FS	0.1387	0.1405	0.1481	0.1524	0.1367	0.3092
FI	0.5656	0.5612	0.5430	0.5329	0.5707	0.2358

Fig. 1 PAN image

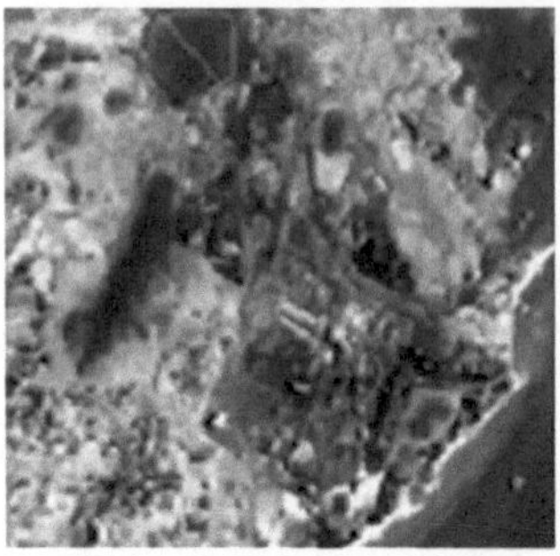

Fig. 2 MS image

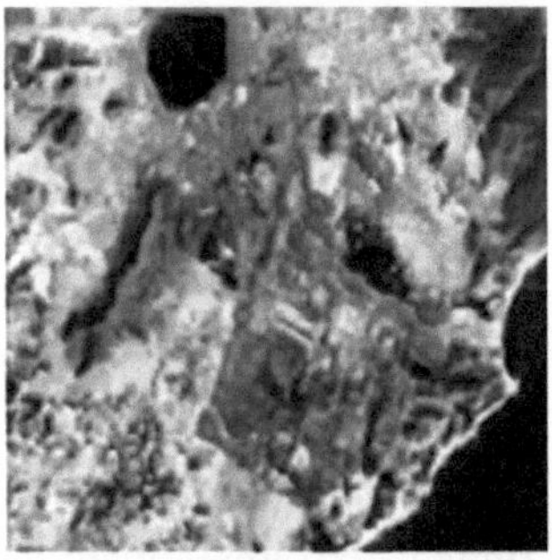

Fig. 3 ORTH

Fig. 4 BIOR

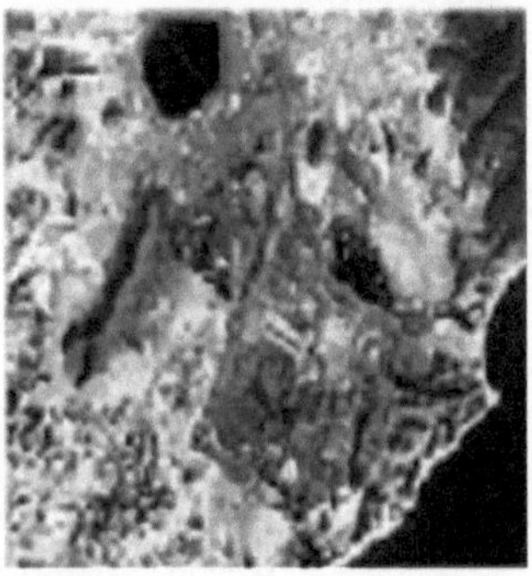

Fig. 5 UORTH

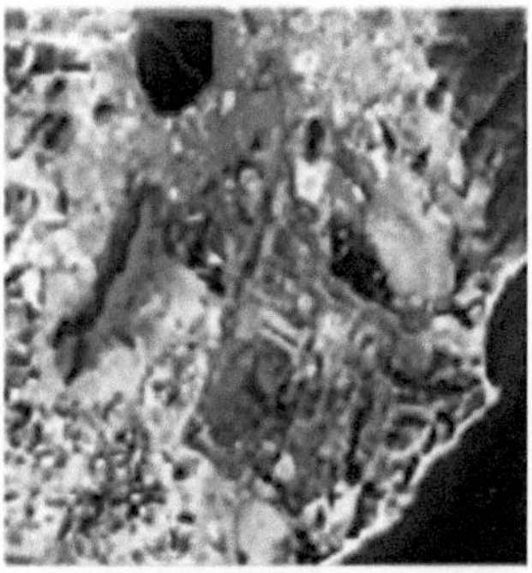

Fig. 6 UBIOR

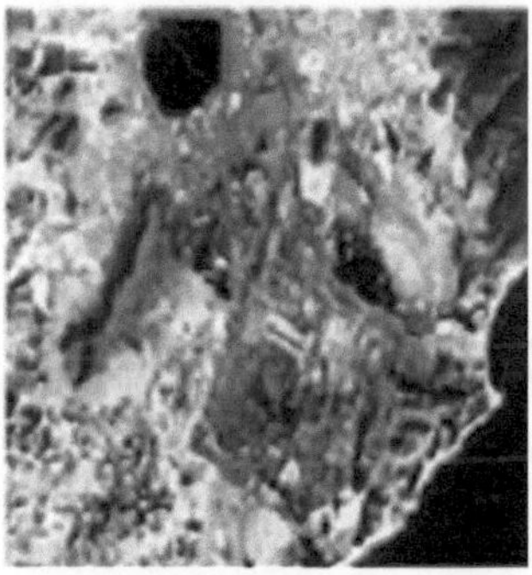

Fig. 7 WIHS

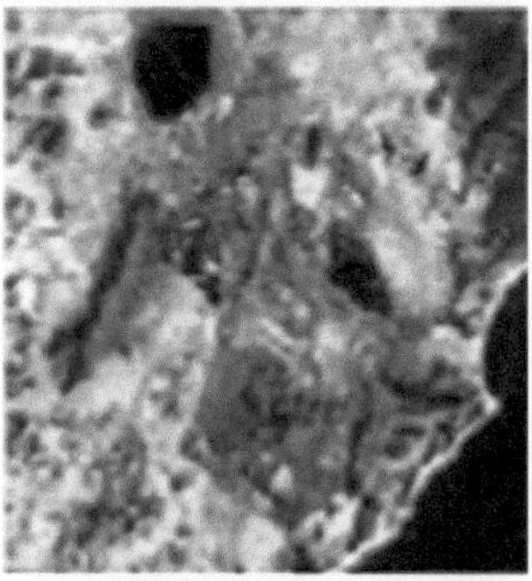

Fig. 8 WPCA

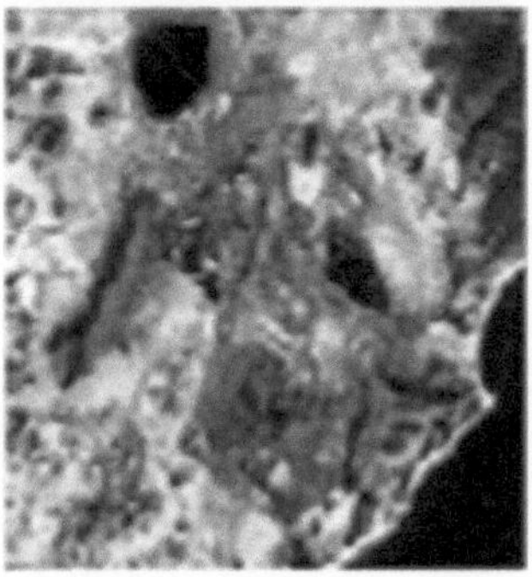

Fig. 9 PAN image

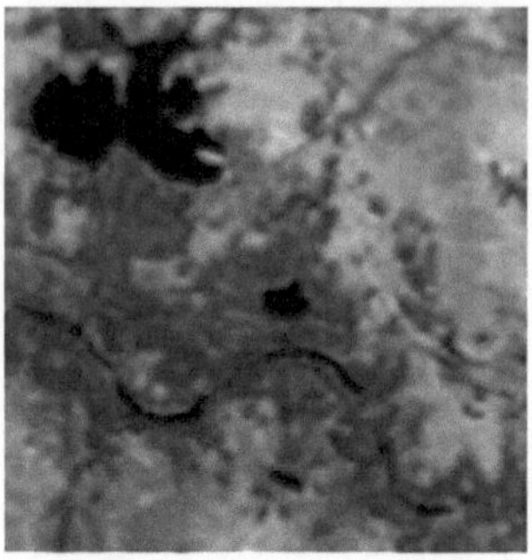

Fig. 10 MS image

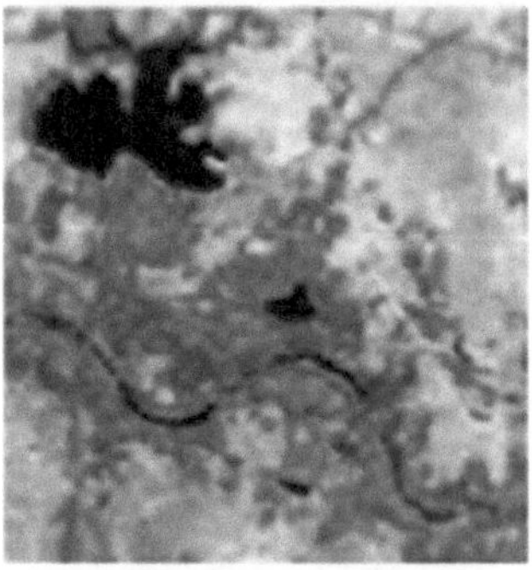

Fig. 11 ORTH

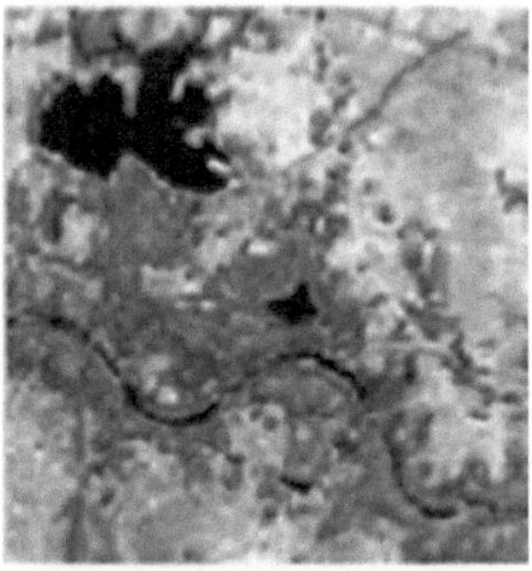

Fig. 12 BIOR

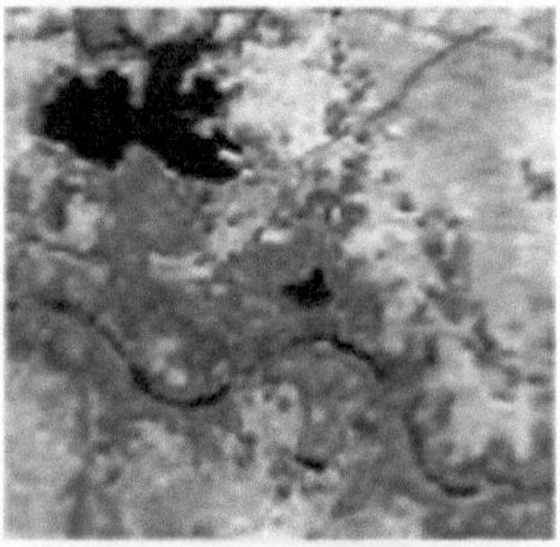

Fig. 13 UORTH

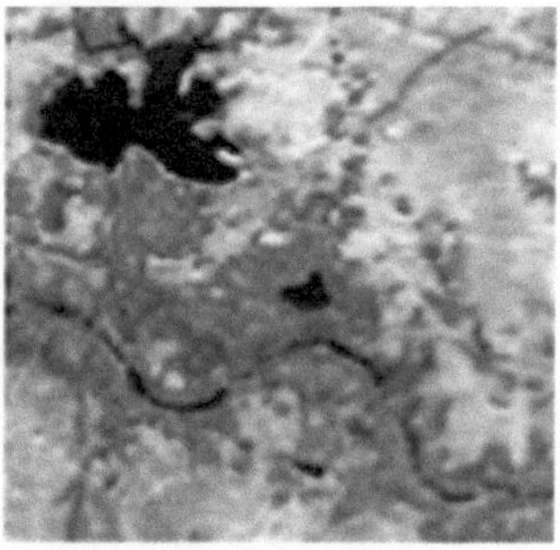

Fig. 14 UBIOR

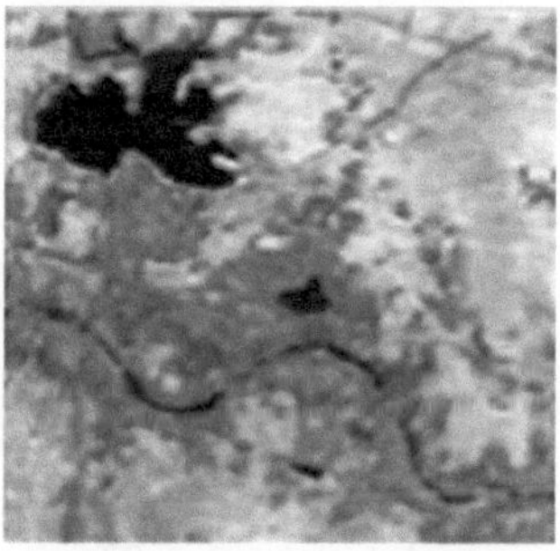

Fig. 15 WIHS

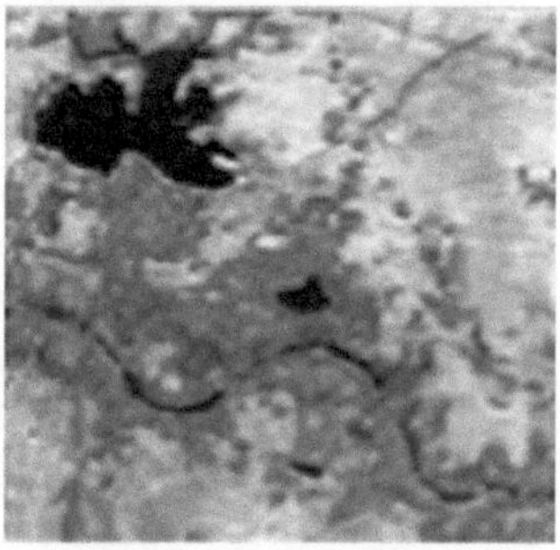

Fig. 16 WPCA

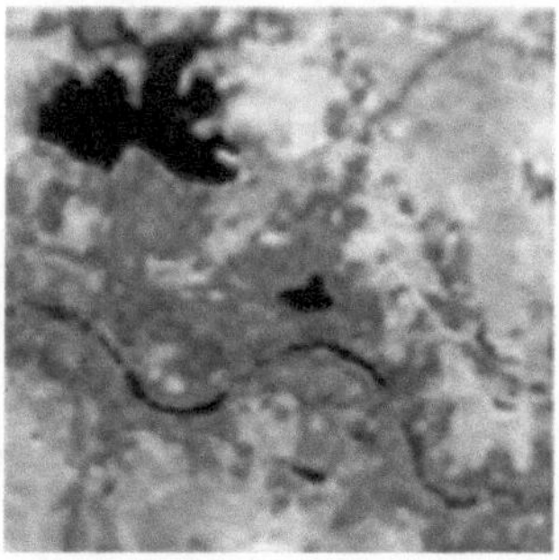

Table 2 Test case 2 for comparison of different metrics using different methods for Hyderabad image

Metrics	ORTH	BIOR	UORTH	UBIOR	WIHS	WPCA
Entropy	7.1814	7.2059	7.1395	7.1559	7.0600	7.8087
CC	0.9640	0.9619	0.9778	0.9787	0.9364	0.9820
MSE	0.0061	0.0062	0.0052	0.0051	0.0068	0.0038
PSNR	70.248	70.173	70.988	71.059	69.778	72.349
RMSE	0.0784	0.0790	0.0720	0.0714	0.0827	0.0615
MAE	0.0617	0.0622	0.0569	0.0564	0.0652	0.0478
MIM	1.1660	1.1420	1.2743	1.2676	1.1778	2.7935
FF	2.3321	2.2839	2.5487	2.5353	2.3556	3.5869
FS	0.0952	0.1000	0.1107	0.1179	0.0633	0.2518
FI	0.6800	0.6666	0.6375	0.6183	0.7753	0.3301

Conclusions

The potentials of Image fusion using wavelet with PCA are explored. The various fusion results are analyzed using quality performance metrics. The higher value for entropy is achieved for wavelet with PCA method. The higher the value of the correlation coefficients, the more similar is the fused image to the corresponding original MS image. The higher correlation between the high-frequency components of the fusion result and the high-frequency component of the PAN image indicates that more spatial information from the PAN image has been injected into the fusion result. The higher value of PSNR implies that the spectral information in MS image was preserved effectively. It also implies that high signal was preserved. The higher values of MM, FS, and FF indicate that symmetry is achieved by retaining spectral information. The smaller value for MSE, RMSE, MAE, and FI were obtained for Wavelet with PCA method. The parameters PSNR, MIM, FF, and FS are maximum for wavelet with PCA method and MSE, RMSE, MAE, and FI are minimum for wavelet with PCA method. Hence, it is ascertained that Wavelet with PCA method has superior performance than Wavelet with IHS method and other different Wavelet Transformation domains. The results are verified for LISS III images and the study can be extended for other types of images (Table 2).

References

1. Lehigh Univ: http://www.eecs.lehigh.edu/SPCRL/IF/image_fusion.htm
2. Li, Z., Jing, Z., Yang, X., Sun, S.: Color transfer based remote sensing image fusion using non-separable wavelet frame transform. Pattern Recogn. Lett. **26** (13), 2006–2014 (2005)
3. Hong, G., Zhang, Y.: Comparison and improvement of wavelet-based image fusion. Int. J. Remote Sens. **29**(3), 673–691 (2008)

4. Chavez, P.S., Kwarteng, A.Y.: Extracting spectral contrast in land sat thematic mapper image data using selective principal component analysis. Photogramm. Eng. Remote Sens. **55**, 339–348 (1989)
5. Pohl, C., Van Genderen, J.L.: Multisensor image fusion in remote sensing: concepts, methods and applications. Int. J. Remote Sens. **19**, 823–854 (1998)
6. Smith, A.R.: Color gamut transform pairs. Comput. Graph. **12**, 12–19 (1978)
7. ACM.: Status report of the graphics standard planning committee. Comput. Graph. **13** (3), 1979
8. Schetselaar, E.M.: Fusion by the IHS transform: should we use cylindrical or spherical coordinates? Int. J. Remote Sens. **19**, 759–765 (1998)
9. Chibani, Y., Houacine, A.: Fusion of multispectral and radar image in the redundant wavelet domain. In: Proceedings of the SPIE, vol. 3500, pp. 330–338, Sept. 1998
10. Gonzalez-Ausicana, M., Saleta, JL., Catalan, RG., Garcia, R.: Fusion of multispectral and panchromatic images using improved IHS and PCA mergers based on wavelet decomposition. IEEE Trans. Geosci. Remote Sens. **42**(6), 2004
11. Shi, W., Zhu, C., Tian, Y., Nichol, J.: Wavelet-based image fusion and quality assessment. Int. J. Appl. Earth Obs. Geoinf. **6**, 241–251 (2005)
12. Li, S., Li, Z., Gong, J.: Multivariate statistical of measures for assessing the quality of image fusion. Int. J. Image Data Fusion (2010)
13. Yuhendra, Wavelet PCA based images fusion techniques and quality assessment

Forensic Investigation Processes for Cyber Crime and Cyber Space

K. K. Sindhu, Rupali Kombade, Reena Gadge and B. B. Meshram

Abstract Computers are an integral part of our life. A significant percentage of today's transactions and processes take place using the computer and Internet. People have readily adopted Internet technology and innocently trust it while using it with the ignorance of the limitations and threats to the system security. With the advance of technology, equally or more advanced form of crimes started emerging. Different types of cyber attacks from various sources may adversely affect computers, software, a network, an agency's operations, an industry, or the Internet itself. Thus companies and their products aim to take assistance of legal and computer forensics. Digital forensics deals with computer-based evidence to determine who, what, where, when, and how crimes are being committed. Computer and network forensics has evolved to assure proper presentation of cyber crime evidentiary data into court. Forensic tools and techniques are an integral part of criminal investigations used to investigate suspect systems, gathering and preserving evidence, reconstructing or simulating the event, and assessing the current state of an event. In this paper we deliberate on two aspects; first, various types of crimes in the cyber space and various sources of cyber attacks, and second, investigation processes for various cyber attacks with the help of digital forensic tools like WinHex [1].

Keywords Cyber crime · Cyber space · Digital forensic · Network forensic · File system forensic · Email forensic

K. K. Sindhu (✉) · R. Kombade · R. Gadge · B. B. Meshram
Computer Department, Veermata Jijabai Technological Institute, Mumbai, India
e-mail: kksindu@gmail.com

R. Kombade
e-mail: rupalikombade@gmail.com

R. Gadge
e-mail: reena.gadge10@gmail.com

B. B. Meshram
e-mail: bbmeshram@vjti.org.in

S. Sathiakumar et al. (eds.), *Proceedings of International Conference on Internet Computing and Information Communications*, Advances in Intelligent Systems and Computing 216,
DOI: 10.1007/978-81-322-1299-7_19,

Introduction

Digital forensics is new in the forensic science and efforts are underway to experiment, explore, discover, enhance, and reconstruct the incidents that work together to make the investigations complete and successful. A digital forensic investigation model is to make the investigation perfect and error free. Each footstep in investigation is decisive and evidence of scrutiny has to surface the facts. Overlooking one step or interchanging any of the steps may lead to incomplete or inconclusive results and erroneous interpretations. A computer crime culprit may escape from light of justice or an innocent suspect may suffer negative consequences. Ultimately, evidence left is in zeros and ones, so forensics investigation can be mislaid by criminal thoughts. Higher levels of concerns are to be on account of a forensics investigation to avoid any misleading and thereby any loop holes to the accused. In this paper we explain the investigation processes of crime committed in cyber space [2].

Cyber space is the electronic space of computer communication or networks. Cyber space was imagined and actually implemented as a borderless new space, transcending physical borders and formal legal rules. Within the past few years a new class of crime scenes has become more prevalent, that is, crimes committed within electronic or digital domains, particularly within cyber space.

Cyber Forensic is the discovery, analysis, and reconstruction of Evidence extracted from and/or contained in a computer, computer system, computer network, computer media, or computer peripheral. A digital forensic investigation is an inquiry into the unfamiliar or questionable activities in a cyber space or digital world. According to the Oxford Dictionary digital forensic science is the systematic gathering of information about electronic devices that can be used in a court of law. Digital forensic science is more popularly called digital forensics and sometimes also called computer forensics as digital forensics is the science of identifying, extracting, analyzing, and presenting the digital evidence that has been stored in the digital electronic storage devices to be used in a court of law.

A cyber crime can be defined as crime committed in the cyber space or crime committed with the assistance of the Internet. In cyber crime externally or internally the computer takes part in the attack. Cyber crime investigations are always difficult because the evidence are very critical, i.e., the life of data are sometimes within fractions of a second. Evidences in the running memory registers are available only in some seconds. Digital evidence [3] at present, the analysis of digital evidence, must depend on forensic tools such as Forensic Toolkit (FTK) of Encase, or WinHex. Most of them are commercial software and are too expensive for the small enterprises or individual. Digital evidence stored in a computer can play a major role in a wide range of crimes, including murder, rape, computer intrusions, espionage, and child pornography as proof of a fact of what did or did not happen. Digital information is fragile in that it can be easily modified, duplicated, restored, or destroyed, etc. In the course of the investigation, the

Table 1 Shows sources of evidence in different types of files in a computer

Sources of evidence in a computer	Description
User created files	Address books, E-mail files. Audio/video files. Image/graphics files. Calendars. Internet bookmarks/favorites. Database files. Spreadsheet files. Documents or text files.
User protected files	Compressed files. Misnamed files. Encrypted files. Password-protected files. Hidden files. Steganography.
Computer–created files	Backup files. Log files. Configuration files. Printer spool files. Cookies. Swap files. Hidden files. System files. History files. Temporary files.
Other data areas	Bad clusters, Computer date, time and password. Deleted files, Free space, Hidden partitions. Lost clusters, Metadata. Other partitions. Reserved areas, Slack space, Software registration information, System areas.

investigator should assure that digital evidence is not modified without proper authorization. The typical goal of an investigation is to collect evidence using generally acceptable methods in order to make the evidence accepted and admitted at court.

The final forensic report must include:

(1) Where the evidence was stored?
(2) Who had obtained the evidence?
(3) What had been done to the evidence?

Any step in the process must be carefully recorded in order to prove that the electronic records were not altered in the investigation procedure (Table 1).

Types of Cyber Crimes

Computers are an integral part of our life. A significant percentage of today's transactions and processes take place using information technology and the future is pregnant with innovations, including nanotechnology, silicon chips, quantum computers, and even biochips. People have readily adopted this technology and have innocently trusted it while performing many tasks, with ignorance about the limitations and threats to their securities. With this advance in technology, an equally advanced form of crimes has emerged. The crimes being committed in cyber space like Internet fraud, business espionage, pornography, sexual assault, online child exploitation, cyber terrorism, and more are on the rise. The following statistical data shows various attacks and their total percentage.

Table 2 Shows the statistical data on different types of attacks reported

Attack	Reported cases (%)
Data theft	33
Email abuse	22
Unauthorized access	19
Data alteration	15
Virus attacks	5
DoS attacks	3
Others	3

Data Theft

Data are precious assets in this modern age of Cyberworld. Data are important raw materials for business organizations, call centers, and I.T. companies. Data have also become an important tool and weapon for companies, to capture larger market shares. Due to the importance of data, in this new age, its security has become a major issue in the I.T. industry. The piracy of data is a threat faced by I.T. players who spend millions to compile or buy data from the market. Their profits depend upon the security of the data. The above statistics reveals that 33 % of cyber crime is data stealing (Table 2).

A case has been reported in Bangalore (9 Crore loss) where some key employees of the company had stolen source code and launched a new product based on stolen source code and mailed to former clients. Social engineering techniques can also be applied for such attacks. For example, a beautiful lady meets the young system admin and collects the username and password.

Email Abuse

Email abuse takes many forms, for example: unsolicited commercial email, unsolicited bulk email, mail bombs, email harassment, and email containing abusive or offensive content. The format for submitting reports to the abuse department regarding abuse of email is always the same whatever the offence.

Unauthorized Access

Unauthorized Access is when a person who does not have permission to connect to or use a system gains entry in a manner unintended by the system owner. The popular term is "hacking".Hacking is the viewing of private accounts, messages, files, or resources, when one has not been given permission from the owner to do so. Viewing confidential information without permission or qualifications can result in legal action.

Data Alteration

Changing /modifying /deleting data causes major losses in the cyber world. In a crime reported in the USA (Cyber murder), a patient file data altered by a criminal caused overdose of medicine and the patient got killed.

Denial of Service

A denial of service (DoS) attack is an incident in which a user or organization is deprived of the services of a resource they would normally expect to have. In a distributed denial-of-service, large numbers of compromised systems (sometimes called a bonnet) attack a single target. Dos causes

- Attempts to "flood" a network, thereby preventing legitimate network traffic.
- Attempts to disrupt connections between two machines, thereby preventing access to a service.
- Attempts to prevent a particular individual from accessing a service.
- Attempts to disrupt service to a specific system or person.

Impact of the DoS is Denial-of-service attacks can essentially disable your computer or your network. Depending on the nature of your enterprise, this can effectively disable your organization. Secondly, some denial-of-service attacks can be executed with limited resources against a large, sophisticated site.

Malicious Codes

Viruses, worms, and Trojans are types of malicious codes which enter into the system without permission of the user and delete, modify, and capture the user files and data.

Sources of Attacks

Cyber crimes such as network intrusion, hacking, virus distribution, denial of service attacks, hijacking (a computer or network), defacing web sites, cyber stalking, and cyber terrorism are included in this category.

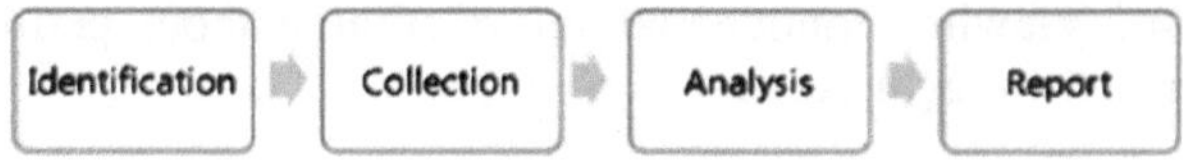

Fig. 1 Shows investigation processes

Basically, the computer itself becomes the "target" as well as "Source" of the crime. This is "unauthorized access" to the targeted system. The transmission of a program, information, code, or command, and as a result of intentionally causes damage without authorization, to a protected computer.

- Program/program source code.
- Disgruntled employees.
- Teenagers.
- Political Hacker.
- Professional Hackers.
- Business Rival.
- Desperados.
- Terrorists.

The Cyber Investigation Process of a Compromised System

Investigation Processes

The entire investigation process can be divided into four phases (Fig. 1).

1. Identification: In this phase it collects the information about the compromised system. System Configuration, software loaded, user profiles, etc [4].
2. Collection Phase: Collects the evidence from the following.
3. Evidence is most commonly found in files that are stored on hard drives and storage devices and media.
4. If file is deleted, recovering data from the deleted files and also collects evidence file deleted files.
5. Analysis phase: Analyze the collecting data/files and find out the actual evidence.
6. Report phase: The audience will be able to understand the evidence data acquired from the evidence collection and analysis phases. The report generation phase records the evidence data found out by each analysis component. Additionally, it records the time and provides hash values of the collected evidence for the chain-of-custody.

Investigating Tools

The goal of the investigator is to find every digital evidence stored in devices, or at least sufficient information for building and supporting a crime logic. Investigating tools are used to collect evidence from the crime scene. Throughout our paper mainly WinHex is using for image creation of compromised disk or folder and image analysis. WinHex [1] can recover deleted files.

Cyber Crime Investigation Steps

1. Assesses the crime scene
2. Evidence collection
    2.1 Select a tool, e.g.: WinHex [1].
    2.2 Create image of the compromised system disk.
        2.2.1 Open WinHex.
        2.2.2 Open particular drive (Tools → open disk).
        2.2.3 Calculate Hash value of the drive/disk (Tools → compute hash) Store hash value in a text file.
        2.2.4 Save disk content as image file extension.img file (Fig. 2).
3. Analyze the Disk image
    3.1 Calculate Hash value of image (Tools → compute hash).
    3.2 Compare the Hash value of original with image. If equal start analysis else acquired data altered.
    3.3 Recover the necessary files and deleted files from the disk image. (Specialist → Interpret as image).
    3.4 Copied into a folder.
    3.5 Start analysis of the content recovered files of files.
    3.6 Image analysis, i.e., hidden data inside an image can be analyzed using steganography tools (Stools).
    3.7 Check header and footer of application file. Copy header and footer and paste into text pad. Sometimes evidence should be present in header and footers.
4. Conclude the investigation and generate report.

Recover Deleted File Using WinHex

1 Open Drive image [1].
2 Make as file view mode.

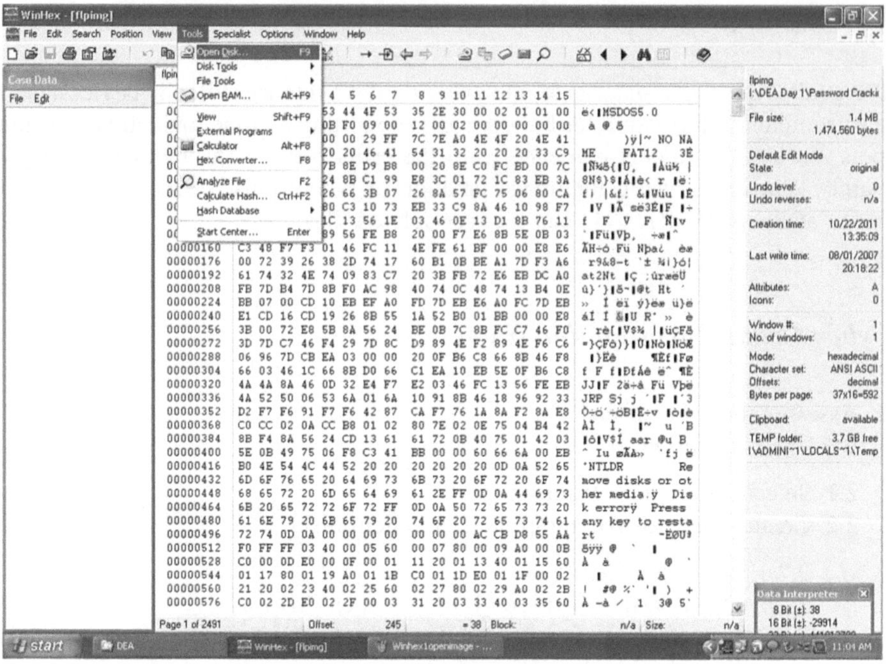

Fig. 2 Shows screenshot of Winhex [1] for creating an image of DISK

3 Recover the necessary files and deleted files from the disk image. (Specialist → Interpret as image).
4 Copy into a folder.
5 Start analysis of the content recovered files.
6 Open drive image–select file and right click.

Investigation on IPR Crime–Source Code Theft

IP crime is generally known as counterfeiting and piracy. Counterfeiting is intentional trademark infringement, while piracy involves intentional copyright infringement. Now, IPR crimes are becoming a big issue in big businesses. However, it is not a new phenomenon but due to globalization and advances in technology counterfeiting and piracy become big business (Fig. 3).

The investigation procedure of a source code theft

1. Assess the crime scene.
2. Find which source code files are thefts.
3. Calculate hash value of the theft files and save it into text file.
4. Find out who are involved in the project.

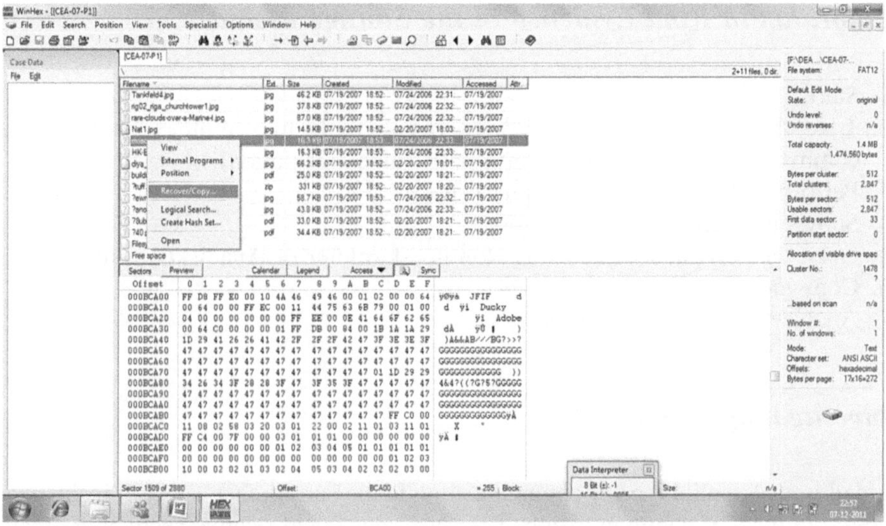

Fig. 3 Screenshot of Winhex tool [1] to recover/copy deleted file

5. Investigate their USBs and PCs–Take an image of all drives using Winhex tool…
6. Investigate their email accounts—take an image of Email folders using Winhex tool.
7. Analyze the drives–Open drive images in Winhex.

   (Tools → open disk).

8. Calculate hash value of the drive/disk.

   (Tools → compute hash).

9. Store hash value in a text file.
10. Find out if source codes are available in drives.
11. If deleted recover all deleted files using winHex (Right click on the files folder and recover).
12. Again calculate the hash value of the recovered files.
13. If the hash value of the recovered source code files are equal, then commit the crime happened by the person who owns the above files.
14. Sometimes source code can hide inside the image.
15. We can find out the hidden data using steganography tools or click the particular image in the winHex and check the content displaying window. If the content contains hidden text data then it will display directly.

Algorithm to Find Evidence in the Storage Media

1. Start.
2. Take the image of the disk drive (using tools like Winhex or Linux dd command) (For integrity, perform hash value calculation).
3. Perform chkdisk of the particular drive.
4. Shows the number of bad sectors.
5. Open the particular drive using Runtime Explorer or Anadisk/BXDR.
6. Copy the content of bad clusters.
7. Analyze using any hexeditors.

Investigating Emails

Email is an essential type of communication in the current fast world. It is the most preferred form of communication. The ease, speed, and relative secrecy of emails have made it a powerful tool for criminals.

The following are the major email-related crimes.

1 Email spoofing.
2 Sending malicious codes through email.
3 Email bombing.
4 Sending threatening emails.
5 Defamatory emails.
6 Email frauds.

Email Investigation Procedure

Email investigation can be done with two methods Email tracking and Email tracing. Email Tracking tells us that tracking down an IP address will give a general idea of what city, state, and other geographical information pertains to the original sender. You can also determine what ISP a computer user is networked with through an IP address lookup tool. www.ReadNotify.com is an online service for email tracking. It tracks an email as to when it was read/reopened/forwarded, and much more. In cases where only an email ID is obtained as clue to track the sender of an email, services like ReadNotify.com can prove helpful. Email Tracing gives other information, such as how many times an email was sent to various servers and is an important method used for determining the original source of an email. By tracing an email you can determine the original sender's IP address, therefore giving you a geographical location of the email sender.

1. Assess the crime scene.
2. Take a copy of compromised email folder.

3. Analyze Email folder.

   3.1.1 Open Email folders–inbox, outbox, spam.
   3.1.2 Open each mail and analyze header of email.
   3.1.3 From header can identify IP address of source.

4. Analyze Body message.

   4.1.1 Copy body message into a notepad (text file).It shows any hidden formatted messages.

5. Download any attachments the particular email has.

If it contains any images, then check with steganography tools. All these analyses give the necessary evidence of the email crime.

Networks Forensic

Network forensics is a subfield of digital forensics where evidence is captured from networks and interpretation is substantially based on the knowledge of cyber attacks. It aims to locate attackers and reconstruct their attacks actions through the analysis of log files and monitoring network traffic.

Log File Analysis

In many cyber crime cases (especially web defacement cases), analysis of FTP and web server logs gives the most crucial evidence—IP address of the suspect [5]. Log files are key informers of web usage. Log files typically keep logs of the requests they receive. Data that are often logged by web servers for each request include Timestamps; IP address, Web server version, Web browser version, and OS of the host making the request; Type of request (GET/POST) read data or write data; the resource is requested and the status code. The response to each request includes a three-digit status code that indicates the success or failure of the request. Log file investigation procedure.

Step1 Collect server log file from the network administrator.
Step2 Open file notepad.
Step3 Searching key words (Given in the case scenario–or IP or admin username) from this file using searching utility of notepad.
Step 4 Copy and Paste those particular lines into another text file.
Step 5 Analyze all fields of logfile, for e.g.: IP, Access Time, file Path, status codes, methods (GET/POST).

Network Traffic Analysis

Network traffic analysis will help investigators to reconstruct and analyze network-based attacks and inappropriate network usage, as well as to troubleshoot various types of operational problems. Network forensic analysis relies on all of the layers. Analysts begin to examine data, likely an IP address, protocol, and port information. Collecting network traffic can create legal issues. Capture (intentional or incidental) of information breaks privacy or security implications, such as passwords or the contents of e-mails.

Sources of Network Traffic

- Firewalls and routers.
- Packet sniffers and protocol analyzers.
- Intrusion detection system.
- Remote access.
- Network forensic analysis tools.
- Other sources are:
- Dynamic Host Configuration Protocol Servers.
- Network Monitoring Software.
- Internet Service Provider Records.
- Client/Server Applications.
- Hosts. Network Configurations and Connections.

Examination and Analysis of Network Traffic

The examination process extracts and prepares data for analysis. The examination process involves data translation, reduction, recovery, organization, and searching. For example, known files are excluded to reduce the amount of data, and encrypted data are decrypted whenever possible to recover incriminating evidence. A thorough examination results in all relevant data being organized and presented in a manner that facilitates detailed analysis. The analysis process involves critical thinking, assessment, experimentation, fusion, correlation, and validation to gain an understanding of and reach conclusions about the incident based on available evidence (Casey and Palmer, 2004). In general, the aim of the analysis process is to gain insight into what happened, where, when, how, who was involved, and why [6].

The first step in the examination process is the identification of an event of interest. There are two types of identification.

1. Someone within the organization–system administrator, network administrator, user, or employee.
2. Some monitoring system like IDS or Firewall alerts showing an incident happened [6].

Child Pornography Investigation Examination Process Would Include

1. Examining all graphics or video files from network traffic.
2. Examining all web sites accessed.
3. Examining all Internet communications such as IRC, Instant Messaging (IM), and email.
4. A search for specific usernames and keywords to locate additional data that may be relevant.
5. Most of the relevant data to the investigation have been extracted from network traffic.
6. Extracted data made readable.
7. They can be organized in ways that help an individual analyze them.
8. Gain an understanding of the crime.

As the analysis process proceeds, a complete picture of the crime emerges often resulting in leads or questions that require the analyst to return to the original data to locate additional evidence, test hypotheses, and validate specific conclusions (Fig. 4).

Flow Reconstruction [6]

Because most networks use TCP/IP to transmit data between hosts each TCP connection is bi-directional, comprising one flow for receiving data and a second flow for sending data. Because TCP breaks data into packets prior to transmission, tools for examining network traffic require some ability to reconstruct flows [6].

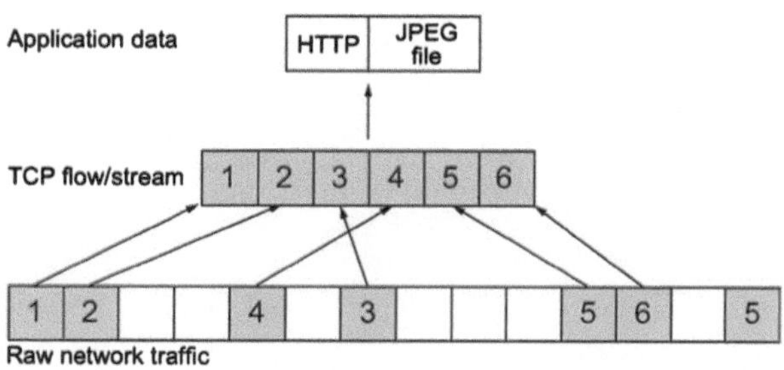

Fig. 4 A conceptual representation of packets in network traffic relating to a single flow being extracted and reconstituted to obtain the data they carry [6]

Conclusion

This paper explains introduction to digital forensics and summarizes different types of attacks and its sources in Sect. "Types of Cyber Crimes". Section "The Cyber Investigation Process of a Compromised System" explains steps in the investigation process practically with WinHex tool. Section "Networks Forensic" explains the network forensics and investigation procedures in the log files as well as in the Network traffic.

References

1. WinHex (http://www.WinHex.com)
2. Carrier, B.D., Spafford, E.H.: Categories of digital investigation analysis techniques based on the computer history model. J. Digit Invest. Sci. 3S, S121–S130 (2006)
3. Di Pietro, R., Verde, N.V.: Digital forensic techniques and tools chapter 17 of security handbook of electronic security and digital forensics
4. Choi, J., Savoldi, A., Gubian, P., Lee, S., Lee, S.: Live forensic analysis of a compromised Linux system using LECT (Linux Evidence Collection Tool) 2008 IEEE
5. Arasteh, A.R., Debbabi, M., Sakha, A., Saleh, M.: Analyzing multiple logs for forensic evidence. Digit. Invest. J. Sci. 4S, S82–S91 (2007)
6. Casey, E.: Network traffic as a source of evidence: tool strengths, weaknesses, and future needs. Digit. Invest. J. 9–148 (2003)

Feasibility Study for Implementing Brain Computer Interface Using Electroencephalograph

S. S. Sridhar and R. Shivaraman

Abstract The purpose of a Brain Computer Interface (BCI) consists of the development of an interface between a human and a computer to allow the control of a device only via brain signals. The conformance of the system and the individual brain patterns of the subject is the major concern while developing the interface. In this paper, we begin with the exploration of variety of brain sensing technologies for detecting the specific forms of brain activity used in HCI research and then analyzing the properties of the most prevalent technology used in HCI research – Electroencephalograph. We have then proceeded with our aim of conducting a feasibility study of this implementation idea and verified the same by clearly determining the requirements and setting up of the appropriate facilities for research. We have also described and discussed the experiments showing the differences in the brain wave patterns for proving the same.

Keywords Brain Computer Interface · Brain sensing technologies · Electroencephalography · Non-invasive interface · Electroencephalogram · 10–20 Electrode placement · EEG wave groups · Task classification

Introduction

A Brain Computer Interface sometimes called the direct neural interface or a brain machine interface is a method of acquiring and analyzing the signals generated in

S. S. Sridhar (✉)
School of Computing, SRM University, Chennai, India
e-mail: sridhar.srm@gmail.com

R. Shivaraman
Nano Technology Research Centre, SRM University, Chennai, India
e-mail: shivi.masti@gmail.com

S. Sathiakumar et al. (eds.), *Proceedings of International Conference on Internet Computing and Information Communications*, Advances in Intelligent Systems and Computing 216,
DOI: 10.1007/978-81-322-1299-7_20, © Springer India 2014

the brain with a goal of creating direct communication pathway between a human or animal brain and an external device. This implies that neural impulses generated by the user's brain are detected, elaborated, and utilized by the machine, approximately in real time, to perform definite tasks. It is one of the most challenging visions to most of the computer manufacturers throughout the world who have an idea of building such a computer that is controlled just by the human's thought process alone.

The research on BCI began in the 1970s but it was not until the 1990s that the first working experimental implants in human appeared. Several human working implants exist and have been designed to restore damaged hearing, sight, and movement. For example, in neural prosthesis, a small computer chip is substituted for the damaged inner ear control organ to enable sound waves to be transformed into electrical signals the brain can interpret. In the case of degenerative disorders like cerebral palsy, BCI is considered to be helpful in their functional recovery by bypassing the damaged area [1]. With advances in technology and knowledge, pioneering researchers are able to make plausible attempts to produce true BCI that augment human functions rather than simply restoring them.

Science fiction movies often present very unrealistic and chilling views of BCI. It is important to dispel this myth that BCI technology provides the ability to interpret minds and decipher thoughts. Rather, it provides a method for users to communicate their intensions to the outside world directly from their thoughts. Such intensions are limited to simple things like whether or not to select a particular icon or a letter from a group of them, or where a cursor should move and in which direction it should move. This ability to convey these very fundamental intentions is sufficient to engineer some very complex and useful tasks such as navigating the desktop, using simple move-and-click functions, using a virtual keyboard, playing simple two-dimensional (2D) games, composing an email, or browsing the internet.

The need for motor movements in computer interfaces is challenging and rewarding, but the full potential of brain sensing technologies as an input mechanism lies in the extremely rich information it could provide about the state of the user. Having access to this state is valuable to HCI researchers because it allows us to derive more direct measures of traditionally elusive phenomenon. Knowing the state of the user as well as the tasks they are performing can provide key information that would allow us to design context sensitive systems that adapt themselves to optimally support the state of the user [2].

In this paper, we explore the different brain sensing technologies that can be applied to HCI research and analyze their properties and limitations. We also study the principle of a non-invasive brain sensing technology Electroencephalograph, which we have used for the initial scanning of the brain wave patterns by conducting experiments and discussing the results. This has led us to the way of studying the feasibility of implementing a Brain Computer Interface by encephalon mapping techniques.

Conceptualization of BCI

The Brain Computer Interfaces can be categorized in different ways, as invasive and non-invasive, synchronous and asynchronous, universal and individual, online and offline BCIs, and BCIs based on the Electroencephalogram (EEG) features [3].

The non-Invasive BCIs are based on the EEG measured with the scalp electrodes. In Invasive BCIs, the electrical activity of the brain is recorded from inside the head with one or more microelectrodes, which can record the activity of a single neuron. Invasive systems are mainly used to experiment with animals.

Brain Computer Interfaces that work in a synchronous mode are in an externally paced mode. The user must produce specific mental states in a predefined time window. The control is system initiated. In an asynchronous mode, the brain activity is analyzed continuously. The user can freely initiate the specific mental task(s) used as the control signal(s) and here the control is user-initiated.

Universal BCI relies on the assumption that by gathering EEG data from few users it is possible to find a classification function that should be valid for everybody. So the BCI is the same for all users. In individual BCI, the fact that no two people are the same, both physiologically and psychologically is taken into account.

Online BCIs are the actual working BCIs. The signal processing, feature extraction, classification, and device control are done in real time. This makes it possible to provide feedback for the user. This is not possible in the offline BCI. The EEG is typically recorded as in online BCI, but using more electrodes. The recordings are then stored and the actual BCI research is done later. This makes it possible for us to examine, for example, different electrode positions, preprocessing and feature extraction methods, classifiers, etc. The results are comparable with the same kind of online BCI, if all the recorded EEG data are used. These offline systems are generally used for the algorithms of classification.

Brain Computer Interfaces are also categorized according to what EEG feature or features they try to detect. These features include α or β rhythm amplitude, P300 and action potential of a single cortical neuron.

The human brain is a dense network consisting of approximately 100 billion nerve cells called neurons. Each neuron communicates with thousands of others to regulate physical processes and produce thought. Neurons communicate either by sending electrical signals to other neurons through physical connections and by exchanging chemicals called neurotransmitters. Advances in brain sensing technologies enable us to observe these electrical, chemical, or blood flow changes as the brain processes information or responds to various stimuli.

Experiments in human, utilizing modern invasive and non-invasive brain imaging technologies as interfaces have been conduced by many HCI research groups. There exist various techniques to monitor the brain activity which have some disadvantage or the other making them unsuitable for the purpose of usage as a crucial component in a Brain Computer Interface. Techniques like Electrocorticography (ECoG) and the Single Neuron Recording (SNR) using microelectrodes are invasive technique for recording the brain wave patterns thus involving surgery. The functional magnetic resonance imaging (fMRI) and magneto encephalography (MEG) are extremely expensive techniques and magnetic shield is required during their usage. Positron emission tomography (PET) and Single Photon Emission Computer Tomorgraphy (SPECT) have a risk of radiation exposure. Optical brain imaging (OBI), magnetic resonance imaging (MRI), and the computed tomography (CT) provide only anatomical data like changes in blood flow or brain images which cannot provide direct information about the brainwave signals. Event-related Optical Signal/Functional Near-Infrared (EROS/fNIR) is a technique still in infancy and is not very cost efficient. The most commonly studied potential interface designed for the humans has been Electroencephalography in lieu of its advantages such as its fine temporal resolution, ease of use, portability, cost efficiency, non-invasive nature, continuous and instantaneous availability of data and free from radiation exposure. Moreover, EEG measures the brain's electrical activity directly, while other methods record changes in blood flow (e.g., SPECT, fMRI) or metabolic activity (e.g., PET), which are indirect markers of brain's electrical activity. However, practical use of EEG as a BCI requires a great deal of user training and is highly susceptible to noise.

The idea of producing a true BCI using EEG is really appreciated as producing a movement either in real or virtual world by imagining is quite different than using one's own hand in doing so [4]. The main challenge in EEG-based BCIs is to identify the particular EEG signal component that can be successfully used as control command as shown in the general architecture of BCI in Fig. 1.

A group of the most important authors in the field of non-invasive BCIs gave the list of goals important for future progress of these systems. The general operation of BCI includes, the placing of EEG Cap on the scalp, making the subject to imagine various movements, use the data from this to extract features that reflect the dynamic properties of the brain signals. The effects of the artifacts such as the electrode movement, eye movement, electrode pop and vascular activity should be removed. The second classifier should be built using advanced machine learning techniques. We can then start the BCI experiment such a mental

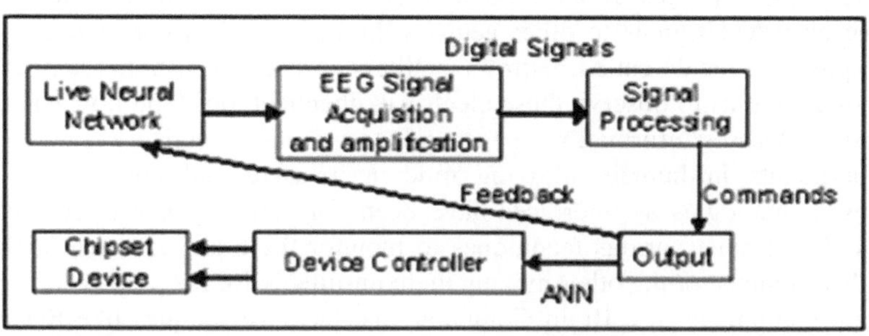

Fig. 1 General architecture of BCI

desktop navigator with real-time feedback. The result is then crosschecked manually by a brief visual inspection. [5].

Thus, the future progress on the development of BCIs will depend on:

- Identification of those signals that users are best able to control.
- Development of training methods for helping users to gain and maintain that control.
- Delineation of the best algorithms for translating these signals into device commands.
- Attention to elimination of artifacts such as electro-myographic (muscle) and electro-oculographic (eye movement) activity.
- Adoption of precise and objective procedures for evaluating BCI performance [6].

The electrical activity of the human nervous system has been recognized for more than a century. It is well known that the variation of the surface potential distribution on the scalp reflects functional activities emerging from the underlying brain. Just as the activity in a computer can be perceived on different levels, from the activity of individual transistors to the function of applications, so can the electrical activity of the brain be described on relatively large scales. Electroen cephalography is the measurement of the surface potential variation that is recorded by affixing an array of electrodes to the scalp, and measuring the voltage between the pairs of these electrodes. This instantaneous data that is obtained from the EEG system is known as the Electroencephalogram.

Electrodes conduct voltage potentials at microvolt level signals and carry them to amplifiers that magnify the signals approximately ten thousand times. The use of this technology depends strongly on the electrodes positioning using the 10–20 Electrode placement system and the electrode conduction as in Fig. 2.

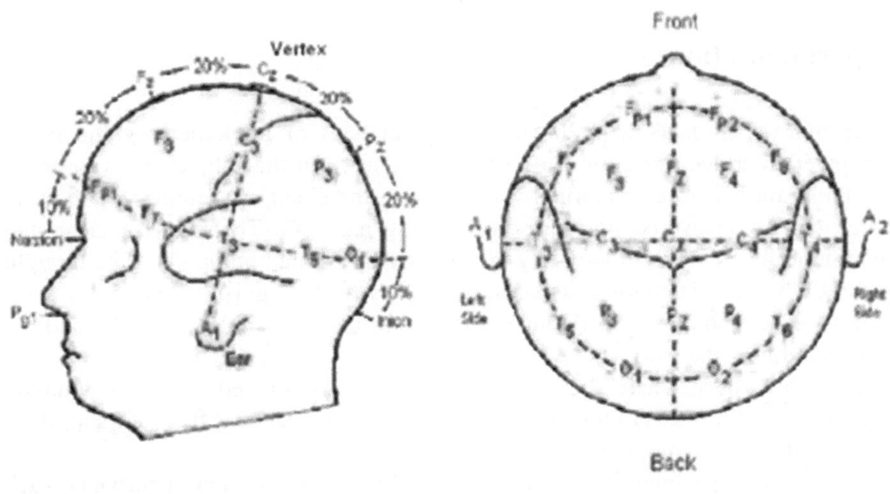

Fig. 2 International 10–20 electrode placement

For this reason, electrodes are usually constructed from conductive materials such as gold or silver chloride, with an approximate diameter of 1 cm, and subjects must also use a conductive gel on the scalp to maintain an acceptable signal-to-noise ratio (the power ratio between a signal - the meaningful information and the background noise).

The set up procedure requires about 10 min. Once the experiment is complete, we can remove the electrodes and subjects could wash of any remaining gel and paste with a brief water rinse.

The electrical activity goes on continuously in every living human's brain. We may sleep one-third of our life times, but the brain never rests. Even when one is unconscious the brain remains active. Unless a person is dead, brain-electrical activity is always present, that is, that person's EEG (voltage vs. time) curve constantly varies. Apart from artifacts because of muscle contractions or eye movement, etc., the signals exhibit contributions from frequencies between 0.5 Hz and about 70 Hz [7]. The analysis of these continuous EEG signals or brain waves is complex, due to the large amount of information received from every electrode.

Different waves, like so many radio stations are categorized by the frequency of the emanations and in some cases by the shape of their waveforms. Although none of these waves are ever emitted alone, the state of consciousness of the individual may make one frequency range more pronounced than others.

Six types are particularly important. They are in the increasing order of their frequencies – Delta, Theta, Alpha, Mu, Beta, and the Gamma waves. Most attempts to control a computer with continuous EEG measurement work by monitoring alpha, beta, and mu waves, because people can learn to change the amplitude of these two waves by making the appropriate mental effort. A person might accomplish this result, for instance, by recalling some strongly stimulating image or by raising his or her level of attention and concentration.

Experimentation

Since this paper deals with the feasibility analysis of implementing the Brain Computer Interface, we conducted some basic tests on the subject so as to prove that the brain generated electrical waves due to the electrical activity in the brain which can be measured non-intrusively using an EEG. There was a clear distinction in the electrical impulses generated by the human brain when the thought process is varied. The emphasis is on the ability to distinguish between the wave patterns generated during different activities and on the complete analysis and the classification of the same.

The work space where these experiments were conducted was very suitable, comfortable with free and uncongested space. It had clean and fresh-air supply. It was also designed completely noise proof.

During the experimentation, we had very little or no external disturbance such as squeaking doors or creaking floors. In the course of experimentation, the subject was asked to perform a number of things to activate his brain in various ways. Each experiment was conducted by asking the subject to concentrate on the task at hand and after around 20 min into the task or after the generated waves stabilized into an observable rhythm the graph generated was frozen to take the reading.

Experiment-1 Meditation

During the first experiment, the subject was asked to listen to a mild and soothing music which involved making the subject to meditate on the music. The environment was controlled to ensure that his meditation was deep and his concentration was directed inward to his self. The graph in Fig. 3 shows an 8 s EEG data that was obtained from the 21 Channel EEG while conducting this experiment. We find a constant stream of low frequency waves and their prolongation was evident in the posterior cortex (parietal region) of the brain.

Experiment-2 Mental Arithmetic

In this phase of experimentation, the subject was asked to perform a series of nearly 20 simple single digit mathematical calculations (like $2 + 2 = ?$) thus making him involve himself in active and busy thinking and concentration. From the graph as shown in Fig. 4, which is a 8 s segment of the 21 channel EEG, we

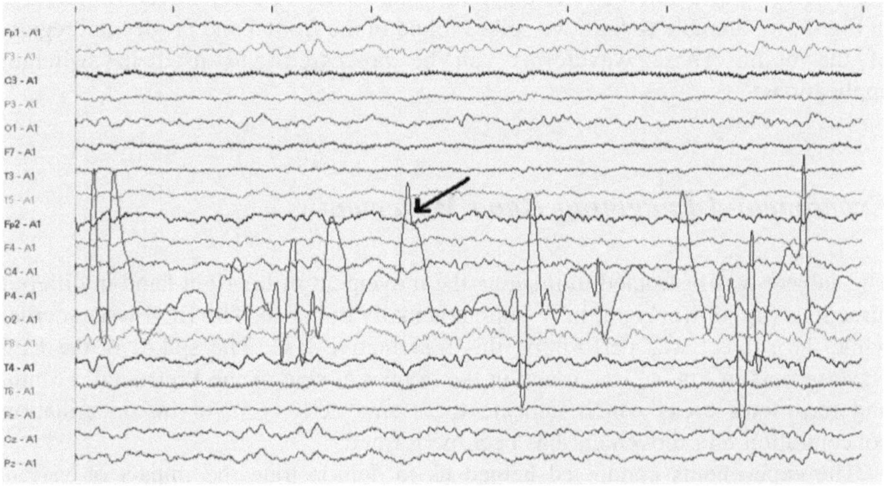

Fig. 3 Meditation

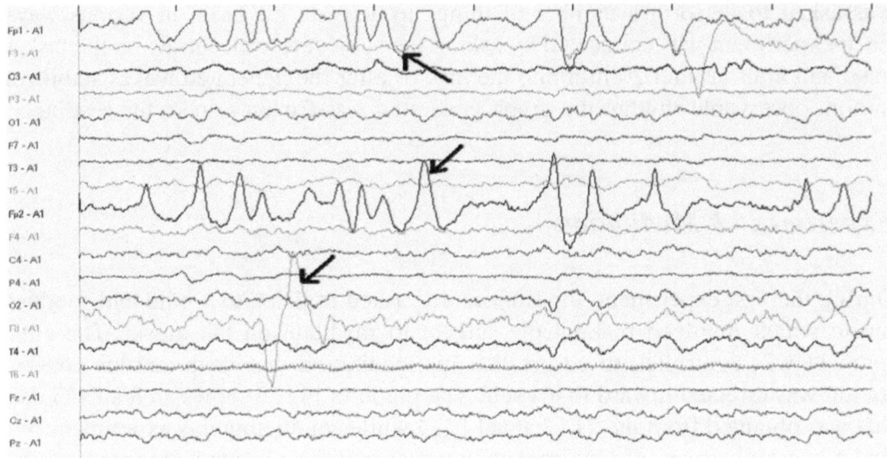

Fig. 4 Mental arithmetic

find persistent high frequency waves in the pre-frontal and the frontal cortex of the brain.

Experiment-3 Open/Close Eyes

This experiment was conducted to analyze the feasibility of BCI to be used in switches (e.g., opening and closing applications, Switching ON and OFF an appliance, etc.). The subject was asked to close his eyes fast and then wait for few seconds and was again asked to open his eyes. From the EEG waveform graph as in Fig. 5, we are able to find two spikes found in the frontal and pre-frontal regions of the brain. These waveforms can be processed and used in switching applications.

Experiment-4 Imagining Hand Movement

The subject was instructed to imagine the movement of his right hand in different directions (up, down, right, left) without actually doing so. The EEG was recorded while the subject was performing this imaginative task. The spikes in the EEG waveform as in Fig. 6, was found in the dominant side of the Prefrontal, Frontal, and the motor areas which indicate that some activity involving imagination/ concentration and movement has been performed.

The experiments conducted helped us to demonstrate the impact of various tasks on EEG-based task classification in a computing scenario. We also found that

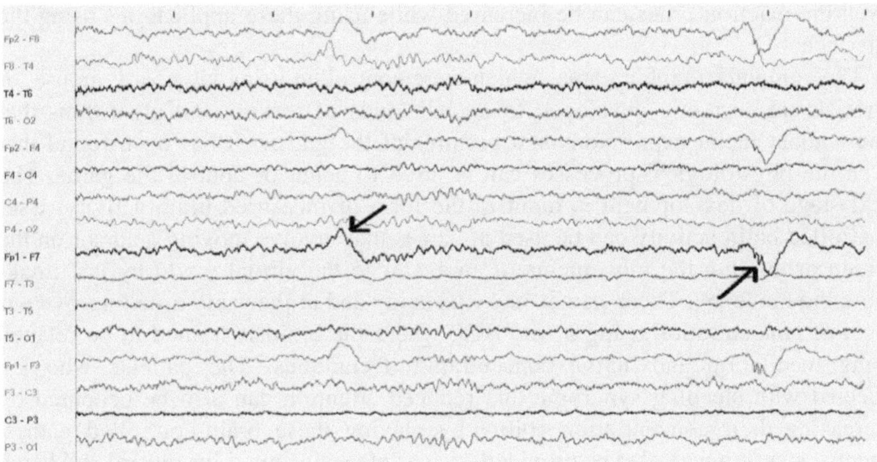

Fig. 5 Open/Close eyes

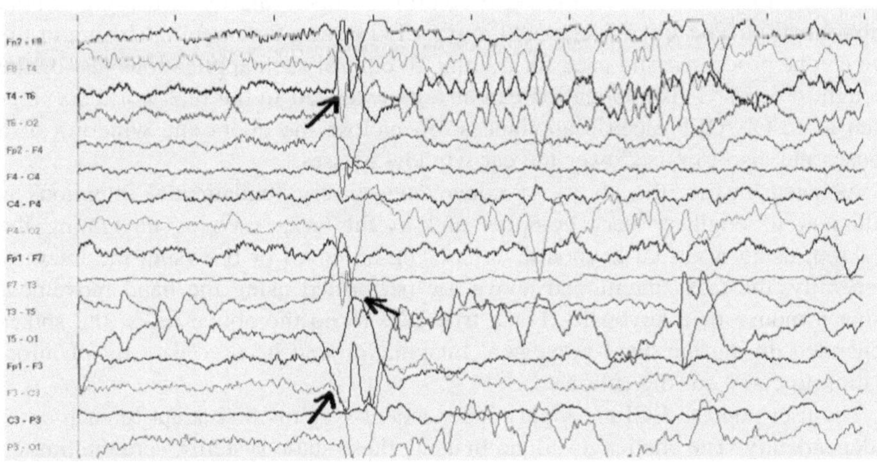

Fig. 6 Imagining hand movement

EEG shows interesting potential as a general physiological input sensor for distinguishing between tasks in a wide variety of computing applications without requiring detailed prior knowledge of the tasks.

Discussion

Having a control over the computer just by thought is the basic aim of a cerebral interface. Several applications of the online BCI include game playing, controlling applications, communication through an email and sensor-assisted navigation.

Even the reaction times can be increased while using these applications using the interface.

The computer graphic games which were controlled using keyboard, mouse, or joystick are currently developed to use all kinds of sensors and algorithms that know about facial expressions and gestures of the gamer. It has been found that even the physiological processes can be used to adapt or control the game. The next stage of development is input in the form of measured brain activity. User controlled brain activity can be used in games that involve moving a cursor on the screen or guiding the movements of an avatar in the virtual world by just imagining the move [8]. These games make us enjoy and at the same time improve our level of concentration using a true BCI, where the brain is trained to be relaxed using meditation and active concentration techniques. The patients who are affected with sleeping syndrome and reduced attention can also be benefited by increasing their concentration ability by playing these brain-controlled games. Feedback system can also be provided such as pleasing sound for reward and harsh tone for punishment while designing such games [9].

Computer applications (a form or a 2D game or a web browser or a word processor) to be used generally have controls to open and close using a simple click function. The spikes generated in the EEG during the opening/closing of the eye can be processed and used for training to control such applications just by the thought process. This application can also be extended to the real world scenario such as to ON/OFF electric appliances, open/close the doors and windows of a house, and also to raise/lower the car window glasses.

As cited earlier, the ability to convey these very fundamental intentions is sufficient to engineer very complex and useful tasks such as navigating the desktop, using a virtual keyboard, composing an email or browsing the internet. Generally, the tasks mentioned above are performed using the hand movement using a mouse or a keyboard. For a true BCI to do the above tasks, the spikes generated during the hand movement imagination can be used as a bit of information to carry out the action.

The user and the BCI are highly independent systems that adapt to each other independently. The challenges found in using these static systems were the limited number of brain activities recorded due the high temporal sampling and limited special resolution of the EEG. The users until trained, find it very complicated to reproduce the mental states sufficiently similar to be correctly classified. Moreover, the preparation of EEG sensors is very time-consuming and it is not suitable to be used daily.

The higher end EEG systems (20 electrode channels or more) benefit the researches by improving their chance in detecting stimulus related brain signals. Extremely high-end EEG systems (256 channels or more) have started exploring the idea of using models of the human brain to perform source locating with respect to the brain. However, this work still remains very experimental. Neuroscience researchers wish to make claims regarding the neurological behavior of the brain and try to get abundant information regarding the brain activity using sophisticated techniques. However, these details can only be used for performing task classification and detection.

Conclusion

While the electroencephalograph was invented centuries ago, it is only recently that researches have begun to apply it to problem outside the medical and neurophysiology domains such as the implementation of a direct neural interface. Applying this technology to these domains depends not only on our understanding the device, but also on the analysis of the feasibility of implementing our creative and innovative BCIs using the same. We believe that this work represents an initial step in exploring how several brain sensing technologies can be applied in a relevant manner to the present-day BCI research.

We also believe that we were able to prove, theoretically as well as with the help of experiments, that the brain generated electrical waves due to electrical activity which can be non-intrusively measured using an EEG. We were also able prove that there is a clear distinction between the patterns generated during various thought process, various levels of concentration and various functioning that can be tapped to create a BCI. Thus, we can conclude that the implementation of a Brain Computer Interface using encephalon mapping techniques using EEG is possible. In the future, efforts can be taken to generate consistent and reliable EEG wave patterns by training the subject that would be helpful in isolating a particular frequency of brain waves. These brain waves can be used to trigger a logical response from a hardware component enabling us to lead towards our ultimate goal of implementing a Brain Computer Interface.

References

1. Friehs, G.M., Zerris, V.A., Ojakangas, C.L., Fellows, M.R., Donoghue, J.P.: Brain machine and brain computer interfaces. Stroke **35**, 2702–2705 (2004)
2. Lee, J.C., Tan, D.S.: Using a Low-Cost Electroencephalograph for task classification in HCI research. Proceedings of the 19th annual ACM symposium on User interface software and technology UIST'06, pp. 81-90, Oct 15–8 (2006)
3. Lehtonen, J.: EEG-based brain computer interfaces. Thesis, Department of Electrical and Communications Engineering, Helsinki University of Technology
4. Popescu, F., Badower, Y., Fazli, S., Dornhege, G., Muller, K.R.: EEG-based control of reaching to visual targets, Dynamical Principles for neuroscience and intelligent biomimetic devices—Abstracts of the EPFLLATSIS Symposium, vol 1–2, pp. 123–124, (2006)
5. Müller, K.-R., Blankertz, B.: Toward noninvasive brain-computer interfaces. IEEE Signal Process. Mag. **23**, 125–128 (2006)
6. Carpi, F., D Rossil.: Non invasive brain-machine interfaces—A Report. University of Pisa, Interdepartmental Research Center "E.Piaggio". http://www.esa.int/act
7. Felzer T: On the possibility of developing a brain-computer interface. Department of computer Science, Technical University of Darmstadt (2001)

8. Nijholt, A., Tan, D.: Playing with Your Brain: Brain-computer interfaces and games. University of Twente, Microsoft Research. Proceedings of the international conference on Advances in computer entertainment technology, 305–306, (ACM) (2007)
9. Dzhafarova, O.A., Donskaya, O.G., Zubkov, A.A.: Biofeedback Computer Games: Development and Current State. Biomedical engineering springer science and business media, Inc.**41**(4), 186–190 (2007)

Scope of Cloud Computing for Multimedia Application

Uttam Vijay and Lalit Kumar Awasthi

Abstract The paper deals with most recent emerging computing technology cloud computing and how it can be used for multimedia application. With so many sources of data having multimedia features, the amount of data needed to be processed have increased significantly and alarmingly. So processing these data on the system itself is becoming very inefficient due to limited resources of the system especially in case of mobile device with limited memory and battery life. So this paper tries to analyze whether the concept of cloud computing can be implemented for processing and storing data related to multimedia application and if yes then what are the difficulties which could arise.

Keywords Cloud computing · Service level agreement · Multimedia on cloud · Media edge cloud computing

Introduction

Cloud computing is a computing mechanism for providing on-demand access to the computing resources such as processor, servers, memory on payment basis. Basically cloud computing is a pay per use model where clients who want to utilize cloud service need to pay for the resource they are using. Cloud computing basically deals with concept of virtualisation [1] where the resources required by the client is provided virtually from the pool of resources cloud service providers have. To provide smooth and coherent running of the cloud system, a very

U. Vijay (✉) · L. K. Awasthi
National Institute of Technology, Hamirpur, India
e-mail: contactuttam.uttam@gmail.com

L. K. Awasthi
e-mail: lalitdec@gmail.com

S. Sathiakumar et al. (eds.), *Proceedings of International Conference on Internet Computing and Information Communications*, Advances in Intelligent Systems and Computing 216,
DOI: 10.1007/978-81-322-1299-7_21,

important component is service level agreement (SLA) which is signed between the cloud service provider and client and violation of the agreement can lead to penalty for the culprit side.

Cloud computing is divided into three layers or classes. They are, namely, Infrastructure as a service (IAAS), Platform as a service (PAAS), and Software as a service (SAAS). IAAS deals with the providing of virtualized resources such as storage, processor to the client. PAAS deals with the programming of cloud itself. It provides an environment where application can be created and deployed according to the need on cloud. SAAS is the top layer which is utilized by end users where services can be accessed using the Internet.

There are three types of models on which cloud can be deployed. They are private, public, and hybrid models. When cloud computing runs within company own data center and infrastructure for its internal use it is said to be private cloud. When cloud service is available to any user through the Internet on per use basis it is said to be public cloud. Public cloud is the basic model of the cloud computing as it is intended that everyone get the benefit of cloud services. But due to certain security issue as public cloud can be accessed by anyone the private model is implemented by certain organization. In private model, infrastructure and other scopes are very limited and thus is not very efficient so to get benefits of both the types, that is, security of cloud and unlimited scalability of public cloud concept of hybrid cloud is integrated in the model.

Why Cloud for Multimedia

In this age, Internet multimedia has emerged as a vast service. Even mobile multimedia has grown very significantly. So to provide rich and effective media services multimedia computing has emerged as a significant technology. These services include to generate, edit, process, and search media contents such as images, audio, video, graphics, etc. In multimedia computing, so many data coming due to more internet and mobile users at same time computation required is very high. So if we can transfer these computations on cloud, it will release a lot of load and thus a need of multimedia cloud computing has come in demand.

Challenges in Implementing Multimedia on Cloud and Need for Multimedia Cloud computing

There are different challenges in implementing multimedia in cloud. It is basically due to the heterogeneity in services, network, devices, and quality of services (QoS) required. These days so many different services available for multimedia like VOIP, photo sharing, streaming, etc., that too happening simultaneously for millions of user and cloud need to accommodate all the processing required. Also different types of network like LAN, wireless, 3G, etc., available these days and cloud need to adapt according to the network being used by the service. Cloud also need to adapt according to the different devices like PC, TV, mobile phones, etc., and also different quality of services required by them. These challenges can't be managed by general cloud. General cloud is based on utility mechanism, that is, it is used for allocating processor and storage on demand but multimedia computing have other requirement like bandwidth delay, jitter apart from storage and processing. So general cloud is not eligible to provide such operation as if we consider a video file with delay in transmission of packets will create problem and thus a low quality of experience (QoE).

How Cloud can be used for Multimedia Application

There are two ways in which cloud can be used for multimedia. First way is from cloud perspective, that is, how an existing cloud can be utilized by multimedia to perform processing and also storage. Second perspective is from multimedia view, that is, how multimedia services and application such as storage, sharing, authoring, mashup, adaption, and delivery can utilize cloud computing resources. For the existing cloud, we need to do parallel processing as amount of data is so vast so cloud system having parallel processing system can only handle the ever growing demand of computation.

How and Why Multimedia Services Should Utilize Cloud Service

Cloud can provide unlimited storage service on pay per use basis so we can store a huge amount of data on cloud and thus would be very beneficial. For sharing a multimedia item cloud can be very beneficial also and is also very important. In normal multimedia sharing both parties need to be active or on during transmission but if we use cloud one can upload its data and other can download it any time. Authoring is the process of editing segments of multimedia contents. Mashup deals with combining multiple segments from different multimedia sources. Authoring and mashup are done both online and offline. Generally it is a very time and space consuming job. Cloud can help a lot in online authoring and mashup. Computation part can be done using cloud. Adaptation and delivery is also very important service of multimedia application. With so many different multimedia devices and also due to heterogeneity in the network the delivery has become very important. So Video adaptation [2] has become very important. Video adaptation is the process of converting video files in the different formats so that it is compatible with different terminals. Video adaptation require lot of processing which can be done on cloud. Video adaptation of non-live videos are relatively simpler and can be achieved through cloud computing. But it is quite difficult to do video adaptation of live videos.

Proposal for Doing Multimedia Computation

It is proposed in the paper [2, 3] to use media edge cloud computing architecture for better performance. In this architecture, basically operations are described in two parts. Firstly the similar types of media services are placed into a cluster of servers based on properties of media services namely Data Hash Table (DHT) [4] for storage and central processing unit (CPU) and graphical processing unit (GPU) for processing. Secondly, processing in CPU and GPU is done in distributed manner. A load balancer is used to schedule task distributedly and for large-scale computation parallel distributed multimedia computing is used. So we can say that load balancing [5] is done at user level while parallel distributed computing is done at task level [3].

We propose some more points which can be used in the model presented in the above referenced model and can be used for a better multimedia cloud computing. Firstly we propose that we can use fast memory (cache memory) for fast data access. DHT we are using is not cache friendly so we need to find a correct algorithm if we want to use the concept. So we propose a processor (CPU) in the storage cluster which mange the memory with caching technique based on locality of reference. Secondly, we propose dynamic load balancing where even at the task level if we can do load balancing for the workloads that can change during computation. Dynamic load balancing must be used very carefully with only proper type of task and via proper algorithm. Thirdly, we want to propose that CPU and GPU clusters are handled at same time by load balancer which we think is unnecessary. GPU clusters are used for graphics and high computation. So we propose two stage load balancing, that is, it should only handle CPU cluster and then if required data should be sent in GPU. For that we need to have a processor in CPU cluster working as load balancer for GPU if required. In our view it will be beneficial in small computation.

Summary

In the paper, we have tried to give the basic concept of cloud computing and then how this cloud computing can be utilized by multimedia services. The area of multimedia cloud computing is very new and there is a scope of lot of new research in the field. In the proposal part of the paper we have presented the mechanism which currently can be used and has been presented in few papers and also proposed our view for improvement of mechanism. In the paper different multimedia services are taken and need for cloud and the benefits which it can get from cloud computing has been discussed briefly. This paper can be used for preliminary knowledge of cloud computing and multimedia computing in combination with cloud computing.

References

1. Voorsluys, W., Broberg, J., Buyya, R.: Introduction to Cloud Computing. In Proceedings 1st International Conference on Cloud Computing, pp. 254–265, Beijing (2009)
2. Wu, D., Hou, Y.T., Zhu, W., Zhang, Y.-Q., Peha, J.M.: Streaming video over the internet: approaches and directions. IEEE Trans. Circuits Syst. Video Technol. **11**(3), 282–300 (2001)
3. Zhu, W., Luo, C., Wang, J., Li, S.: Multimedia cloud computing. IEEE Signal Process Mag. 59–69 (2011). doi:10.1109/MSP.2011.940269
4. Trajkovska, Salvachúa, J., Velasco, A.M.: A novel P2P and cloud computing hybrid architecture for multimedia streaming with QoS cost functions. In Proceeding of ACM Multimedia, pp. 1227–1230 (2010)
5. Golub, G.H., Van Loan, C.: Dynamic load balancing for distributed memory multiprocessors, J. Parallel Distrib. Comput. **7**(2), 279–301 (1989)

Jena with SPARQL to Find Indian Natural Plants Used as Medicine for Diseases

Vadivu Ganesan, Hopper S. Waheeta and H. Srimathi

Abstract The voluminous information available on the Web necessitates the integrative approach which retrieves sequence and process the information. The Semantic Web technology promises to store, share, and retrieve information with meaningful relationships. Semantic Web technology also allows to link heterogeneous data model into graphical form without changing the existing data storage. This makes the document-centric idea of current Web to more fine-grained semantic structures. Our paper focuses on Ontology construction of Natural Food Resources related with diseases, with the focus of context-based retrieval. Jena framework is used to build Semantic Web applications with the ontology representation of RDF (Resource Description Framework) and OWL (Web Ontology Language) form. SPARQL (Simple Protocol And RDF Query Language) is used to retrieve various Query Patterns. Thus using SPARQL in Jena it is possible to retrieve more specific and semantically related resources those are useful can identified without affecting the existing data models.

Keywords Semantic Web · RDF (Resource description language) · OWL (Web ontology language) · DL (Description logic) · SWRL (Semantic web rule language) · SPARQL (Sparql query language)

V. Ganesan (✉)
Department of IT, SRM University, Chennai, India
e-mail: vadivu.g@ktr.srmuniv.ac.in

H. S. Waheeta
Department of BioInfo, SRM University, Chennai, India
e-mail: Waheeta.hopper@ktr.srmuniv.ac.in

H. Srimathi
MCA Department, SRM University, Chennai, India
e-mail: Srimathi.h@ktr.srmuniv.ac.in

S. Sathiakumar et al. (eds.), *Proceedings of International Conference on Internet Computing and Information Communications*, Advances in Intelligent Systems and Computing 216,
DOI: 10.1007/978-81-322-1299-7_22, © Springer India 2014

Introduction

The widely used traditional data models are relational using structured query languages (SQL) and hierarchical languages such as Extensible Markup Language (XML). In relational model, data are grouped in tables and ensure integrity constraints. In hierarchical data model, data represented in the form of tree structures, where each entity has one parent entity except the root. There is limited options of representing relationships in the above two said models. However, the new data model such as graph model (see Fig. 1) permits to include or establish links among various entities.

Semantic Web makes use of this graph data model for representation. Semantic Web is about explicitly declaring the knowledge embedded in many Web-based applications, integrating information in an intelligent way, providing semantic-based access, and extracting information from texts. Though the current Web is rich of content, it is very difficult to link the existing resources to retrieve more useful information from the large set. This consumes more human effort. The current Web is hampered while handling advanced applications such as processing, understanding, and semantic interoperability of information contained in several Web documents. Semantic Web tries to represent information such that it can be used by machines, not just for display purposes, but for automation, integration, and reuse across applications [1]. Semantic Web is considered to be able to solve all such problems presented by the current web using Ontologies. Ontology is considered as a formal description of the concepts and relationships. RDF is basic building block of Web semantics ontological representation in the form of graphic triples, where OWL can be called as superset of RDF with more granular representation. The approach of Semantic Web facilitates resource annotation, integration, and querying of data with little or no human intervention from existing web pages.

Semantic Web is increasingly used in the area of bioinformatics especially in health informatics to get the integrated knowledge. The study of relationships among natural food and medicinal properties of the natural plants used for diseases are discussed in this paper. The name of the natural food and medicine is mentioned with the reference context of Indian Ayurvedic context. The physiological process of digestive system is a base and compliments the whole body health. The food habit and medicines intake of the person decide his health and any deficiency in this

Fig. 1 Graph data

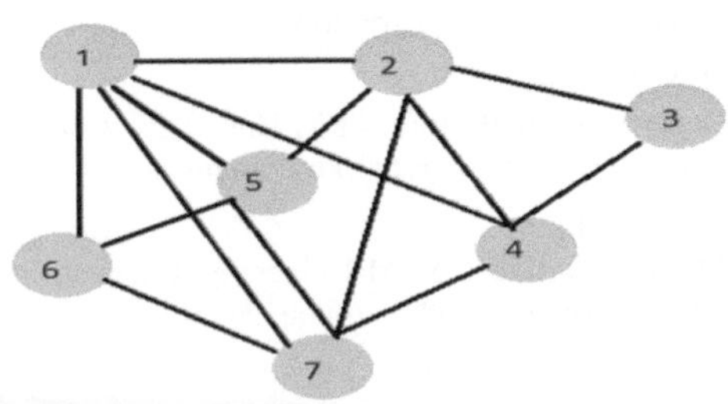

lead to chronic diseases. However, the balanced intake of natural foods with rich in nutritional factor has high impact on curing chronic diseases such as cancer, diabetes, asthma, and hypertension, etc. It is very difficult to represent in hierarchical or relational table form of relationship among the entities such as natural food, disease, and physiological process [2]. Ontology representation is suggested and used for the identified bioinformatics problem to utilize the listed benefits of Semantic Web. Our study attempted Ontology development using semantic relationships to the domains of diseases and natural food sources for the better querying and linking of information. The paper is structured as follows: Section "Choice of Ontology Language and Tools" describes the choice of ontology language; Sect. "Construction of Knowledge Base" describes ontology/knowledge construction; and Sect. "Querying Using SPARQL and Jena" describes Jena with SPARQL for querying the datasets.

Choice of Ontology Language and Tools

Ontology is defined as explicit and formal specification of conceptualization. Ontology comprises a set of knowledge terms, including the vocabulary, the semantic interconnections, simple rules of inference, and logic for some particular topic. The Semantic Web architecture lists the underlying machine understandable languages for knowledge representation: XML, RDF, and OWL.

XML with XML namespace and XML schema definitions makes sure that there is a common syntax used in the Semantic Web. However, XML has no provision on representing semantic interoperability. Using XML, it is not possible to specify more relationships among the data. On top of XML is the RDF, which represents data in a graph form. RDF, follows binary predicates with the format <Subject, Predicate, Object> used to create the relationship in the form of triples. For example, "Ocimum Sanctum used as Medicine for Cold", <OcimumSanctum, usedAsMedicineFor, Cold> is the triple form. RDF Schema (RDFS) defines the vocabulary of RDF model. It provides a mechanism to describe domain-specific properties and classes of resources to which those properties can be applied, using a set of basic modeling primitives (class, subclass-of, property, subproperty-of, domain, range, type). However, both RDF and RDFS are rather simple and still do not provide exact semantics of a domain. OWL is the next layer used in Semantic Web Architecture, is having more features compared to RDF. OWL is a set of XML elements and attributes, with well-defined meaning, that are used to define terms and their relationships (e.g., Class, equivalentProperty, intersectionOF, unionOF, etc.). Reasoning tasks like verification of ontology consistency, computing inferences, and realizations can be easily executed with the OWL representation. OWL is the preferred language for ontology representation.

In practice, ontologies are often developed using integrated, graphical, ontology authoring tools, such as Protégé, OILed, and OntoEdit. Protégé facilitates extensible infrastructure and allows an easy construction of knowledge rich domain

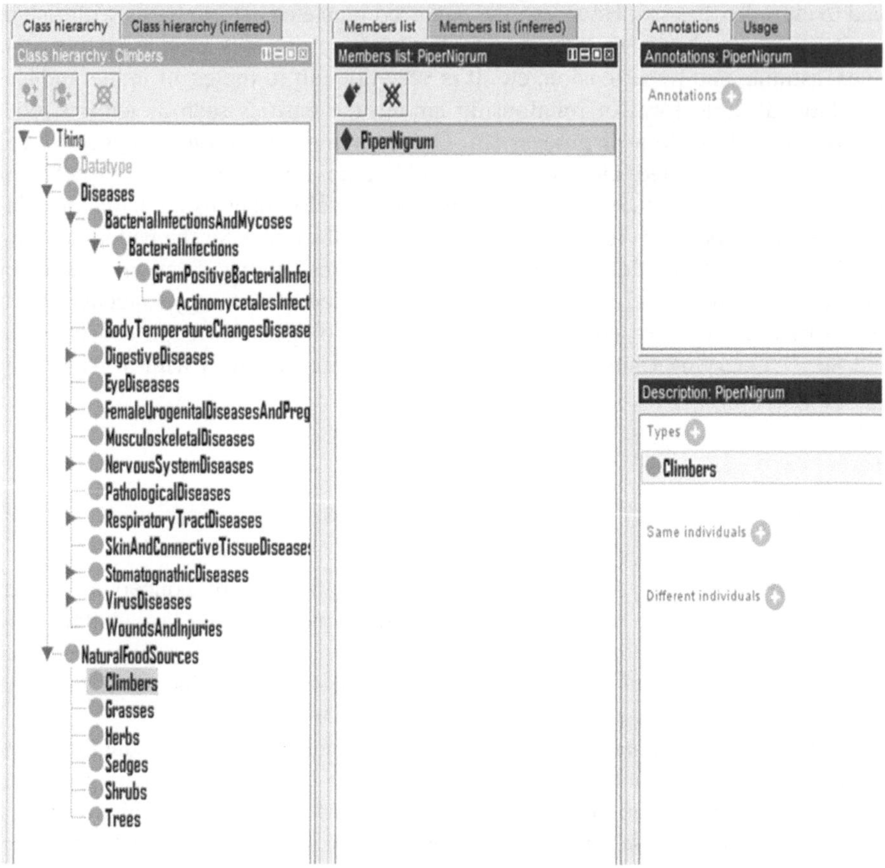

Fig. 2 Class, subclass, individual, and their properties representation in protégé tool

ontologies. Protégé tool is used to develop domain ontology and querying since it has adopted the recent recommendation of the W3C, i.e., OWL standard. Protégé is also a knowledge-based editor and it is open source Java tool that allows the easy construction of Ontologies.

Most data in Web will continue to be stored in either hierarchical or relational databases [3]. To work with these databases in ontology-based applications, tools and techniques that bridge these models are required. Mapping all the existing hierarchical and relational data to ontology instances is often not practical so dynamic data access approaches are typically employed. The OWL's rule language SWRL with these systems presents an opportunity to employ optimization techniques that can significantly reduce the amount of data transferred from databases. The SWRLTab is a development environment for working with SWRL rules in Protégé-OWL [4]. It supports the editing and execution of SWRL rules. It also provides mechanisms to allow interoperation with a variety of rule engines

```
<owl:Class rdf:ID="Diseases"/>
<Microcephaly rdf:ID="HeartDisorder">
<curedBy rdf:resource="#Tulsi"/>
<withMESHID rdf:datatype="&xsd;string">C537544</withMESHID>
<withReference
rdf:datatype="&xsd;string"
http://ctd.mdibl.org/detail.go?type=disease&amp;db=MESH&amp;acc=C5
37544#top</withReference>
<withSynonym rdf:datatype="&xsd;string"
>MentalRetardation</withSynonym>
<withSynonym
rdf:datatype="&xsd;string">Microcephaly</withSynonym>
<withSynonym
rdf:datatype="&xsd;string">Seizures</withSynonym>
<withSynonym rdf:datatype="&xsd;string"
>SkeletaAbnormalities</withSynonym>
<withSynonym rdf:datatype="&xsd;string"
>congenitalHeartDisease</withSynonym>
</Microcephaly>
```

Fig. 3 OWL code representation of < subject, predicate, object> triple

and the incorporation of user-defined libraries of methods that can be used in rules. Several built-in libraries are provided; include collections of mathematical, string, and temporal operators, in addition to operators than can be used to effectively turn SWRL into a query language. This language provides a simple but powerful means of extracting information from OWL ontologies.

Construction of Knowledge Base

The effort is taken to provide an explicit specification of the conceptual model for the identified bioinformatics domain. Initially, the identified components of natural food and diseases are represented in the form of class hierarchy using protégé tool, since the ontology development focuses on developing design hierarchy toward the explicit specification of relationships. The partial list of created class hierarchy, individuals, and their properties is shown in Fig. 2. The equivalent RDF/OWL code representation is given in Fig. 3. The ontology representation in Protégé is mapped on different dimensions. Semantic link is used to connect different datasets. The semantic linking provides more meaningful navigational paths to the users. There are two different types of links are used in our study. Explicit links and relationships are given by the developers to find the relationship among the entities. Implicit relationships among the data cannot be automatically identified by the machines; developers have to find the internal relationships to assist the machine [5, 6–8]. This type of sharing and finding implicit information is more important and also critical. The dynamic hyperlink updating is useful for inferring and retrieving information from the existing Web resources. However, it is possible only for smaller set of datasets, and it results cumbersome for larger datasets.

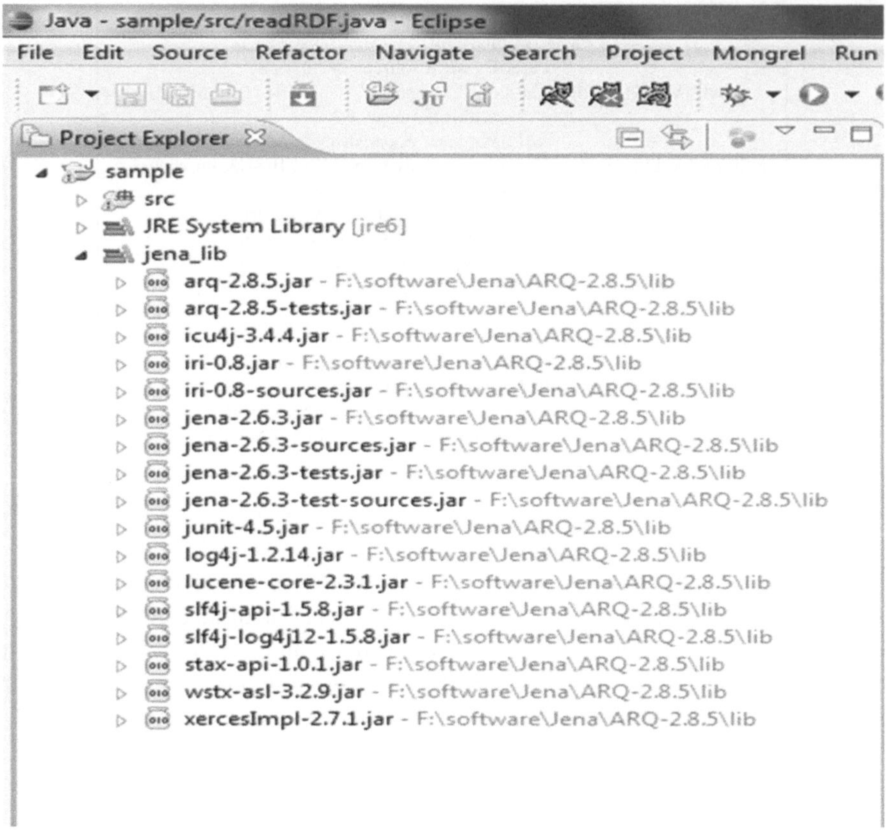

Fig. 4 Including Jena library files in Java eclipse interface

The Semantic Web performance can be improved through linking of related entities from different datasets and establishing relationships automatically [9].

Querying Using SPARQL and Jena

The application development of Ontology-based knowledge querying is made simple by using Jena programming toolkit. Jena aims to provide a consistent programming interface for ontology application development with the base of Java Programming. The inclusion of Jena Library in Java Platform using Eclipse interface is shown in Fig. 4. 'OntClass' is used to represent OWL class or RDFS class. 'OntModel' extends support for the kinds of objects expected to be in an ontology: classes (in a class hierarchy), properties (in a property hierarchy), and individuals [10–12]. In Java, ontology models are created through the Jena ModelFactory. The simplest way to create an ontology model is as follows:

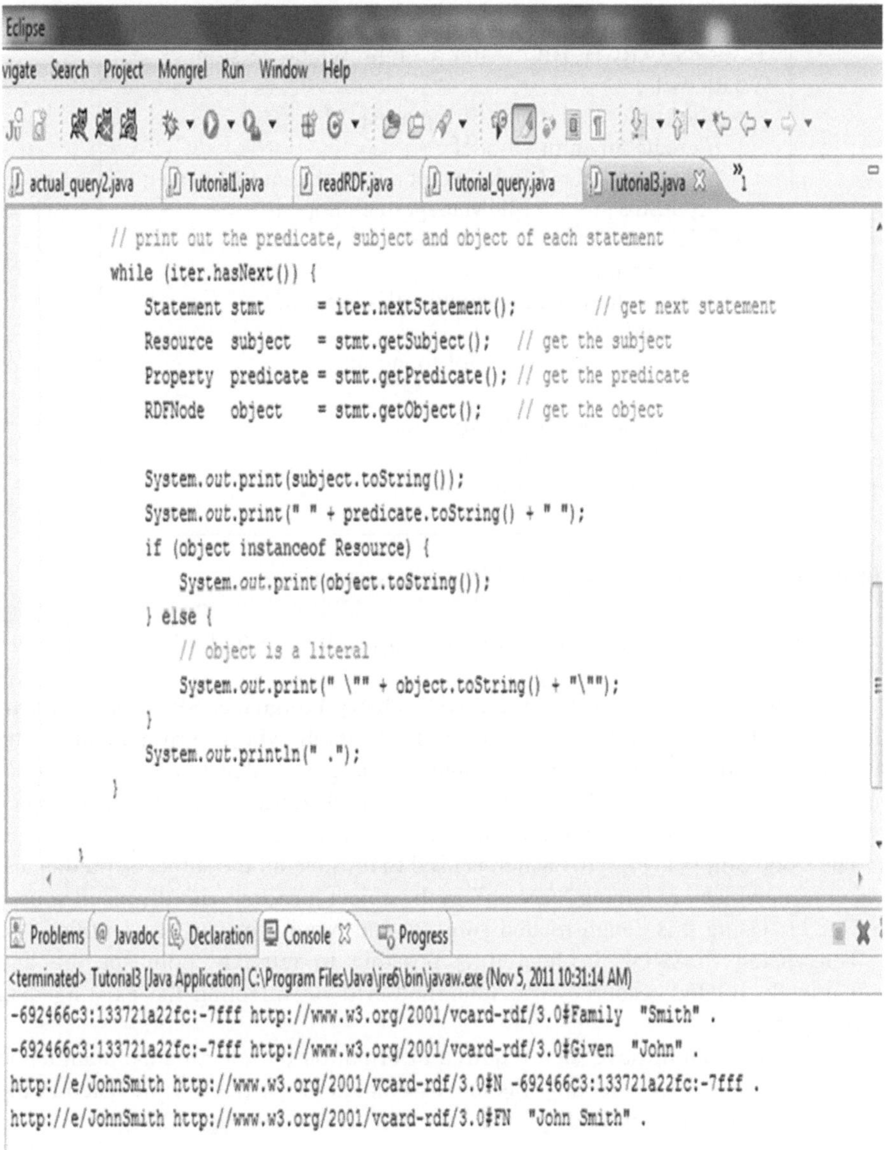

Fig. 5 <subject, predicate, object> triple display using Jena

OntModel m = ModelFactory.createOntologyModel();

Jena supports a simple notion of ontology representation of <subject, Predicate, Object> triple, which essential base building block of ontology [13] in statement as given in Fig. 5.

```
static final String inputFileName =
"E:\\RESEARCH_BIO\\After_Compre\\data_version
4\\data.owl";
   public static void main (String args[]) {
      // create an empty model
      Model model = ModelFactory.createDefaultMod el();
      InputStream in = FileManager.get().open(
inputFileName );
      if (in == null) {
         throw new IllegalArgumentException(
"File:"+inputFileName + "
                              not found");
      }
         // read the RDF/XML file
      model.read(in, "");
      // write it to standard out
      model.write(System.out);
```

Fig. 6 Jena code snipped to retrieve the RDF/OWL resources

The Jena code snippet execution of Fig. 6 resulted to find all the valid RDF/OWL statements as given in Fig. 7.

SPARQL is a Simple Protocol and RDF Query Language. SPARQL is a syntactically SQL-like language for querying RDF graphs via pattern matching. The language's features include basic conjunctive patterns, value filters, and optional patterns. The SPARQL code snippet (Fig. 8) which retrieves all the three triples given in [14] the OWL File and results are shown in Fig. 9.

The Code Snippet (Fig. 10) which is used to retrieve all the subjects, predicates, and objects which are having "Cough" in the object location with the result shown in Fig. 11. Using this cough-related subjects and predicates can be identified.

Thus using SPARQL in Jena it is possible to retrieve more specific and semantically related resources can identified without affecting the existing data models [15].

Naïve Users can retrieve their required information by selecting the plant name or disease name. After selecting this, the associated properties will be listed in the list box. The output will be displayed in a Text Box as in Fig. 12.

```
  </rdf:Description>
  <rdf:Description rdf:about="http://www.owl-ontologies.com/Ontology1320072228.owl#Fever">
    <curedBy rdf:resource="http://www.owl-ontologies.com/Ontology1320072228.owl#Adathoda_Vasica"/>
    <rdf:type rdf:resource="http://www.owl-ontologies.com/Ontology1320072228.owl#Diseases"/>
  </rdf:Description>
  <rdf:Description rdf:about="http://www.owl-ontologies.com/Ontology1320072228.owl#Chemicals">
    <rdf:type rdf:resource="http://www.w3.org/2002/07/owl#Class"/>
  </rdf:Description>
  <rdf:Description rdf:about="http://www.owl-ontologies.com/Ontology1320072228.owl#Plants">
    <rdf:type rdf:resource="http://www.w3.org/2002/07/owl#Class"/>
  </rdf:Description>
  <rdf:Description rdf:about="http://www.owl-ontologies.com/Ontology1320072228.owl#curedBy">
    <rdfs:range rdf:resource="http://www.owl-ontologies.com/Ontology1320072228.owl#Plants"/>
    <rdfs:domain rdf:resource="http://www.owl-ontologies.com/Ontology1320072228.owl#Diseases"/>
    <rdf:type rdf:resource="http://www.w3.org/2002/07/owl#ObjectProperty"/>
  </rdf:Description>
  <rdf:Description rdf:about="http://www.owl-ontologies.com/Ontology1320072228.owl">
    <rdf:type rdf:resource="http://www.w3.org/2002/07/owl#Ontology"/>
  </rdf:Description>
  <rdf:Description rdf:about="http://www.owl-ontologies.com/Ontology1320072228.owl#Diseases">
    <rdf:type rdf:resource="http://www.w3.org/2002/07/owl#Class"/>
  </rdf:Description>
</rdf:RDF>
```

Fig. 7 Display all the resources

```
model.read(in, "");
String sparqlQueryString1= "SELECT ?subject ?predicate
?object WHERE { ?subject ?predicate ?object }";
Query query = QueryFactory.create(sparqlQueryStrin g1);
QueryExecution qexec =
QueryExecutionFactory.create(query, model);
try  {
ResultSet results = qexec.execSelect();
   for(;results.hasNext();)
    {QuerySolution soln = results.nextSolution();
          //RDFNode x = soln.get("Staff#");
    ResultSetFormatter.out(System.out, results,
          query);
            }
      }
```

Fig. 8 Code snippet to retrieve all <subject, predicate, object>

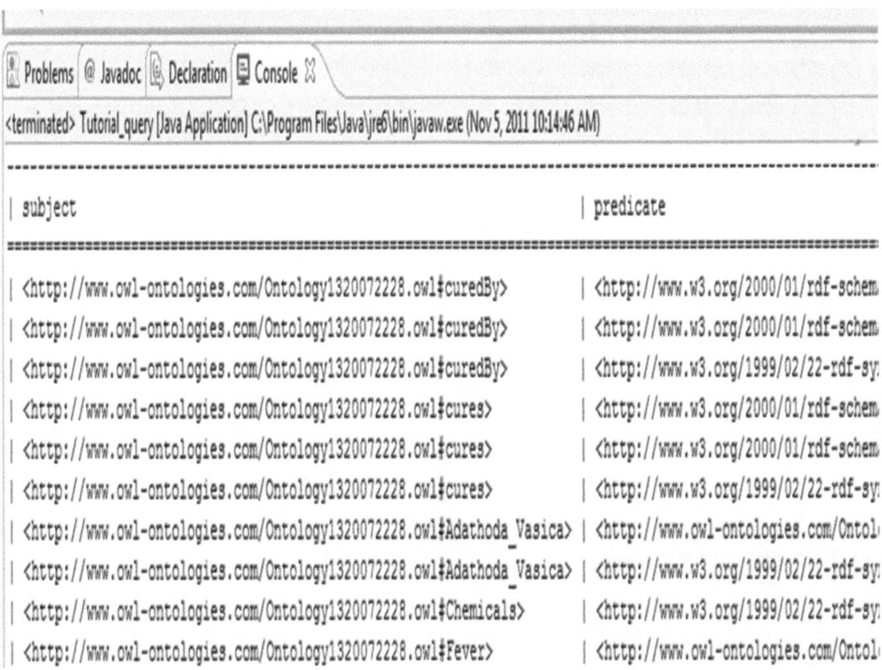

Fig. 9 Partial result of code snippet of Fig. 8

```
Model model = ModelFactory.createDefaultModel();
                    InputStream in = FileManager.get().open(
inputFileName );
if (in == null) {
        throw new IllegalArgumentException( "File: " +
inputFileName + " not found");
      }
model.read(in, "");
      String sparqlQueryString1= "SELECT ?subject ?object
WHERE {?subject ?predicate ?object FILTER
(regex(str(?predicate),\"^?usedAsMedicineFor?\")) filter
(regex(str(?object),\"^?Cough?\")) }";

      Query query =
QueryFactory.create(sparqlQueryString1);
      QueryExecution qexec =
QueryExecutionFactory.create(query, model);
 try
 {
          ResultSet results = qexec.execSelect();
          for(;results.hasNext();)
```

Fig. 10 SPARQL code snippet to retrieve "Cough" in object

```
-----------------------------------------------------
| subject           | predicate          object   |
===================================================
| #OcimumSanctum>  | usedAsMedicineFor #Cough>  |
| #Asparagus>      | usedAsMedicineFor #Cough>  |
```

Fig. 11 Output of the code snippet given in Fig. 10 to retrieve "Cough" in object

Conclusion

In this paper, the semantic representation and querying the data of bio-informatics is discussed using Semantic technologies and tools such as Protégé, Jena, SWRL, and SPARQL. The study highlighted the possibilities of information extraction, automatic, and semi-automatic generation of meta-data for web information, generic, and heuristic reasoning methods for the web, knowledge retrieval, and content-based information retrieval in the existing large data set.

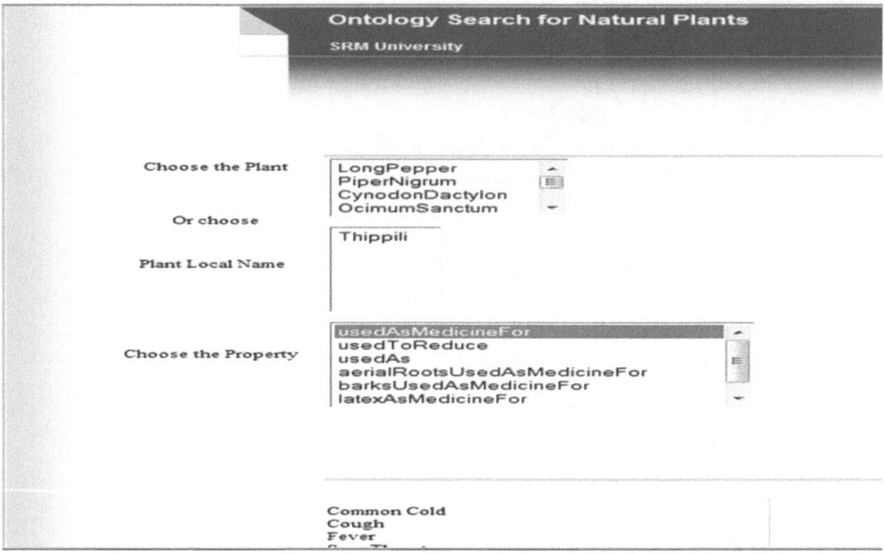

Fig. 12 Result of SPARQL with filter option of "long pepper"

References

1. Berners-Lee, T., Hendler, J., Lasilla, O.: The Semanticweb. Sci. Am. http://www.SemanticWeb.org/knowmarkup.html (2001)
2. Chen, H., Wang, Y., Wang, H., Mao, Y., Tang, J., Zhou, C., Yin, A., Wu, Z.: Towards a semantic web of relational databases: a practical semantic toolkit and an in-use case from traditional chinese medicine, fifth international semantic web conference, China. (2006)
3. Mendes, P.N., Kapanipathi, P., Cameron, D., Sheth, A.P.: Dynamic associative relationships on the linked open data web, Web Science Conference. (2010)
4. O'Connor, M.J., Shankar, R.D., Tu, S.W., Nyulas, C., Parrish, D.B., Musen, M.A., Das, A.K.: Using semantic web technologies for knowledge-driven querying of biomedical data. 11th conference on artificial intelligence in medicine (AIME 07). Amsterdam, Netherlands, (2007)
5. Lehmann, J., Schuppel, J., Auer, S.: Discovering unknown connections-the DBpedia relationship Finder. (2007)
6. Aleman-Meza, B., Halaschek-Wiener, C., Budak Arpinar, I., Ramakrishnan, C., Sheth, A.: Ranking complex relationships on the semantic web. (2005)
7. Bizer, C., Lehmann, J., Kobilarov, G., Auer, S., Becker, C., Cyganiak, R., Hellmann, S.: DBpedia: a crystallization point for the web of data. Web Semantics: science, services and agents on the world wide web. (2009)
8. Horridge, M., Jupp, S., Moulton, G., Rector, A., Stevens, R., Wroe, C.: A practical guide to building OWL ontologies using Prot′eg′e 4 and CO-ODE tools. (2007)
9. Cregan, A.M.: Linked open data: a new resource for eResearch. (2009)
10. Heim1, P., Lohmann2, S., Stegemann, T.: Interactive relationship discovery via the semantic web. (2010)
11. Heath, T., Hausenblas, M., Bizer, C., Cyganiak, R., Hartig, O.: How to publish linked data on the web. 7th European Semantic Web Conference, ESW 2010, Heraklion, Crete, Greece, May 30-June 3. (2009)
12. http://www.w3.org/Submission/SWRL/

13. http://protege.cim3.net/cgi-bin/wiki.pl?SWRLTab
14. Farooq, A., Arshad, M.J., Shah, A.: A layered approach for similarity measurement between ontologies. J. Am. Sci. (2010)
15. Horridge, M., Knublauch, H., Rector, A., Stevens, R., Wroe, C.: A practical guide to building OWL ontologies using the prot′eg′e-OWL plugin and CO-ODE tools, Edition 1.0 (2004)

Early Detection of Diabetic Retinopathy by CRA (Cluster Rejection Approach)

S. Vijayalakshmi Karthiga, T. Sudalai Muthu and M. Roberts Masillamani

Abstract Diabetic Retinopathy (DR) is a major public health issue, since it can lead to blindness in patients with diabetes. Since large number of existing method of detecting DR had given importance to robust modeling of MA (Microaneurysm) by explicit segmentation of optic disk and vessels. Existing methods of detecting MA involve complex modeling results in high computation cost and time consuming process. In order to improve efficiency of the system, proposed a new approach for detecting DR based of cluster rejection methodology for detecting MA from retinal image. The proposed technique involves cluster separation of retinal images and selecting candidate sets based on simple threshold and rejection of candidates from DR affected retinal image. Our proposed methodology for detecting MA results in easy computation of fungus image and less time consuming process.

Keywords Diabetic retinopathy · Microaneurysm · Cluster rejection · Retinal image · Candidate selection · False positive

Introduction

Diabetic retinopathy (DR) is a widespread retinal complication associated with diabetes. It is a main cause of blindness in middle as well as older age groups. Early detection of DR by analyzing MAs is the best practices in scientific research.

S. Vijayalakshmi Karthiga (✉) · T. Sudalai Muthu · M. Roberts Masillamani
School of Computing Science and Engineering, Hindustan University,
Padur, Chennai, Tamilnadu, India
e-mail: vij.karthiga@gmail.com

T. Sudalai Muthu
e-mail: sudalaimuthut@gmail.com

M. Roberts Masillamani
e-mail: deancs@hindustanuniv.ac.in

S. Sathiakumar et al. (eds.), *Proceedings of International Conference on Internet Computing and Information Communications*, Advances in Intelligent Systems and Computing 216,
DOI: 10.1007/978-81-322-1299-7_23,

In the early stages of DR, patients are usually asymptomatic, but in more advanced stages of the disease patients may experience symptoms that include floaters, distortion, and/or blurred vision. MAs are the earliest clinical sign of DR [1–4].

Background

DR is the foremost cause of new blindness in persons aged 25–74 years in the United States. About 700,000 persons in the United States have proliferative DR, with an annual occurrence of 65,000. A recent estimate of the occurrence of DR in the United States showed a high prevalence of 28.5 % among those with diabetes aged 40 years and older. Patients with diabetes often develop ophthalmic complications, such as corneal abnormalities, glaucoma, iris neovascularization, cataracts, and neuropathies. The most widespread and potentially most blinding of these complications, however, is DR. In the preliminary stages of DR, patients are generally asymptomatic, but in more advanced stages of the disease patients may experience symptoms that include floaters, distortion, and/or blurred vision. MAs are the earliest clinical sign of DR. Workup for DR includes fasting glucose and hemoglobin A1C measurements.

Retinal disease, as evidenced by proteinuria and eminent BUN/creatinine levels, is an excellent predictor of retinopathy; both conditions are caused by DM-related microangiopathies, and the presence and severity of one reflects that of the other. Aggressive treatment of the nephropathy may slow progression of DR and neovascular glaucoma. According to the diabetes control and difficulties trial controlling diabetes and maintaining the HbA1c level in the 6–7 % range can significantly reduce the progression of DR. One of the leading aspects in the management of DR is patient education. Create awareness among patients because they play an integral role in their own eye care.

Proposed System

The Proposed system consists of several steps to process the retinal image in various situations. The following steps are:

- Preprocessing
- Clustering
- Feature Extraction
- Candidate Selection and Matching

Steps to Process Retinal Image

Robust representations for retinal image recognition must be invariant to changes in the size, position, and orientation of the patterns. Retinal image from different people may be captured in different sizes and, even for retina from the same eye; the size may change due to illumination variations and other factors. Segmentation of image requires image segmentation in homogeneous regions; image regions generally have homogeneous characteristics (e.g., intensity, texture). Mathematical morphology as a tool for extracting retinal image components that are useful in the representation and description of region shape, such as boundaries, skeletons, etc. We are also interested in morphological techniques for pre- and postprocessing, such as morphological filtering, thinning, and pruning. Morphological representation of retinal image processing based on set theory for reflection and translation of gray-scale images. The two main processes are those of dilation and erosion (Fig. 1).

These processes involve a special mechanism of combining two sets of pixels. Usually, one set consists of the image being processed and the other a smaller set of pixels known as a structuring element or kernel [5, 6]. In dilation, every point in

Fig. 1 System overview

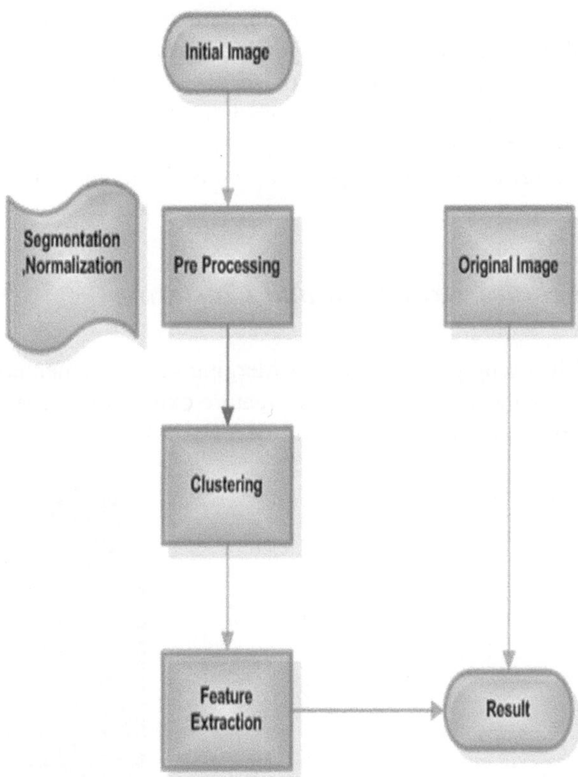

the image is superimposed onto by the kernel, with its surrounding pixels. The resultant effect of dilation is of increasing the size of the original object. Erosion is an inverse procedure in which an image is thinned through subtraction via a structuring element or kernel. The kernel is superimposed onto the original image and only at locations when it fits entirely within its boundaries will a resultant central pixel be accepted. The algorithms of opening and closing are based upon these processes. Opening consists of erosion followed by dilation, and tends to smooth an image, breaking narrow joints, and removing thin protrusions.

$$(A)_z = \{c | c \in a + z, \quad for\, a \in A\}$$

$$\hat{B} = \{w | w \in -b, \quad for\ b \in B\}$$

Set *A*, structuring element *B*, and erosion of *A* by *B*. Boundary of set *A*. Set of values of *z* such that satisfy the reflection and transformation (Figs. 2, 3, and 4).

Normalization

The method is based on the estimation of the luminosity and contrast variability in the background part of the image and the subsequent compensation of this variability in the whole image. Normalizing retinal images involve extraction on estimation of luminosity and drift contrast If the intensity range of the image is 50–180 and the desired range is 0–255, the process entails subtracting 50 from each of pixel intensity, making the range 0–130 which should result in eliminating the noisy data in the pixel region of segmented images.

Feature Extraction and Clustering

Clustering generally allows merging set of segmented retinal images as cluster and the same should be used for feature extraction. Feature extraction clustered images give us effective output and which would give best result when compared to other

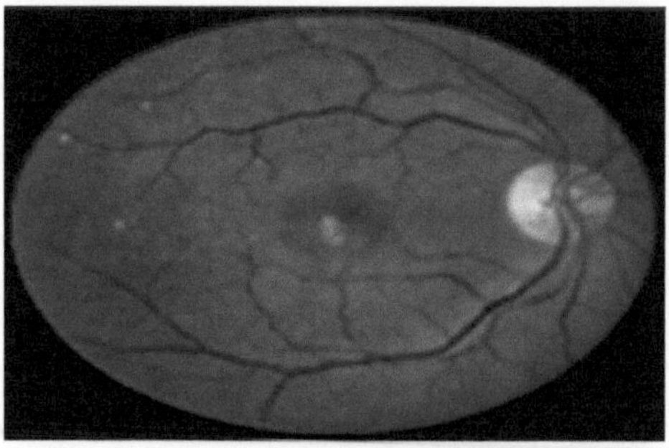

Fig. 2 Fungus image

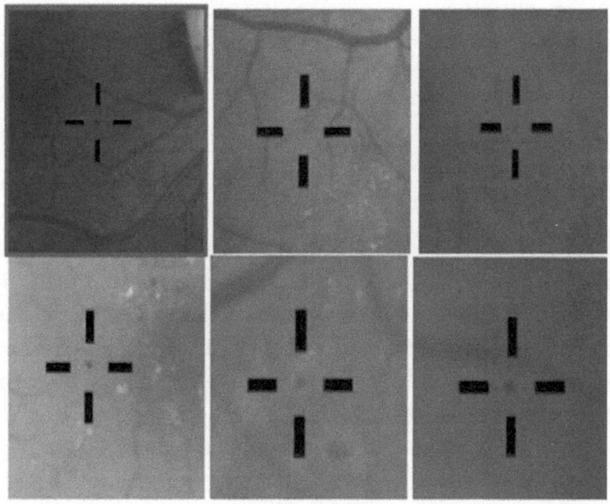

Fig. 3 Segmented image

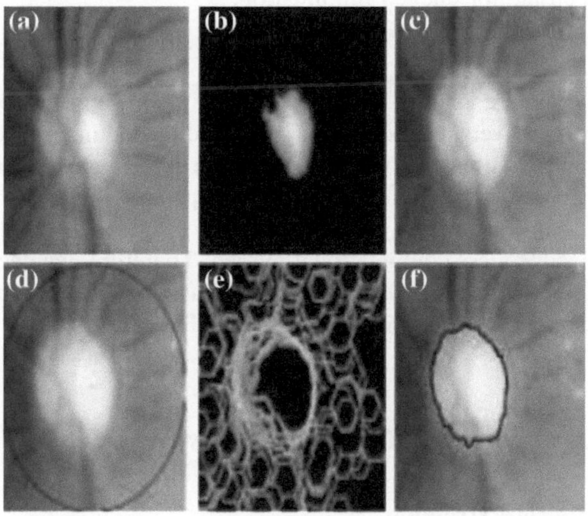

Fig. 4 Preprocessed retinal image

types of feature extraction. Cluster rejection of irregular behavior in fundus images provides a way to remove the unwanted vessel in retina.

Retinal images are acquired with a digital fundus camera, which captures the illumination rejected from the retinal surface. Despite the controlled conditions, many retinal images suffer from nonuniform illumination given by several factors: The curved surface of the retina, pupil dilation (highly variable among patients), or presence of diseases, among others. The curved retinal surface and the geometrical configuration of the light source and camera, lead to the fact that the peripheral

Fig. 5 Luminosity component

part of the retina appears darker than the central region. In many works, such as, the local contrast enhancement method is used for equalizing uneven illumination in the intensity. Most approaches are designed for gray-scale images. Attempts to extend them to color images tend to produce hue-shifting related artifacts, given by the introduction of new colors to the image. In this section, we implement a color retinal image enhancement based on the knowledge of the retina geometry and imaging conditions. We determine the strengths of this approach and discuss its further improvement [7, 8]. The general idea is that the image can be enhanced by estimating the background luminosity and contrast distribution in order to compensate for uneven illumination.

$$U(x; y) = I(x; y) - L(x; y)/C(x; y)$$

Where I is the original image, C and L are the contrast and luminosity drifts, respectively. C and L can also be understood in terms of gain and offset. They have to be estimated by sampling the original image. The sampling approach of the whole image divides into small squares, whereas in they use a more intuitive sampling scheme based on the knowledge of the illumination distribution that leads to less computational burden. Therefore, we decided to use a similar type of nonuniform sampling grid shown in figure. The sampling is coarse in the central region and dense in the periphery (Figs. 5, 6, and 7).

Detection of Microaneurysms and Hemorrhages

Vessel Extraction

Retinal images have pathological noise and backgrounds with varied texture, which lead to problems in vessel extraction. In order to remove the noise, gray opening operation and an opening by reconstruction with a linear structuring element are employed to the original image at various orientations Second, the Top-Hat transform combined with reconstruction opening and closing operations are proposed to strengthen the smoothed image by iteratively filtering the image (Fig. 8).

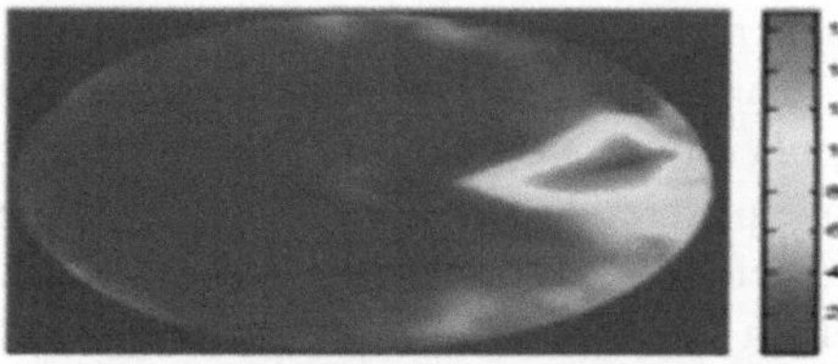

Fig. 6 Contrast component

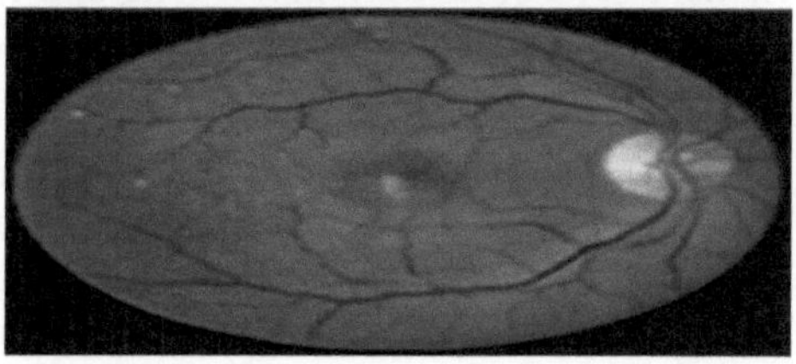

Fig. 7 Enhanced image

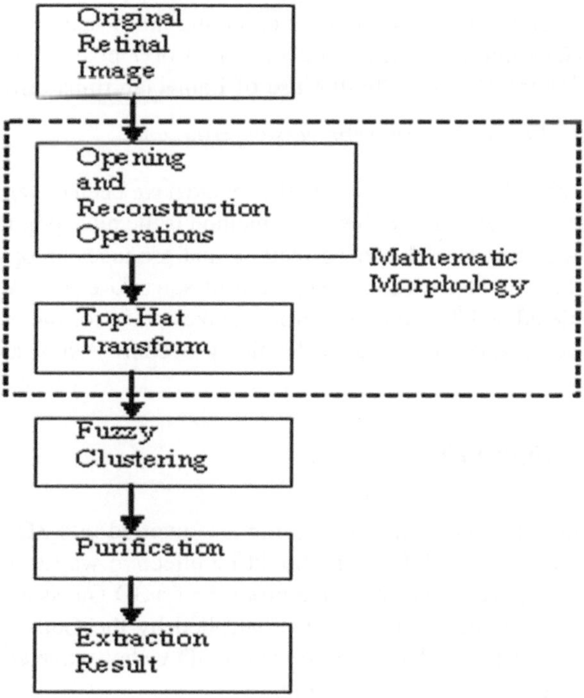

Fig. 8 Detection of microaneurysms

1. *Smooth image and remove noise*

The image is smoothed by gray mathematical morphology by selecting the structuring element. Here, linear structuring elements based on the line type property of vessels have been used. However, it is important to note that an opening operation used by a linear structuring element will remove a vessel or some parts of it, when the vessels in the image have orthogonal directions or the structuring element is longer than the vessel *International Journal of Computer Applications width*. On the contrary, when the structuring element and the vessel have parallel directions, the vessel will never be changed, hence linear structuring element at different orientations has been used to get the maximal response. The length of the structuring element is close to the diameter of the largest vessel. Since the diameter of the largest vessels is approximately 6 pixels, the structuring element is taken to be 7 pixels long and 1 pixel in wide. Structuring elements are applied to perform a smoothening operation first by erosion and then dilatation.

2. *Strengthening of vessels and removing background*

After smoothening the image, the Top-Hat transform is applied to strengthen the vessels by selecting appropriate structuring elements. Here, the Top-Hat transform is applied to the smoothed image at 12 directions, and the computational results of the 12 directions are summed up to increase the gray difference between the vessels and the background. The corresponding formula is as follows:

However, most images containing noisy data need to be further smoothed by a Gaussian filter, which has the width of 7 pixels, and further strengthen the curve feature of vessels by the use of Laplacian transform.

3. *Fuzzy clustering for vessel extraction*

After the retinal vessels in the image have been enhanced, the next step is to extract the vessels. Fuzzy clustering method is the most popular tool to extract the vessel. It is an iterative clustering method that produces an optimal partition by minimizing the weighted within group sum of squared errors. When the partitioning is completed, a defuzzification process takes place for the conversion of fuzzy partition to a crisp one. Finally, a purification procedure is done and final extraction is achieved.

Gabor Filter

Retinal image after processing is converted into 3D image to obtain z index while selecting candidates, it should be effective way to select candidates from fungus image. A 3D Gabor is the product of a 3D Gaussian and a 3D harmonic function. The length of the axes is controlled by the Gaussian and the frequency is controlled by the harmonic function. 3D Gabor wavelets are used for spatio-temporal analysis of 3D objects (Fig. 9).

Similarity Measure Computation (L)

The rejector cascade outputs a set *C2* of candidates which is expected to be true MAs. This final module assigns a numerical similarity score to each sample in *C2*.

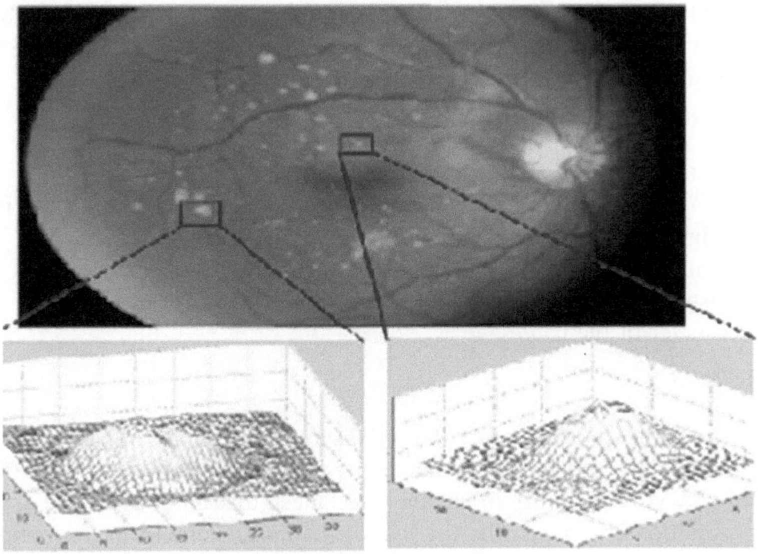

Fig. 9 Conversion of 2D image into 3D using gabor filter

It is indicating the chance of being a true lesion. We choose to perform the score assignment by considering the signed distance of a sample from the optimal hyper plane of a two-class SVM, in feature space. A full representation for true MA is obtained by considering features from the previous rejection stages in addition to the features encoding context and structure symmetry information which are explained next.

Context features: The following set of context features is derived. It considers the pixels within the candidate, and a context neighboring it.

- Variation in mean value of the candidate region and its surround of size (49 × 49) pixels computed in 4 spectral bands: red, green, blue, and hue.

$$msdj = meanj(cand) - mean(surround),\ \text{where}\ j = \{red, green, blue, hue\}$$

- The response of the candidate to a center-surround binary filter with off-center. This is used as a rough descriptor of candidate computed on an image patch centered at local minima.
- The perimeter p of the candidate, establish as the number of pixels in the level curve at lc (defined in $FS2$).
- Mean response and standard deviation of derivative of Gaussian filter bank: *gx, gy, gxx, gyy, gxy* at pixels within the candidate (5 filters at 4 scales each, resulting in 20 features; scales used are $\sigma = \{1, 2, 4, 8\}$) pixels.

Symmetry features: By filtering with rotated Haar-like wavelets, a set of 8 features is obtained at each candidate. To get 16 filters, the vertical 2D nonstandard Haar wavelet is rotated in 16 orientations (each separated by $\pi/8$). The axially antisymmetric feature pairs capture symmetry of the candidate along different axes, and the ratio of the pair responses is used as features.

A. *Feature Set-3*

- *dtrue, A*($l1$), *A*(lc), Γ, Ω, v: Taken from feature set-2.
- *Mdr, mdg, mdb*: Difference in mean values within the candidate and its surrounding region for red, green, and blue color plane.
- *mdh*: Same as previous, in hue plane
- *c.s*: Response of center-surround binary filter
- *p*: Perimeter of the candidate

B. *Classifier-3*

Here, we calculate approximately the similarity score for a sample based on its distance from the optimal hyper plane of a support vector machine (SVM). Strength of SVM is its ability to handle imbalanced distributions of true and false samples. Additionally, it permits the use of nonlinear kernel transformations, to overcome hyper plane linearity assumption. The similarity score ψ (a function of x) obtained is, such that it models a posterior probability of the two-class SVM assigning a label "true-MA" to xq, given its feature values that is $\psi(xq) = p(yq \leftarrow true|xq)$. In our trialing, we parameterized this probability score to get detection sensitivities at different false positive per image (fppi) rates.

Datasets

For the purpose of assessment, three datasets were considered: Two are the publicly obtainable datasets namely, the DIARETDB1 and *ROCd* datasets; and a custom-built dataset called CRIAS. Images in each dataset are divided into two, training and testing sets. An image in each dataset gives a range of image sizes (768 × 586 to 1500 × 1100), resolution, etc. The detailed specifications of the selected datasets are given next. DIARETDB1 consists of 89 images, of which 5 images do not contain any DR-indicative lesions. From a screening program, the images were collected and taken under a fixed imaging protocol. The medical experts selected the images, but their distribution does not correspond to any typical population. The ground truth supplied with this dataset is a soft map consisting of regions indicating expert consensus level information averaged from multiple experts. A bright region thus indicates high consensus about the presence of MA. According to the guiding principle given with the dataset, evaluation of the presented method is done on a 75 % consensus level (relative to maximum) as the ground truth. A total of 182 MAs are obtained at 75 % consensus level (i.e., majority vote among 4 experts). A test set of 68 images is formed and remaining 21 images were used in training. *ROCd* consists of 50 training images with associated ground truth, and a test set of 50 images, whose ground truth is retained by the organizers of a competition. The images are taken from a DR screening program across numerous sites, and hence captured with different cameras, fields of view, and resolution. These images are mainly collected for clinical documentation and patient profiling. These images are of diabetic patients who already have been diagnosed with DR. Therefore, these images have high pathology occurrence and laser marks. Ground truth on these images was obtained from two experts and merged using the OR rule; a location is considered to have MA if at least one expert has marked it. This dataset contains a total of 1436 MAs based on the above criteria, which is far higher compared to the two public datasets.

Conclusion

In this work, we formulated MA detection as target *detection in cluster* problem and proposed a successive cluster rejection solution for MA detection. The rejection stages are formulated based on the occurrence frequency and discriminability of the underlying cluster. Retinal image from different stages is collected and tested to get the effective resultant MAs. The proposed methodology to detect MAs in retinal image gives us the expected rested result when compared to other methods.

Future Enhancement

Thus, the proposed system works well in all kinds of situations for retinal image, but the systems need to be enhanced for detecting retinopathy for new born babies because the normal behavior of eye formed only after 6 months. Other than this situation, the system functions well in detecting DR by analyzing MAs in retinal image.

References

1. Pettigrew, J.D.: Is there a single, most-efficient algorithm for stereopsis? In: Blakemore, C. (Ed) Vision: Coding and Efficiency. Cambridge University Press, Cambridge (1990)
2. Jain, K.: Fundamentals of Digital Image Processing. Prentice-Hall, Englewood Cliffs (1989)
3. Marr, D., Poggio, T.: A computational theory of human stereo vision. Proc. R. Soc. Lond. B **204**, 301–328 (1979)
4. Grimson, W.E.L.: A computer implementation of a theory of human stereo vision. Philos. Trans. R. Soc. Lond. B **292**, 217–253 (1981)
5. Agarwal, A., Triggs, B.: Recovering 3D human pose from monocular images. IEEE Trans. Pattern Anal. Mach. Intell. **28**(1), 44–58 (2006)

6. Andriluka, M., Roth, S., Schiele, B. Pictorial structures revisited: people detection and articulated pose estimation. In: Proceedings of the IEEE conference on computer vision and pattern recognition (CVPR) (2009)
7. Balan, A., Black, M. An adaptive appearance model approach for model-based articulated object tracking. In: Proceedings of the IEEE conference on computer vision and pattern recognition (CVPR) (2006)
8. Balan, A. O., Sigal, L., Black, M. J., Davis, J. E., Haussecker, H. W. Detailed human shape and pose from images. In: CVPR 1–8 (2007)

A Review Paper on IEEE 802.11 WLAN

Vikram Singh and Lalit Kumar Awasthi

Abstract IEEE 802.11 wireless networks are becoming omnipresent nowadays, providing mobility as well as flexibility to the users accessing the information. Presently, it acts as an alternative to the wired network, but soon it may replace the wired network completely. The protocol is based on multiple access where a node competes with other nodes to get access to the communication medium and to transmit the data. A major aspect of wireless technology is roaming, which is defined as the ability to seamlessly change from one wireless AP to another. TCP used in wired networks is not appropriate for wireless networks because TCP assumes all loses as a result of congestion, which is not the case always in wireless links. Also, wireless networks are more vulnerable to security threats as compared to wired networks.

Keywords IEEE 802.11 WLAN · Wireless networks · Wireless technology · Multiple access

Introduction

Wireless LANs (WLANs) are becoming quite popular nowadays as they provide a way to connect a device to the network without using wires. These networks utilize radio waves to transfer the information. A WLAN connects the computers and other components to the network using an Access Point (AP). Each device is

V. Singh (✉) · L. K. Awasthi
Department of Computer Science and Engineering, National Institute of Technology, Hamirpur 177005 Himachal Pradesh, India
e-mail: vikram87in@yahoo.com

L. K. Awasthi
e-mail: lalitdec@yahoo.com

S. Sathiakumar et al. (eds.), *Proceedings of International Conference on Internet Computing and Information Communications*, Advances in Intelligent Systems and Computing 216,
DOI: 10.1007/978-81-322-1299-7_24,

equipped with a wireless network adaptor with the help of which the device connects to the AP. The AP is also connected to a wired Ethernet LAN via an RJ-45 port. An AP usually has a coverage area of up to 100 m. This coverage area is called cell. The user is allowed to move around with the device within the cell.

IEEE 802.11 is based on multiple access where a node competes with other nodes to get access to the communication medium. For that Carrier Sense Multiple Access is used with Collision Avoidance (CSMA/CA). The conventional CSMA/CD which is used in wired networks does not work here because it is not possible to detect the collisions at the sender side. Two more problems are there: Hidden node and Exposed node problem. For this RTS-CTS scheme can be used.

WLAN, which is also known as Wi-Fi, has various versions like IEEE 802.11a, 802.11b, 802.11 g, 802.11n. These operates on two frequency bands, 2.4 GHz and 5 GHz. Higher frequency means it has a shorter range and faces more difficulty in penetrating walls and other obstructions. Let's see those versions briefly:

IEEE 802.11. This was the first WLAN standard developed in 1997. It supported data rate of not more than 2 Mbps, which was too low for most of the applications. So it is not in much use now [1].

IEEE 802.11b. It was developed in 1999 and it supported data rate up to 11 Mbps. It operates on 2.4 GHz.

IEEE 802.11a. This too was developed in 1999 along with IEEE 802.11b. Supports data rate up to 54 Mbps. Operates on 5 GHz. It has greater data rate and cost as compared to IEEE 802.11b but shorter range signal, which is easily obstructed.

IEEE 802.11g. Came in 2002–2003, which tried to combine the best of both 802.11a and 802.11b. Data rate up to 54 Mbps. Operates on 2.4 GHz for greater range. Also, it is backward compatible with 802.11b.

IEEE 802.11n. This is the newest standard. It uses MIMO technology. That means it utilizes multiple antennas and multiple signals instead of one. The data rate depends upon the number of antennas used.

A table is shown with comparisons between 802.11a, b, g, and n showing the frequencies they operate on, approximate data rates, and approximate range [2] (Table 1).

Table 1 Comparison of various wireless standards

Standard	Frequency (GHz)	Data rate (Mbps)	Range
802.11a	5	54	120
802.11b	2.4	11	150
802.11g	2.4	54	150
802.11n	2.4/5	220 (Four antennas)	250

Table 2 Modulation and coding techniques used in various wireless standards [2]

Technique	802.11a	802.11b	802.11g	802.11n
PSK	✔	✔	✔	
FSK				
ASK				
CCK		✔		
OFDM	✔		✔	✔
SS		✔		

PSK phase-shift keying
FSK frequency-shift keying
ASK amplitude-shift keying
CCK complementary code keying
OFDM orthogonal frequency-division multiplexing
SS spread spectrum

Modulation and Signal Coding

To transmit the data over a wireless network, digital to analog conversion is needed. Similarly, at the receiving end analog-to digital conversion is to be done. Various modulation techniques are there having different characteristics in terms of power consumption, noise resistance, etc. A table is shown for different standards (Table 2).

Roaming

Roaming is said to take place when a user moves, from being under the range of one AP to another AP. Whenever a mobile device moves away from an AP, the strength of the signal received from that AP reduces. Similarly, signal strength increases on moving closer to an AP [3]. That mobile device periodically and continuously monitors the signal strength from the APs which are nearby to see which is providing the strongest signals. For roaming to take place, few points to note [4]:

1. All the wireless devices and the APs must be on the same IP subnet.
2. When a WLAN client moves from the range of one AP to another, the wired network has to relearn the client's location so that the data could be sent to it.
3. After selecting the new AP, association and authentication need to be done with that AP. This may take up to one or even more seconds. Actually, this will not create a problem in applications like email, etc. but the problem arises if voice application is being used.

TCP in Wireless Networks

We know that TCP is a protocol that is used in wired networks. TCP is widely used in the current Internet as the reliable end-to-end transport protocol. For delivering the error-free and in-order data, it uses flow control, sequence numbers, acknowledgements, and timers. But the same protocol is not appropriate [5] for wireless networks. The reason is that TCP assumes all the losses as a result of congestion, which is not the case always in wireless links.

In wired networks, packet loss is assumed because of congestion mainly. But in wireless networks, the reason for the packet loss may be congestion or link-failure [6]. So if the packet is lost or delayed, in case say when topology changes, TCP might misinterpret the reason as congestion and reduce the transmission rate. This may lead to performance degradation, which may lead to a lower performance throughput of TCP over wireless link. So, TCP needs to be modified to understand the reason of packet loss or delay and the difference between link-failure and the congestion so that the performance could be improved.

Several solutions can be there for replacing the conventional TCP. Snoop protocol [5] can be an option. In this, there is a wired link and a wireless link. The packets are cached or buffered at the Base Station (BS) and local retransmission is performed across the wireless links by a snoop agent. Another option may be Fast Retransmission [5]. Whenever the mobile nodes are on a new link, TCP should go into fast retransmission mode instead of going into congestion mode where slow start condition is started. This will cause the mobile nodes to transmit unacknowledged segments as soon as they are on a new link. Split TCP approach can also be used in which the connection is segmented or divided into two parts. One wired and one wireless. This approach allows the TCP in the wireless link to be more aggressive when errors are met without influencing the fixed or wired connection. That means local recovery is done with the help of local retransmission. This approach provides separate flow control for each part. The connection between the fixed node and the mobile node goes through the BS.

$$\mathrm{FH} - \mathrm{MH} = \mathrm{FH} - \mathrm{BS} + \mathrm{BS} - \mathrm{MH}$$

Another approach to improve performance is Selective-Acknowledgment. It is used where the information about the segments that have been received is provided. It may happen that the packets may not have arrived sequentially. So only those segments are sent which have not been received in spite of the sequence (Fig. 1).

Fig. 1 Split-TCP approach

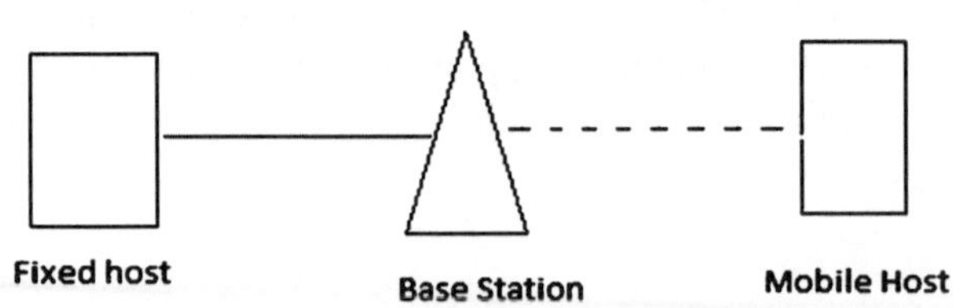

Security

It should be kept in mind that the communication in wireless network is broadcast in nature. That means all the devices within the range will receive the transmission. That makes wireless network more vulnerable to security threats. We will see various protocols defined for this purpose.

WEP. Designed in 1999. It uses 64 bits RC4 cipher to provide privacy. Two types of authentication can be done. Open System and Shared Key authentication. For data integrity, 32 bits Integrity Check Value (ICV) field, which is a simple 32-bit CRC, is used. 24 bits Initialization Vector (IV) is used which can provide 16,777,216 different RC4 cipher streams and this will allow easy decryption of data [7]. 40-bit WEP keys are also not adequate.

IEEE 802.1x. Came into existence in 2001. It is a port-based [6] standard and it is mainly designed to use Extensible Authentication Protocol (EAP) to provide strong authentication, access control, and easy key management. The 802.1x authentication protocol is used between the wireless clients and the AP, while RADIUS operates between the AP and the Authentication Server. The actual authentication is provided by EAP, not by 802.1x itself. That means 802.1x is only a standard for using EAP over the network.

WPA. Came into existence in 2003. Wi-Fi Protected Access (WPA) was developed to eliminate the shortcomings of WEP. It is actually subset of 802.11i standard. For privacy, it used Temporal Key Integrity Protocol (TKIP) [6] along with RC4 cipher. For data integrity, Message Integrity Check (MIC) is used instead of CRC which uses Michael Algorithm itself. The size of IV has been increased to 48 bits. Thus it is more secure than WEP.

IEEE 802.11i. It is also known as WPA2. For privacy it uses three algorithms: WEP, TKIP, and Counter-mode/CBC-MAC Protocol (CCMP). WEP and TKIP are based on RC4 algorithm, and CCMP is based on Advanced Encryption Standard (AES) which is one of the most secured encryption standards. For data integrity, it uses CBC-MAC protocol and MIC. IV size is 48-bit.

Conclusion

We have seen various aspects of WLANs in brief. This technology is omnipresent nowadays. A lot of work is being done to improve the wireless technology to the extent such that it supersedes the wired network. The time is not far when we may see the wires getting disappeared completely from networks, but still a lot of work is yet to be done.

References

1. The 802.11 family explained, http://compnetworking.about.com/cs/wireless80211/a/aa80211standard.htm
2. Tjensvold, J.M.: Comparison of the IEEE 802.11, 802.15.1, 802.15.4 and 802.15.6 wireless standards, (2007)
3. Kille, B.: 802.11 Wireless implementation – existing network infrastructure issues
4. WLAN Roaming – The Basics, http://features.techworld.com/mobile-wireless/435/wlan-roaming–the-basics/
5. Eshak, N., Dani, M.: Improving TCP performance in mobile ad hoc networks
6. Baghaei, N., Hunt, R.: IEEE 802.11 wireless LAN security performance using multiple clients
7. Walker, J.R.: IEEE P802.11 Wireless LANs: unsafe at any key size: an analysis of the WEP encapsulation, 2000.

Compression of Color Image by Using Dual Tree Complex Wavelet Transfrom

Vimal Kumar and Surendra Kumar Agarwal

Abstract In this paper, we explore the use of Dual tree Complex wavelet Transform which is nearly shift invariant and directionally selective in two and higher dimensions. The multidimensional dual tree CWT is a nonseparable and is based on computationally efficient, Separable filter bank (FB). This paper describes how the complex wavelet transform with directional properties is designed and use of it in image compression. When we take the dual tree complex wavelet transform then many wavelet coefficients are close to zero and have intra-subband dependency. We further evaluate the performance of SPIHT coding schema for coding of those coefficients. The result of proposed schema gives higher rate of compression and lover MSE compared to the schema based of DWT. Dual tree complex wavelet transform-SPIHT schema outperform DWT based schema at lower bit rates.

Keywords Image compression · Complex wavelet transform · Image texture · Dual tree · SPIHT

Introduction

Since last 20 years the discrete wavelet transform (DWT) has witnessed great success for image compression. In 2-D DWT based compression technique, two 1-D transform is used in which one 1-D transform is used for vertical Directions and another for horizontal directions [1]. The conventional 2-D discrete wavelet transform (DWT) efficiently captures point singularities of image, but fails to capture line singularities because edges and contours in images are not aligned

V. Kumar (✉) · S. K. Agarwal
Department of Electronics and Communication, Sobhasaria Engineering College, Sikar, Rajasthan, India
e-mail: vimalelectric@gmail.com

S. K. Agarwal
e-mail: skagarwal5@rediffmail.com

S. Sathiakumar et al. (eds.), *Proceedings of International Conference on Internet Computing and Information Communications*, Advances in Intelligent Systems and Computing 216,
DOI: 10.1007/978-81-322-1299-7_25, © Springer India 2014

with the horizontal or vertical direction of image. That's why there is requirement of new transform which remove the drawbacks by filtering the image in edge or contours direction.

According to the property of DWT the energy of image is spread across sub-bands if the edges and contours are not aligned vertically and horizontally. That's why there is requirement of directional transform so that energy cannot spread in sub-bands. Attempts on orientation adaptive transform can be classified into two categories: one category analyses an image along a predetermined set of directions (e.g., [2]), whereas the other category adapts the directional analysis itself to the oriented features of image [3–5].

Lifting structure based, several adaptive wavelet transforms, which adapt the filtering directions to the orientations of edges and textures, have been proposed [3, 4]. Different types of direction selection method are given in [5–9], are based on minimizing the prediction error.

Kingsbury and Reeves have already discussed that Dual tree-DWT gives higher PSNR at same Bit-per pixel level [10]. Image coding with Dual tree DWT is also reported in [11].

In this paper we evaluate the result of compression schema based Dual tree CWT and DWT with SPIHT coding schema (bior 4.4 filter coefficient). By analysing the result of compression schema we see that proposed schema outperform than DWT schema at lower bit rate.

This paper is organised as follows: Section "2-D Dual Tree CWT" gives the detailed presentation of 2-D Dual tree Complex wavelet Transform. Section "Compression Algorithm and Image Coding" gives Image Compression algorithm and Image coding sect. "Image Coding Performance Analysis and Experimental Result" shows Performance and Experimental Result. Conclusion and future work is given in sect. "Conclusion".

2-D Dual Tree CWT

Discrete Wavelet Transfrom is widely used schema for signal processing task, but the performance of DWT is limited by following reason: (i) not able to select higher dimensions, (ii) cause aliasing when signal is downsampled, (iii) energy division in high frequency band etc. The Complex wavelet transform provide analytic or quadrature wavelets which remove the problem of aliasing [12].

1D Dual Tree Wavelet

As we know that the DWT multiresolution schema provide perfect reconstruction of input signal, Dual tree CWT schema also provide perfect reconstruction [13–17] of input signal but it use analytical filter instead of real filter for remove the deficiency of DWT schema. The dual tree CWT schema use two DWT trees of real filter coefficient. The analytic function $\psi c(t)$ is composed of two real wavelets $\psi real(t)$ and $\psi imag(t)$, which are orthogonal to each other. Similar to wavelet functions there are two scaling functions $\phi r(t)$ and $\phi i(t)$, are relationship between them are:-

$$\psi(t) = \sqrt{2} \sum_{n} h1(n)\phi(2t - n) \tag{1}$$

$$\phi(t) = \sqrt{2} \sum_{n} h0(n)\phi(2t - n) \tag{2}$$

The scaling filter $h0$ and $g0$ (used for tree 1 and 2) are HT pair and follow the half sample delay condition:

$$g0(\mathrm{n}) = h0(\mathrm{n} - 0.5) \tag{3}$$

and in frequency domain the magnitude and phase condition is given by:

$$\left|G0\left(\mathrm{e}^{jw}\right)\right| = \left|H0\left(\mathrm{e}^{jw}\right)\right| \tag{4}$$

$$\angle G0\left(\mathrm{e}^{jw}\right) = \angle H0\left(\mathrm{e}^{jw}\right) \tag{5}$$

In this schema two standard DWT operating in parallel, one is known as real tree (tree-a) and other is known as imaginary tree (tree-b). The conjugate filter for tree-a are represented by ha = $\{h0, h1\}$, and for tree-b are gb = $\{g0, g1\}$. The filter for first is different from filter of other level. Similarly the filter pair $\left\{\widetilde{h0}, \widetilde{h1}\right\}$ and $\left\{\widetilde{g0}, \widetilde{g1}\right\}$ are used for synthesis purpose (Fig. 1).

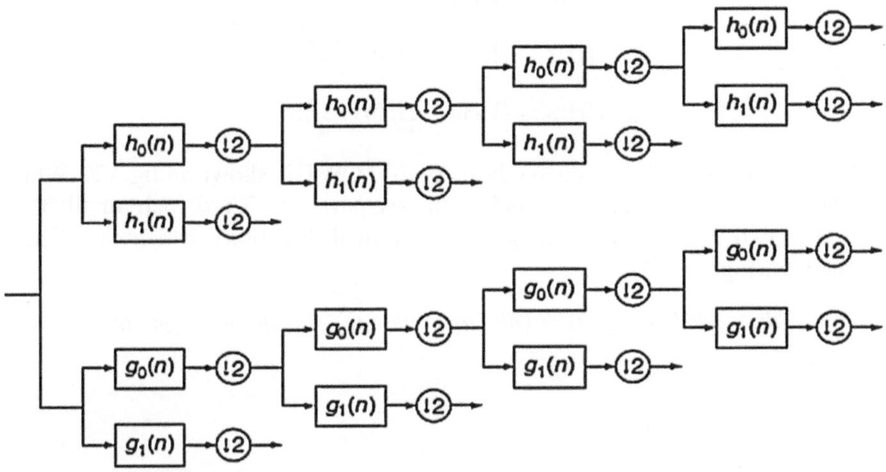

Fig. 1 Analysis filter bank for 1-D dual tree CWT

If the two real DWTs are represented by the square matrices Fh and Fg, then the dual tree CWT can be represented by the matrix:

$$F = \begin{pmatrix} Fh \\ Fg \end{pmatrix} \tag{6}$$

And for the synthesis filter:

$$F^{-1} = \frac{1}{2}[\mathrm{Fh}^{-1}\mathrm{Fg}^{-1}] \tag{7}$$

2-D Dual Tree Wavelet

The implementation of Dual Tree CWT is very straight forward. Firstly the input image is decomposed by two set of filters, both satisfying the PR conditions and every set contains both low pass and high pass filters, filtering the image horizontally and vertically similar two conventional 2-D DWT schema. These two filters are jointly designed such that the overall transform is approximately analytical.

Then six high pass subband (HLa, LHa, HHa, HLa, LHb, and HHb) and two low pass subband are generated at each level of decomposition. These sub-bands are decoded in form of:

$$\begin{aligned} \mathrm{LHs} &= (\mathrm{LHa} + \mathrm{LHb})\Big/\sqrt{2}, \\ \mathrm{LHm} &= (\mathrm{LHa} - \ \mathrm{LHb})\Big/\sqrt{2}, \\ \mathrm{HLs} &= (\mathrm{HLa} + \mathrm{HLb})\Big/\sqrt{2}, \\ \mathrm{HLm} &= (\mathrm{HLa} - \mathrm{HLb})\Big/\sqrt{2}, \\ \mathrm{HHs} &= (\mathrm{HHa} + \mathrm{HHb})\Big/\sqrt{2}, \\ \mathrm{HHm} &= (\mathrm{HHa} - \mathrm{HHb})\Big/\sqrt{2}. \end{aligned} \tag{8}$$

The filter bank analysis and synthesis of 2-D CWT is shown in fig. (2). Which consists four trees for analysis and synthesis purpose? These pair of filters is applied in two dimensions (horizontal and vertical direction), which can be represented as (Figs. 3, 4):

$$(hh + jgh)(hv + jgv) = (hh \cdot hv - gh \cdot gv) + \mathrm{j}(hh \cdot gv + gh \cdot hv) \tag{9}$$

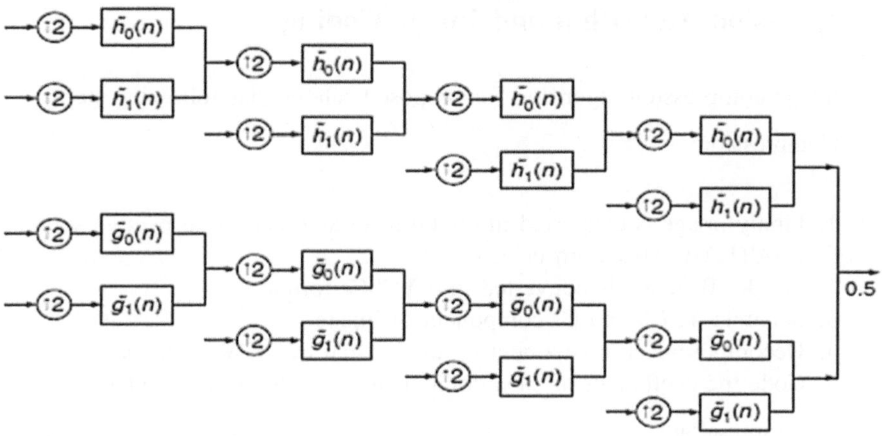

Fig. 2 Synthesis filter bank for 1-D dual tree CWT

Fig. 3 Filter bank structute for 2-D dual tree CWT

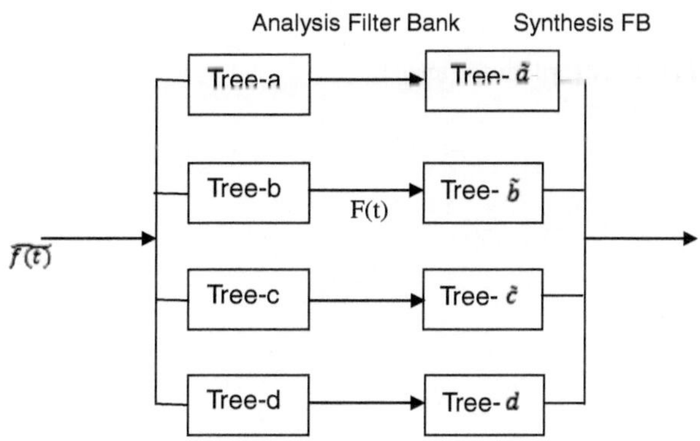

Fig. 4 Filter bank structure of tree-a

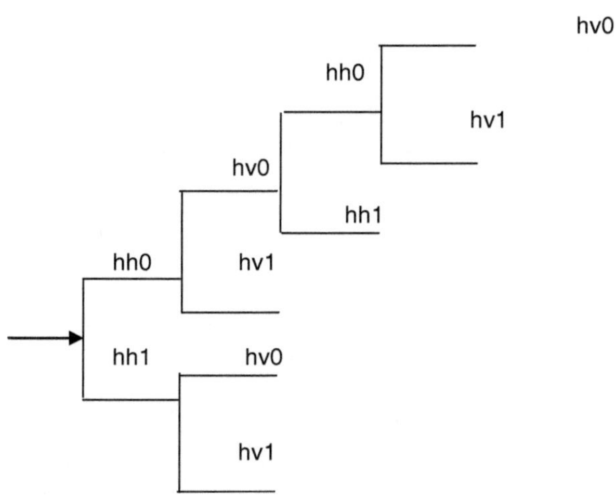

Compression Algorithm and Image Coding

The image compression algorithm for proposed schema has following steps:

(A) Compression:

1. Firstly image is converted in digital form and read by respective software (MATLAB (That I am using)).
2. The RGB image is converted into YCbCr format.
3. Separate Y, Cb and Cr component of image.
4. Decompose each component by using Dual tree CWT schema.
5. Code the coefficient of each component by using SPIHT coder.

(B) Decompression:

1. Read the coded image.

Fig. 5 Algorithm for image compression

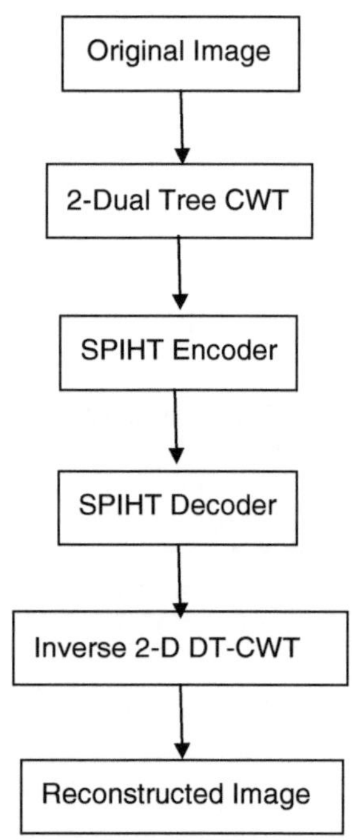

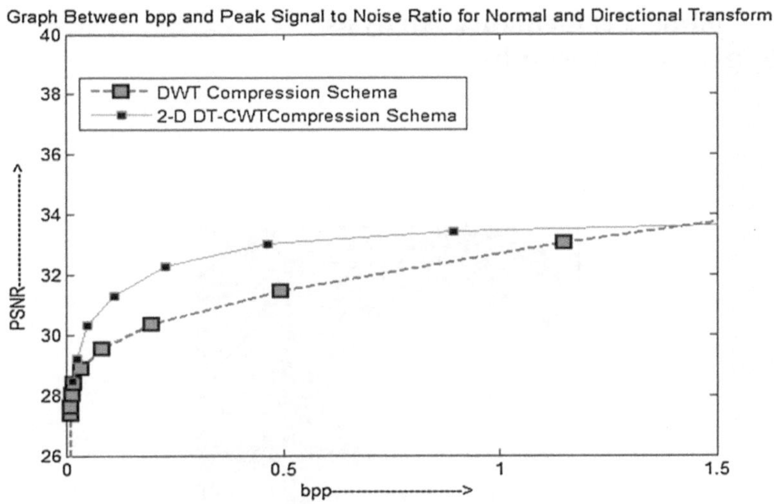

Fig. 6 Performance analysis using JPEG 2000 and proposed compression schema for my daughter image

Table 1 Comparison in performance for different image compressed by Jpeg 2000 and proposed schema

Image	Size	bpp	JPEG 2000	Proposed schema
My daughter image	256*256	0.1	27.22	27.32
		0.25	30.43	30.49
		0.5	31.04	32.25
		1	31.79	33.24
Lena	256*256	0.1	28.34	28.94
		0.25	33.74	34.30
		0.5	37.01	37.40
		1	40.6	40.77
Baboon	256*256	0.1	21.34	21.46
		0.25	23.06	23.29
		0.5	25.48	25.78
		1	29.01	29.47

2. Decode the coded image by using SPIHT encoder.
3. Pass the decoded image through inverse DT-CWT.
4. Covert the image from YCbCr to RGB format.
5. Measure MSE and PSNR for the image (Fig. 5).

Image Coding Performance Analysis and Experimental Results

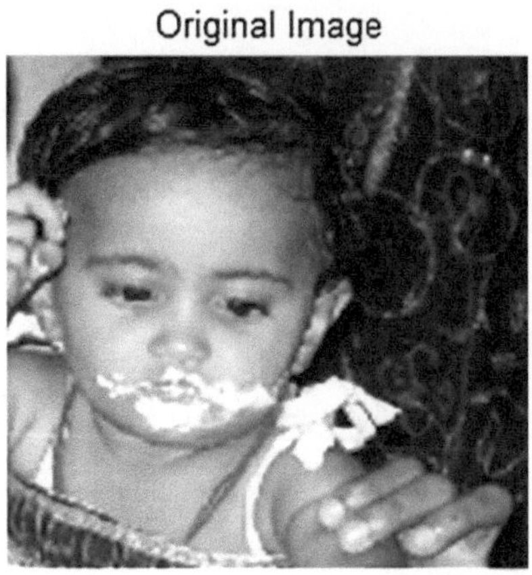

Fig. 7 Original image my daughter

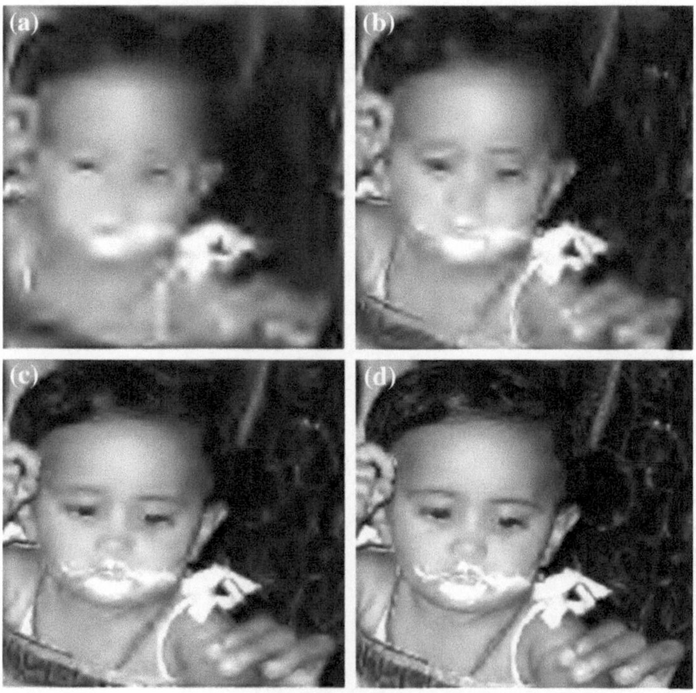

Fig. 8 Image reconstructed at **a** 0.1bpp, **b** 0.25bpp, **c** 0.5 bpp and **d** 1 bpp

The image coding results are compared in this section between JPEG 2000 and given 2-D DT-CWT schema. Because the given transform has capability of direction selection it outperforms the DWT based compression schema Here SPIHT coding schema is utilised to organize the compressed bit stream in the compression scheme. The compression ratio is set as the input of the compression system. The experimental results include four different ratio PSNR values for DWT-SPIHT and for given method (Fig. 6).

In shown figure we see that the proposed schema is better than the old DWT-SPIHT compression schema.

The Comparison in performance (in decimals) for different image is shown in Table 1 (Figs. 7, 8).

Conclusion

In this paper the 2-D dual tree CWT have been presented. The proposed transform covert all the coefficients near to zero which reduce the no of bits required to encode the image. That's why image compression result (MSE and PSNR) of by proposed transform is better than the JPEG 2000 Schema at low bit rate. Our future work includes applying this schema for wavelet based video coding at low computational complexity.

References

1. Taubman, D., Marcellin, M.: JPEG2000: Image Compression Fundamentals, Standards, and Practice. Kluwer, Norwell (2001)
2. Taubman, D., Zakhor, A.: Orientation adaptive subband coding of images. IEEE Trans. Image Process. **3**(4), 421–437 (1994)
3. Taubman, D.: Adaptive nonseparable lifting transforms for image compression. In: Proceedings IEEE International Conference Image Process, Kobe, Japan, 772–776, 3 Oct 1999
4. Claypoole, R.L., Davis, G.M., Sweldens, W., Baraniuk, R.G.: Nonlinear wavelet transforms for image coding via lifting. IEEE Trans. Image Process. **12**(12), 1449–1459 (2003)
5. Gerek, O.N., Cetin, A.E.: A 2-D orientation-adaptive prediction filter in lifting structures for image coding. IEEE Trans. Image Process. **15**(1), 106–111 (2006)
6. Ding, W., Wu, F., Wu, X., Li, S., Li, H.: Adaptive directional lifting-based wavelet transform for image coding. IEEE Trans. Image Process. **16**(2), 416–427 (2007)
7. Chang, C.L., Maleki, A., Girod, B.: Adaptive wavelet transform for image compression via directional quincunx lifting. In: Proceedings IEEE Workshop Multimedia Signal Processing, Shanghai, China, Oct 2005
8. Chang, C.L., Girod, B.: Direction-adaptive discrete wavelet transform via directional lifting and bandeletization. In: Proceedings IEEE International Conference Image Processing, Atlanta, GA, Oct 2006
9. Dong, W, Shi, G. Member, IEEE, and JizhengXu, Member, IEEE," Adaptive Nonseparable Interpolation for Image Compression With Directional Wavelet Transform", in IEEE SIGNAL PROCESSING LETTERS, VOL. 15, 2008

10. Kingsbury, N.G., Reeves, T.H.: Redundant representation with complex wavelets: how to achieve sparsity. In: Proceedings International Conference Image Processing, Barcelona, Sept 2003
11. Reeves, T.H., Kingsbury,N.G.: Overcomplete image coding using iterative projection-based noise shaping. In: Proceedings International Conference Image Processing, Rochester, NY, Sept 2002
12. Wang, B., et al.: An investigation of 3D dual-tree wavelet transform for video coding. In: Proceedings International Conference Image Processing, Singapore, Oct 2004
13. Wang, B., et al.: Video coding using 3-D dual-tree wavelet transforms. In: Proceedings International Conference on Acoustics, Speech, and Signal Processing, Philadelphia, Mar 2005
14. Liu, J., Moulin, P.: Information-theoretic analysis of interscale and intrascale dependencies between image wavelet coefficients. IEEE Trans. Image Process. **10**(11), 1647–1658 (2001)
15. Figueras, R.M., Ventura, I., et al.: Low-rate and flexible image coding with redundant representations, IEEE Trans. Image Process. 15(3), Mar 2006
16. Said, A., Pearlman, W.A.: A new, fast and efficient image codec based on set partitioning in hierarchical trees. IEEE Trans. Circuits Syst. Video Tech. **6**, 243–250 (1996)
17. Taubman, D.: High performance scalable image compression with ebcot. IEEE Trans. Image Process. **9**, 1158–1170 (2000)

Review of Software Quality Metrics for Object-Oriented Methodology

Suresh Yeresime, Jayadeep Pati and Santanu Ku Rath

Abstract This paper presents a review of metrics used in object-oriented programming, it includes a small set of the most well-known and commonly applied traditional metrics which could be applied to object-oriented methodology and a set object-oriented metrics (specifically applicable to object-oriented programming) for software development. The need for such metrics is notably more when an organization is keen on adopting such metrics to develop good quality software. The demand has increased for new or improved metrics for software development and the most prominent being object-oriented methodology.

Keywords Cyclomatic complexity · CK metric suite · Class · Inheritance · Polymorphism · Cohesion

Introduction

Several metrics related to object-oriented systems have been proposed, there are metrics still being introduced and even several books representing such metrics exist, such as Fenton's [1], Shepperd's [2], and others. Even though many metrics have been applicable to programming languages such as object-oriented language, these metrics are extracted from the concept of structured programming and

S. Yeresime (✉) · J. Pati · S. K. Rath
Department of Computer Science and Engineering, National Institute of Technology, Rourkela, India
e-mail: suresh.vec04@gmail.com

J. Pati
e-mail: jayadeeppati@gmail.com

S. K. Rath
e-mail: skrath@nitrkl.ac.in

S. Sathiakumar et al. (eds.), *Proceedings of International Conference on Internet Computing and Information Communications*, Advances in Intelligent Systems and Computing 216,
DOI: 10.1007/978-81-322-1299-7_26, © Springer India 2014

adjusted such measurements so as to match the needs of object-oriented programming. But the point of concern for a researcher or a practitioner is to select a suitable metric, to meet the needs of software project from a large number of available object-oriented metrics.

The Metrics

The metrics presented and evaluated in this paper are both traditional and object-oriented metrics and where in some traditional metrics can be applied to object-oriented programming. The objective of this paper is not to mention or explain all the existing metrics but just to create awareness of the existing metrics to the readers for the future references and research.

Traditional Metrics

Traditional Metrics are generally applied to the methods where in operations related to the class are needed.

Metric 1: Cyclomatic Complexity

Cyclomatic Complexity [3], is used to evaluate the complexity of an algorithm in an method. Thomas J McCabe proposed the concept of CyclomaticComplexity, where it gives insight to the number of decision points which contribute to the complexity of the method. It gives the count of number of test cases (lower bound) need to test the method completely. The formula used for computing the value is $V(G) = e - n + 2$, where V(G) represents the cyclomatic complexity value, e is the number of edges and n is the number of nodes, which are all determined from the control flow graph drawn for the particular algorithm.

Cyclomatic complexity cannot be directly used to measure complexity of the class because of inheritance. Hence complexity of individual methods can be summed up together with other measures to evaluate the complexity of class.

Metric 2 : Size

Size of the class is used to estimate the ease with which the maintainers understand the code written by developers. There are different ways to determine the size of the class viz., physical lines of code, the number of statements, the number of blank lines and the number of commented lines. Lines of Code (LOC) counts all the lines, usually referring to non-commentary lines, meaning pure whitespace and lines containing only comments are not included in the metric. Non-comment Non-blank (NCNB) lines are sometimes referred to as Source Lines of Code, which count all lines that are not commented and are not blank. Since lines of code only estimate the volume of code, we can only use it to compare software programs that use same language.

Metric 3 : Comment Percentage

The line counts used to compute the size metrics can be expanded to include a count of the number of comments, both on-line (with code) and stand-alone.

Comment percentage is given by the ratio as total number of comments divided by the total lines of code minus the number of blank lines. The Software Assurance Technology, a department in NASA has found a comment percentage of about 30 % to be most effective. Since these metrics boost the developers and maintainers to easily evaluate the attributes of understandability, Reusability and Maintainability.

Object-Oriented Metrics

As the present day development practices are based on object-oriented concepts, these object-oriented metrics help in understanding the product quality and help to assess the process effectiveness. The major benefit of using object-oriented metrics is to improve quality of the work performed at the project level. The following are some of the important metric suites:

Chidamber and Kemerer Metric Suite

Metric 4 : Weighted Method per Class (WMC)

WMC is included in the suite because it has been proven to be useful in predicting maintenance and testing effort [4]. This method has its own flaws, but it can be combined with other metrics such as cyclomatic complexity to contribute vital information about a class complexity.

Consider a Class C1, with method M1,….,Mn that are defined in a class. Let c1,….,cn be the complexity of the methods, then the WMC is given as follows: WMC $= \sum$ ci. For i $=$ 1 to n. Where ci is the complexity of the class i'th method, if all the method complexities are considered to be unity, then the value of WMC will be n, the number of methods.

WMC relates directly to the definition of complexity of a methods, since methods are properties of object classes and complexity is determined by the cardinality of its set of properties. The number of methods is therefore a measure of class definition as well as being attributes of a class, since attributes correspond to properties. The number of methods and the complexity of methods involved is a predictor of how much time and effort is required to develop and maintain the class.

Metric 5 : Depth of Inheritance Tree (DIT)

Depth of Inheritance of the class is the DIT metric for the class. In cases involving multiple references, the DIT will be the maximum length from the node to the root of the tree.

DIT relates to notion of the scope of properties. DIT is a measure of how many ancestor classes can potentially affect this class. The deeper a class will be in the hierarchy the greater the possibility to inherit more number of methods which makes it difficult to assess the behavior of the class. On the other side DIT value imply that many methods may be reused.

Figure 1 shows class hierarchy example in which the value of Depth of Inheritance Tree (DIT) is 4.

Metric 6 : Number of Children (NOC)

NOC is termed as the number of immediate sub-classes subordinate to a class in the class hierarchy. NOC is a measure of how many subclasses are going to inherit the methods of the parent class.

As the NOC value increases, reuse increases, but it is also true that as NOC value increases the abstraction represented by the parent class can be diluted i.e., there is a possibility that some of the children are not really appropriate members of the parent class. As NOC increases the effort required for testing also increases. For the Fig. 1, the NOC value for Class 'C' in the hierarchy is three, the children (subclasses) being E, F & G.

Metric 7 : Coupling between Object Classes (CBO)

CBO for a class is a count of the number of other classes to which it is coupled. CBO relates to the approach that an object is coupled or bound to another object if one of them acts on the other. As we know since object of same class have same properties, two classes can be coupled when methods declared in one class use methods or variables defined by the other class.

Fig. 1 A class hierarchy

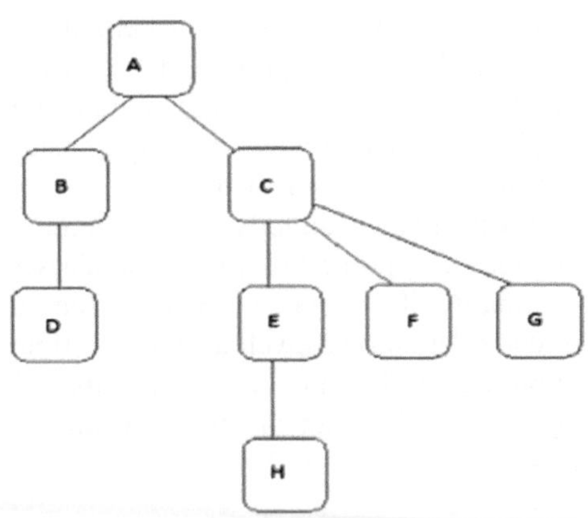

Excessive coupling between object classes is detrimental to modular design and prevent reuse of class and its properties. A measure of coupling is useful to determine how complex the testing of various parts in the design is likely to be. The higher the inter-object class coupling, the more rigorous the testing needs to be.

Metric 8 : Response for a Class (RFC)

The response set of class is a "a set of methods that can potentially be executed in response to a message received by an object of that class" [4]. RFC is defined as the number of methods in response set.

Basically RFC can be represented as RFC = | RS |, where RS is the response set of the class. RS = {M}∪} all in i {Ri}, where {Ri} = set of method called by the method i and {M} = set of all methods in the class. The cardinality in this set is the amount of attributes of the object in the class. Since it clearly includes methods called from outside the class, it is also a measure of the potential communication between the class and other classes.

As RFC increases, the effort required for testing also increases and the overall complexity of the class increases which is because of the test sequence grows. The worst case value for possible responses will assist in appropriate allocation of testing time.

Metric 9 : Lack of Cohesion in Methods (LCOM)

Consider a Class C1 with n methods M1, M2,.....,Mn. Let {Ij} = set of instance variables used by method Mi. There are n such sets {I1},....,{In}. Let A = {(Ii, Ij) | Ii ∩ Ij = ∅} and B = {(Ii, Ij) | Ii ∩ Ij = ∅},If all n sets {I1},...,{In} are ∅ then let A = ∅. LCOM = |A| - |B|, if |A| > |B| = 0.

Example: Consider a Class C1 with three methods M1, M2, and M3. Let {A} = {a, b,c,d,e} and {B} = {a,b,e} and {C} = {x,y,z}. {A} ∩ {B} is non-empty, but {A} ∩{C} and {B}∩{C} are null sets. LCOM is the (number of null intersections-number of non-empty intersections), which in this scenario is 1. Where {A}, {B} are instance variables of methods M1 and M2.

If LCOM is high, methods may be coupled to other methods via attributes, which increases the complexity of class design. In a conventional way high values of LCOM imply that the class might be better designed by breaking it into two or more separate classes where in the concept of modularization comes into act,so as to reduce the complexity and increase the understandability of the class implemented in object-oriented systems.

Metrics for Object-Oriented Design

The object-oriented design metrics [5], are a set of metrics that relevant to basic structural mechanism of the object-oriented paradigm such as the inheritance, polymorphism, message passing which are expressed in the form of quotients.

The set includes the following metrics:

Metric 10 : Method Hiding Factor (MHF)

MHF is defined as the ratio of the sum of the invisibilities of all methods that are defined in all the classes to the total number of methods defined in the system under consideration. The invisibility of a method is the percentage of the total classes from which this method is not visible.

Metric 11 : Attribute Hiding Factor (AHF)

AHF is the ratio of the sum of the invisibilities of all attributes defined in all the classes to the total number of attributes defined in the system under consideration.

Metric 12 : Method Inheritance Factor (MIF)

MIF is defined as the ratio of the sum of the inherited methods in all the classes of the system under consideration to the total number of available methods (i.e., methods that are locally defined including the inherited methods) for all classes.

Metric 13 : Attribute Inheritance Factor (AIF)

AIF is defined as the ratio of the sum of inherited attributes in all he classes of the system under consideration to the total number of available attributes (i.e., methods that are locally defined including the inherited methods) for all classes.

Metric 14 : Polymorphism Factor (PF)

PF is defined as the ratio of the actual number of possible different polymorphic situation for class Ci to the maximum number of possible distinct polymorphic situations for class Ci.

Metric 15 : Coupling Factor (CF)

CF is defined as the ratio of the maximum possible number of couplings in the system to the actual number of couplings not imputable to inheritance.

Balasubramanian's Metric Suite

Balasubramanian has proposed two improved versions of old metrics and a completely new metric [6]. The advanced metrics are Class Complexity (CC) and Cohesion Ratio, and the new metric is the Weighted Method Send Out.

The two improved metrics, Class Complexity and Cohesion Ration, have been influenced by the Chidamber and Kemerer work [4].

Balasubramanian states that, in this Class Complexity (CC) metrics, that the class complexity is the sum of the number of instance variables in a class and the sum of weighted static complexity of a local method in the class. And to measure the static complexity, Balasubramanian used the traditional metric, the cyclomatic complexity proposed by Thomas J McCabe [3].

Cohesion Ratio is the ratio of number of method pairs with similarities divided with the sum of method pairs with and without similarities. Cohesion Ratio is least when none of the method pairs share their instance variables and the value of cohesion ratio is zero and value is one, when the methods share their instance variables.

Objects do communicate using message passing. The Weighted Method Send Out metric uses the number of parameters in each message to weigh the method. This admits that not only more number of message connections but also the more number of parameters increases the coupling between the classes.

Balasubramanian also states that his improved metrics of the Chidamber and Kemerer metric suite yield better results at the early tests, but at the same time he agrees that thorough testing needs to be done before using these new metrics in the object-oriented methodology.

Additional Metrics Commonly Used

In the following section we will give a brief description of the additional metrics calculated, such as Number of Operations Overridden by a subclass (NOO), Specialization Index (SI), Instability.

Number of Operations Overridden by subclass (NOO)

There are instances when a subclass replaces an operation inherited from its parent class with a specialized version for its own use, this is called Overriding. Larger values of NOO imply a design problem.

If NOO is high, it implies that abstraction has been violated from parent class by the designer which leads to weak class hierarchy and the software can be difficult to test and modify.

Specialization Index

The specialization index provides a rough indication of the degree of specialization for each of the subclass in an object-oriented system. Specialization index can be achieved by adding or deleting operations or by overriding.

If the value of specialization index is more, then it implies that the class hierarchy has classes which do not conform to their parent class.

Percent public and protected (PAP)

Public attributes are inherited from other classes and hence are visible to those classes, whereas protected area specialization and private to specific classes. This metric is an indication of the percentage of class attributes that are public. If the value of PAP is more then there is a likelihood of side effects among classes.

Public access to data members (PAD)

This metric is an indication of the number of methods that can access another class's attributes. If the value of PAD is high, then it leads to side effects among other classes.

Number of scenario scripts (NSS)

The scenario scripts are nothing but the Use Cases, which is directly proportional to the number of classes required to meet the requirements, the number of states for each class and the number of methods, attributes and collaborations. NSS is strong indication for program size.

Metric Evaluation Criteria

While it is important to come up with certain metrics for object-oriented system, the metrics worth is in their application to programs- how these metrics suit the developers to improve the quality of programs? While there are many guidelines to interpret the metrics, we list a few preferable values for various metrics in Table 1, for the metrics described in the previous sections.

Results

Gathered Results

Two systems viz., Library Management System and Chess game software were analyzed to obtain the traditional and object-oriented metrics values. These systems were analyzed using Eclipse IDE and some metrics calculated by C code, the results and analysis is shown in the following section.

Table 1 Metric evaluation guidelines

Metric	Objective
Cyclomatic complexity(CC)	Low (CC < 10)
Lines of code/executable statement	Low
Comment percentage	~20–30 %
Weighted method per class (WMC)	Low
Depth of inheritance (DIT)	Less than six
Number of children (NOC)	Low
Coupling between objects (CBO)	Low
Response for Class (RFC)	Low
Lack of cohesion among methods (LCOM)	Low
Cohesion of methods	High

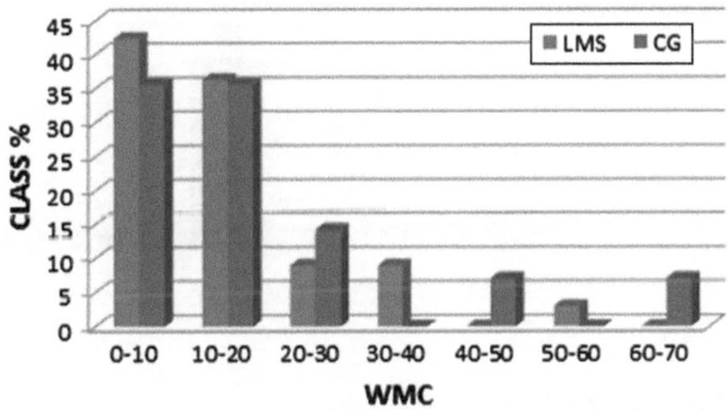

Fig. 2 Comparison of WMC for two projects

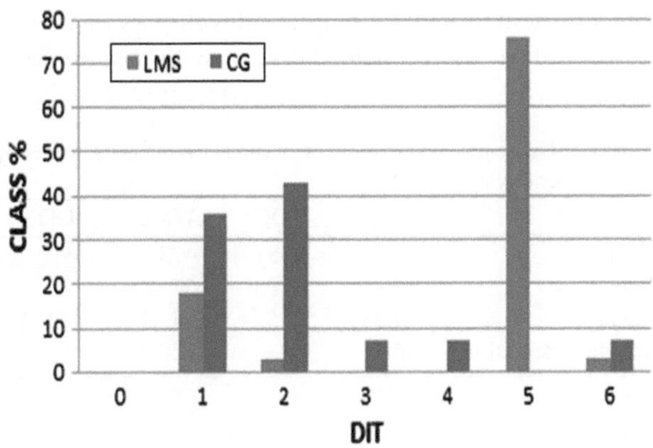

Fig. 3 Comparison of DIT for two projects

Figures 2, 3, 4 and 5 depict the histogram comparison of various object-oriented metrics statistics evaluated for the gathered results from Table 2 and Table 3.

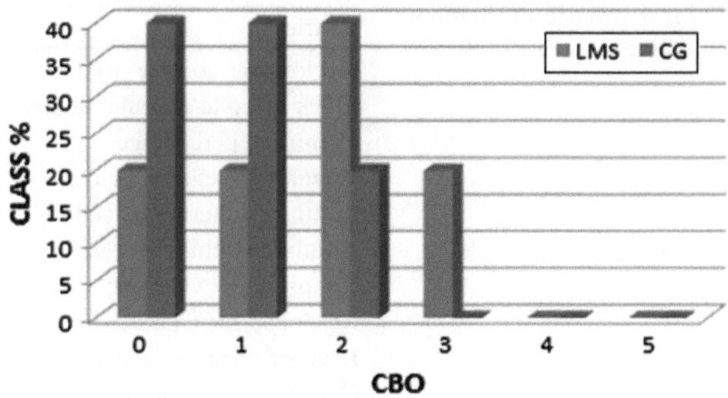

Fig. 4 Comparison of CBO for two projects

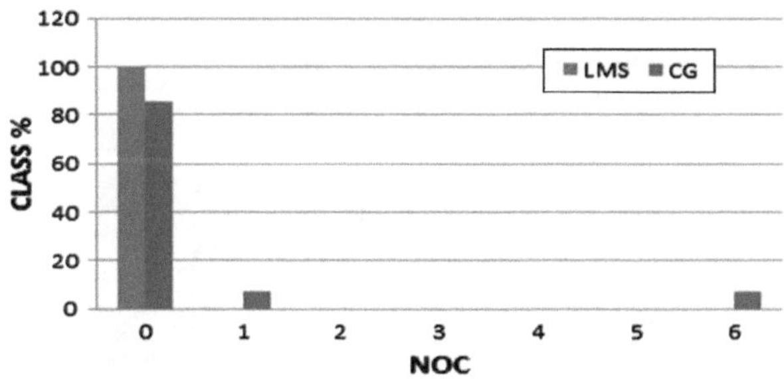

Fig. 5 Comparison of NOC for two projects

Interpretation of Results

Object-oriented metrics help the developers or maintainers to assess the required testing effort, to estimate the ease with which the maintainers can understand,

Table 2 Descriptive statistics of various traditional and object-oriented metrics for library management system (LMS)

	CC	LOC	WMC	DIT	LCOM	NOO	SI
Total		3431	473	6		3	
Max	55		57		1	2	
Mean	4.38		14.33	4.121	0.47	0.91	0.076
Standard deviation	6.31		12.19	1.64	0.43	0.37	0.351

Table 3 Descriptive statistics of various traditional and object-oriented metrics for chess game (CG)

	CC	LOC	WMC	DIT	NOC	LCOM	NOO	SI
Total		1181	269		7		3	0
Max	36		69	6	6	0.87	2	1
Mean	3.58		19.21	2.14	0.5	0.23	0.21	0.08
Standard deviation	6.35		17.33	1.35	1.54	0.32	0.55	

maintain and reuse code. The information tabularized and the bar graphs show the analysis of various metric values. For a given scenario, the objective of cyclomatic complexity is to have lower values (preferably below 10), which in turn reduces the testing efforts, increases the ease of understandability and maintenance of the code for developers and maintainers.

Almost all the metrics, except cyclomatic complexity, don't have a solid theory foundation. The current metrics of the software complexity are somehow, based upon the experience [7].

Future Work

Research needs to be carried out in the field of modularization, as from the results we can interpret that the values for Weighted method per class which in turn uses cyclomatic complexity has higher values. So there is work need to be done in reduction of these values in order to design the object-oriented systems in modular fashion, reduce in complexity of the system helps in ease of understand ability and maintainability and also minimize the testing efforts. This also will help researchers in predicting the failure of fault prone systems and to estimate the testing effort required to analyse the system.

Conclusion

Being able to obtain metrics is important for object-oriented methodology, as it will assist researchers to evaluate the software. There are many metrics available as mentioned in the above sections, like the CK Metric, Balasubramanian Metric suite to determine a systems complexity. There is a necessity to take full advantage of the metrics and tools used to sort out the statistics of analyzed characteristics of classes and systems in object-oriented programming.

The existing tools do concentrate on mining metric values of the present object-oriented systems; further research needs to be carried out to come up with new metrics to evaluate the object-oriented systems design. It is dubious to design a universally acceptable quality metrics and model, so that the metric will satisfy all the programming languages in their development environments and for different platforms and domains. So there is need to properly identify and validate the use of metric to suit the particular development environment. It should be noted that metric are a set of guidelines but not rules [5].

References

1. Fenton, N., Pfleeger, S.L.: Software Metrics : A Riforous and Practical Approach, 2nd edn. International Thomson Computer Press, London (1997)
2. Shepperd, M.J., Ince, D.: Derivation and Validation of Software Metrics. Clarendon Press, Oxford (1993)
3. McCabe, T.J.: A complexity measure. IEEE Trans. Softw. Eng. **2**, 308–320 (1976)
4. Chidamber, S.R., Kemerer, C.F.: A metric suite for object oriented design. IEEE Trans. Softw. Eng. **20**(6), 476–493 (1994)
5. Brito e Abreu, F.: The MOOD metric suite. In: ECOOP'95 Workshop on Metrics, Denmark, Aug (1995)
6. Balasubramanian, N.V.: Object-oriented metrics. In: Asia-Pacific Software Engineering Conference, Seoul, South Korea, pp. 30–34 (1996)
7. Sheng, Y., Shijie, Z.: A survey on metric of software complexity. In: IEEE International Conference on Multimedia & Expo, Chengudu, China pp. 352–356 (2010)

RRTS: A Task Scheduling Algorithm to Minimize Makespan in Grid Environment

Sanjaya Kumar Panda, Sourav Kumar Bhoi and Pabitra Mohan Khilar

Abstract Task scheduling is one of the major issues of grid environment. This is an essential process in utilizing the resources efficiently by reducing the completion time. The performance of the grid can be enhanced by using efficient task scheduling algorithms. In this paper, we have proposed a new technique called Round Robin Task Scheduling (RRTS) for minimizing the Makespan by using concept of Round Robin. The idea of the approach is to execute the tasks by using Dynamic Time Slice (DTS). Our experimental analysis shows better results than other task scheduling algorithms (Minimum Execution Time (MET), Minimum Completion Time (MCT), Min–Max, and Max–Min) in terms of Makespan and Average Resource Utilization.

Keywords Round robin task scheduling · Makespan · Grid computing · Task scheduling · Dynamic time slice

Introduction

Task scheduling is a most essential integrated component in grid environment. Task is a set of related instructions or an executable program which provides a single function [1]. It is an integrated component of computing which makes the

S. K. Panda (✉) · S. K. Bhoi · P. M. Khilar
Department of Computer Science and Engineering, National Institute of Technology, Rourkela, India
e-mail: sanjayauce@gmail.com

S. K. Bhoi
e-mail: souravbhoi@gmail.com

P. M. Khilar
e-mail: pmkhilar@nitrkl.ac.in

S. Sathiakumar et al. (eds.), *Proceedings of International Conference on Internet Computing and Information Communications*, Advances in Intelligent Systems and Computing 216,
DOI: 10.1007/978-81-322-1299-7_27,

resources busy and balance the load to avoid idle time of resources [2–5]. Task scheduling algorithms are the solutions to schedule the tasks in an optimal manner. Many researchers proposed their solutions to the task assignments [6]. In grid environment, there are heterogeneous machines to schedule the tasks in these machines optimally is a NP problem [1, 6–10]. It leads to minimum completion time and minimum Makespan (both Grid and nonGrid) [11]. The problem of allocating tasks to processors in a distributed system to minimize execution and communication costs [10]. To solve the scheduling problem many heuristics [12] have been developed like RASA [13], rule-based algorithm [14], selective algorithm [6] etc. The main goal of their algorithms is to minimize the throughput, maximize resource utilization, and satisfying the user demands. Heterogeneous machines are the machines having different architecture, operating systems [15], and speeds.

We must consider a dynamic environment in which jobs arrive over the time and remove from the task queue at their completion time [14]. Grid task scheduling is not limited to resource utilization but can be extended to the security, central control in administrative domains, real-time scheduling, and quality of service [2, 8, 11, 16, 17]. Real time has a time limit on computation [18]. Grid resource broker is responsible for allocation of a task to a particular resource which takes less time. It is also responsible for splitting the job into various tasks [3].

Grid applications are divided into number of interdependent subtasks in real applications. Subtask can be processed concurrently to reduce the task execution time [7]. Grid environment consists of many clusters. So the processors are not only heterogeneous but also the communication is larger [19]. We cannot guarantee the optimum solution but always find solution which is close to optimum [20].

RR is a pre-emptive scheduling algorithm [21]. It means the processor released the tasks in the middle of the execution. The tasks which do not have interdependency among them are called Meta tasks. As there is no interdependency, we can execute two tasks in parallel. In other words, if there are two processors in the grid environment then two tasks can be scheduled simultaneously.

The rest of paper is organized as follows: Section “Related Work” presents the related work, Sect. “Preliminaries” presents the preliminaries which discuss about notations use in this paper. It also includes assumptions and the existing scheduling algorithms. Section “Proposed Algorithm” proposes the RRTS algorithm with the performance metrics. Section “Experimental Results” elaborates the illustration and experimental analysis. We conclude this study in Sect. “Experimental Results”.

Related Work

In recent years, many algorithms have been designed to schedule the task efficiently in grid environment. As we know, task scheduling is a NP problem; it is difficult to find an optimal solution. Etminani et al. and Parsa et al. proposed a new algorithm based on the two existing algorithms Min–Min and Max–Min [6, 13]. It chooses the two existing algorithms based on standard deviation of the expected CT [6]. Rasooli et al. introduced a rule-based scheduling which contains two rules for resource selection and three rules for job queue [14]. Parsa et al. designed a task scheduling algorithm called RASA which selects Min–Min strategy to execute small task first and selects Max–Min strategy to execute large task first [13]. It seems to no starvation to the tasks. Xiaoshan et al. proposed a Min–Min Heuristic. It is based on adaptive scheduling heuristics which includes Quality of Service guidance [2]. Sun et al. developed a priority-based task scheduling (P-TSA). Tasks are sorted based on the priority and assign to processor by comparing P-TSA with the existing grid scheduling algorithms [7]. Zhang et al. introduced a new measurement called effective aggregated computing power (EACP) that strongly improves the performance of schedulers [19]. Navimipour et al. used genetic algorithm (mutation and new approach of crossover) in linear genetic representation to overcome demerits of previous methods [20]. Mansouri et al., Tang et al., and Abdi et al. proposed combine scheduling strategy in which several replication strategies and their performance are evaluated [5, 22, 23].

Preliminaries

A Notations

RRTS: Round Robin task Scheduling
RET: Remaining execution time
MET: Minimum execution time
MCT: Minimum completion time
DTS: Dynamic time slice
TQ: Task queue
T_X: Task number X
R_Y: Resource number Y
ATAT: Average turn around time
AWT: Average waiting time
ART: Average response time
ARU: Average resource utilization
FCFS: First come first served
CT: Completion time
ET: Execution time
CM: Current makespan
i, j: Loop variable
X: Total number of tasks
Y: Total number of resources
M: Makespan

Mod:	Modulus Function
MaxET:	Maximum execution time
MinET:	Minimum execution time
ETS:	Execution time spent
Max =	Find maximum value
T_{iY} =	Task i in resource Y

B Assumptions

All the experiments are performed in a heterogeneous environment. It consists of a number of computers (resources) with different operating systems and architecture. All the tasks have same priority. Attributes like execution time, number of computers (resources), and number of tasks are known before submitting to the environment. All the tasks are CPU bound. We assume that no tasks are I/O bound. Communication time is assumed to be negligible. DTS is taken in seconds. Tasks are divided into instructions. A task can be executed by switching it from one resource to other resources. We have taken switching time to be negligible. All the tasks are similar types of ET. ET of the tasks should not differ by large value. Skewness should be avoided.

C Scheduling Algorithms

There are many task scheduling algorithms in grid computing like MET, MCT, Min–Min, Max–Min, and RASA. We discuss it by using an example [13] shown in Table 1.

(1) *Minimum Execution Time*: It follows FCFS order. It finds the resource which takes less execution time. One of the major demerits of this algorithm is load imbalance. Time complexity is O(X) to assign a given task to a resource [6]. The gantt chart of MET is shown in Fig. 1. It gives a Makespan of 11.

Table 1 Execution time of tasks

	Resources	
Tasks	R_0	R_1
T_0	3	10
T_1	2	13
T_2	5	15
T_3	1	12

(2) *Minimum Completion Time*: It also follows FCFS order. It finds the resource which takes less completion time. CT is the sum of ET and CM. Time complexity remains same as MET. The gantt chart of MCT is shown in Fig. 2. It gives a Makespan of 11.
(3) *Min–Min*: It does not follow FCFS order. It follows shortest task first mechanism. It means the task having MET is to be scheduled first. First Min indicates MET whereas second Min indicates MCT. Time complexity is O (X^2Y) [6]. It leads to starvation to the larger tasks. The gantt chart of Min–Min is shown in Fig. 3. It gives a Makespan of 11.
(4) *Max–Min*: It does not follow FCFS order. It follows largest task first mechanism. It means the task having Maximum Execution Time is to be scheduled first. In Max–Min, Max indicates Maximum Execution Time whereas Min indicates MCT. It is very similar to Min–Min scheduling algorithm. Time complexity remains same as Min–Min. It leads to starvation to the small tasks. The gantt chart of Max–Min is shown in Fig. 4. It gives a Makespan of 11.

Fig. 1 Gantt chart for MET

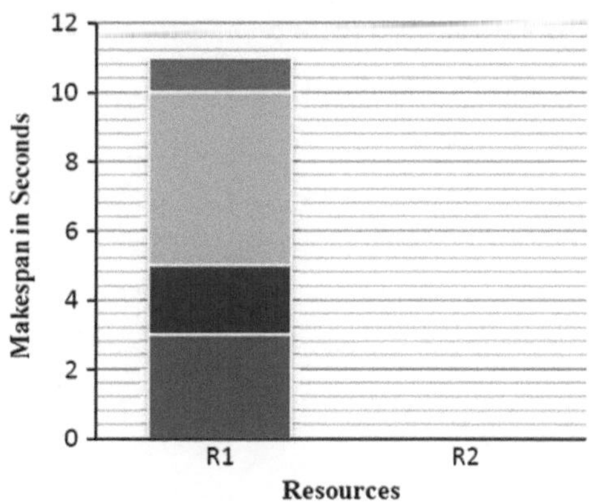

Fig. 2 Gantt chart for MCT

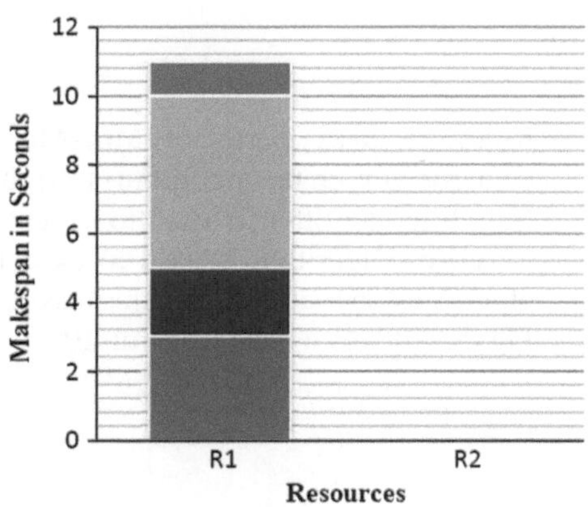

Fig. 3 Gantt chart for Min–Min

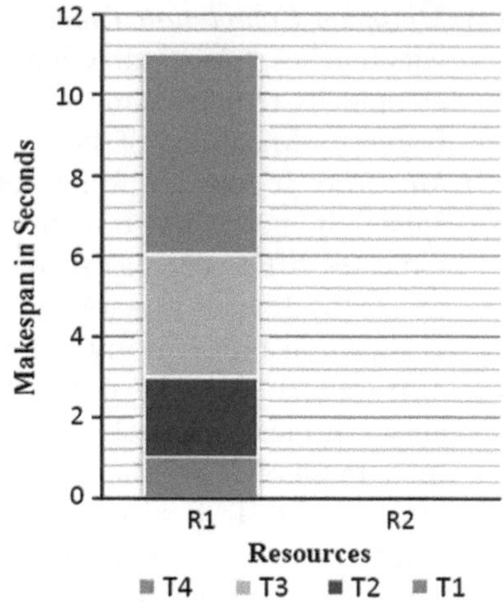

Fig. 4 Gantt chart for Max–Min

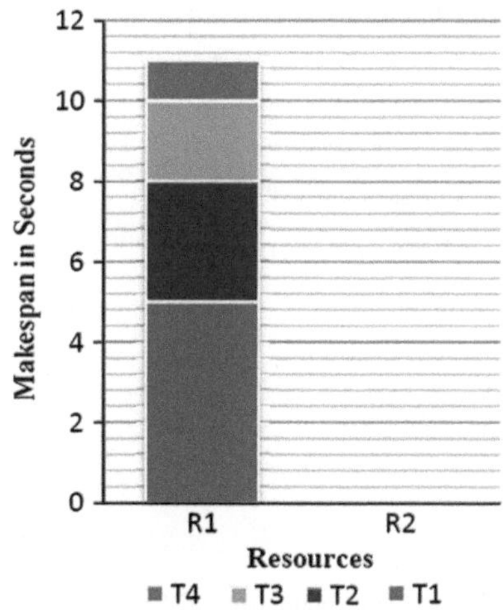

(5) *RASA*: It takes merits of Min–Min and Max–Min algorithm and avoids demerits of both the algorithms [13]. It is a hybrid approach. In this algorithm, Min–Min algorithm is used to execute small tasks rather than large one and Max–Min algorithm is used to execute large tasks rather than small one. It eliminates starvation from the tasks. The gantt chart of RASA is shown in Fig. 5. It gives a Makespan of 10.

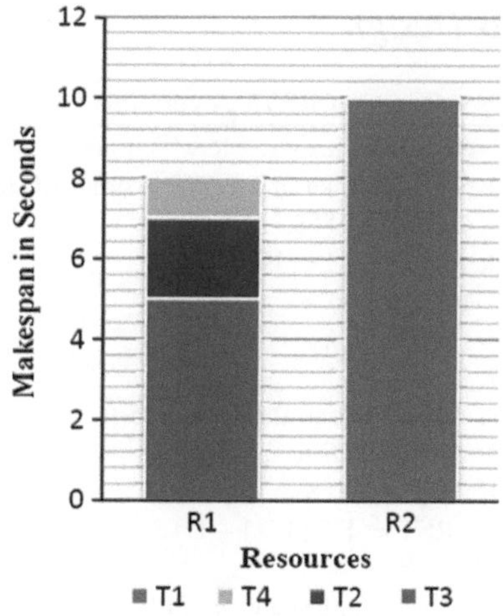

Fig. 5 Gantt chart for RASA

Proposed Algorithm

A Description

In our algorithm, tasks are present in the TQ and then sorted according to the fastest processor's execution time. DTS can be calculated using a formula shown in Eq. 1.

$$\text{DTS} = (\text{MaxET} - \text{MinET}) / X \quad (1)$$

DTS is assigned to the tasks present in task queue. Resources are assigned to the tasks based on the concept of round robin. Tasks can be switched between resources to minimize the completion time. Fastest processor remains 100 % busy in our approach. Task scheduling in RRTS is shown in Fig. 6.

Fig. 6 Task scheduling in RRTS algorithm

T_X/R_Y	T_0	T_1	T_2	T_3
R_0	51	33	27	47
R_1	123	97	54	112

↓ Sorted order

T_X/R_Y	T_2	T_1	T_3	T_0
R_0	27	33	47	51
R_1	54	97	112	123

↓ Calculate Dynamic Time Slice

Assign Processor

B Design

Pseudocode of RRTS Algorithm

1. Select the resource R which takes less ET for all tasks.
2. Sort the tasks in ascending order of their ET. (Rest processors tasks are sorted accordingly)
3. Calculate DTS.
4. while (TQ ! = NULL)
5. for $i = 0$ to X
6. i = i mod X
7. j = i mod Y
8. Assign TQ_i to the Resource R_j
9. Assign DTS to task TQ_i
10. TQ_i -> DTS
11. RET = ET [TQ_i] – DTS
12. if (RET == 0)
13. Task TQ_i has successfully executed.
14. swap ();
15. else if (RET > 0)
16. Pre-empt the task and re-schedule it to end of the TQ.
17. Update the rest resources RET.
18. else if (RET < 0)
19. Task has successfully executed before DTS expires.
20. swap ();
21. end if
22. end for
23. Update TQ and X.
24. end while
25. Calculate Makespan, ATAT, AWT, ART, and ARU.

1. swap()
2. {
    a. if (TQ == NULL && R == NULL)
    b. Pre-empt the task from next resource which takes less ET after R and re-schedule it to R.
    c. else
    d. return 0;
    e. end if

   }

C Performance Metrics

Our algorithm shows better results than other task algorithm scheduling in terms of Makespan, Resource Utilization (RU), Average Waiting Time (AWT), Response Time (RT), and Average Turn Around Time (ATAT).

(1) *Makespan*: Makespan is a performance measure of the throughput of the heterogeneous grid environment. It is a maximum time that all processor taken. Makespan can be calculated using a formula shown in equation 2.

$$M = \max(\text{ETS}) \tag{2}$$

where $\text{ETS} = (\sum_{i=1\text{to}X} \text{ETS}(T_{i1}), \sum_{i=1\text{to}X} \text{ETS}(T_{i2}), \ldots, \sum_{i=1\text{to}X} \text{ETS}(T_{iY})$

(2) *Turn Around Time (TAT)*: The time interval between submissions of a task to completion of that task. ATAT can be calculated using a formula shown in equation 3.

$$\text{ATAT} = (1/X) \sum_{i=1\text{to}X} (\text{Finish time } (T_i) - \text{Arrival time } (T_i)) \tag{3}$$

(3) *Waiting Time (WT)*: The time interval between start time and arrival time of a task. AWT can be calculated using a formula shown in Eq. 4.

$$\text{AWT} - (1/X) \sum_{i=1\text{to}X} (\text{Start time } (T_i) - \text{Arrival time } (T_i)) \tag{4}$$

(4) *Response Time (RT)*: The time interval between first response and arrival time of a task. ART can be calculated using a formula shown in Eq. 5.

$$\text{ART} = (1/X) \sum_{i=1\text{to}X} (\text{1st Response } (T_i) - \text{Arrival time } (T_i)) \tag{5}$$

(5) *Resource Utilization (RU)*: It is the percentage of time that the resource is busy. RU and ARU are calculated using a formula shown in Eqs. 6 and 7, respectively.

$$\text{RU } (R_Y) = \sum_{i=1\text{to}X} \text{ETS } (T_{iY}) \tag{6}$$

$$\text{ARU} = \left(\sum_{i=1\text{to}Y} \text{RU } (R_Y) \right) / Y \times 100 \tag{7}$$

Experimental results

A Illustration

Let us consider a problem having four tasks T_0, T_1, T_2, and T_3 and two resources R_0 and R_1. It shows that $X = 4$ and $Y = 2$. Table 1 shows the execution time of the tasks [13].

Table 2 Execution time of sorted tasks

Tasks	Resources	
	R_0	R_1
T_3	1	12
T_1	2	13
T_0	3	10
T_2	5	15

Step 1: Select the resource *R* which takes less ET for all tasks i.e. *R*1.
Step 2: Sort the tasks in ascending order of their ET. Table 2 shows this scenario.
Step 3: Calculate DTS.
DTS = (5−1)/4 = 1
So, DTS is 1 for all the tasks.
Step 4: TQ contains T_3, T_1, T_0 and T_2, respectively.
Step 5: Initially, *i* value is 0.
Step 6: New value of $i = i \bmod X = 0$.
Step 7: Similarly, $j = i \bmod Y = 0$.
Step 8: Task 3 is assigned to R_0.

TQ	T_3	T_1	T_0	T_2
	0	1	2	3

Step 9: Assign DTS = 1 to task T_3.
Step 10: T_3 has ET = 1.
Step 11: RET = 1−1 = 0.
Step 12: The condition for RET = 0 is satisfied.
Step 13: Task 3 has successfully executed.
Step 14: Call swap function. As TQ not equal to NULL, it returns 0. Go to Step 5.

So we continue our iteration until TQ is empty. Final gantt chart is shown in Fig. 7. It gives a Makespan of 9.04. T_{ij} shows task *i* in *j*th iteration.

B Experimental Analysis

The experimental analysis shows that our proposed algorithm RRTS is more efficient than the other task scheduling algorithms in terms of Makespan, ARU, and ART. We have considered three cases in which we have compared our algorithm with other four algorithms. Performance metrics of all the algorithms for three different cases are shown in Figs. 8, 9, 10, 11, 12, respectively.

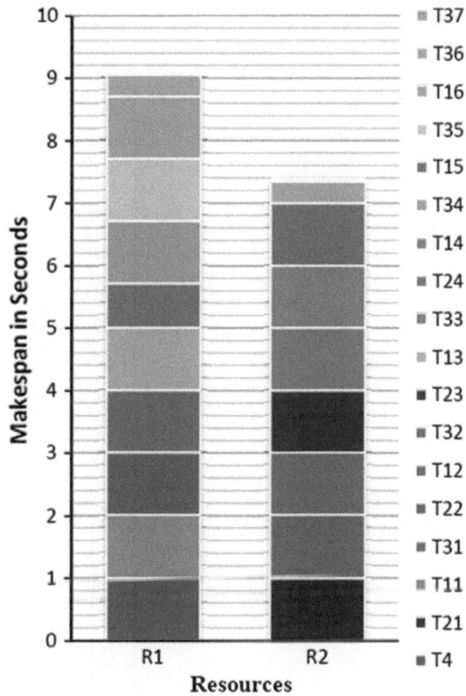

Fig. 7 Gantt chart for RRTS

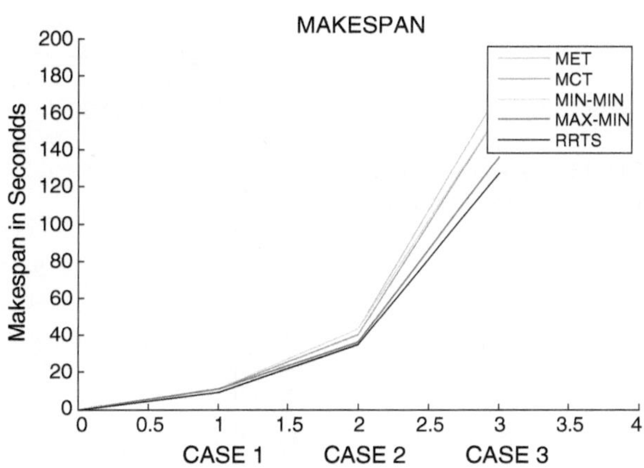

Fig. 8 Comparison of makespan for case 1, case 2 and case 3

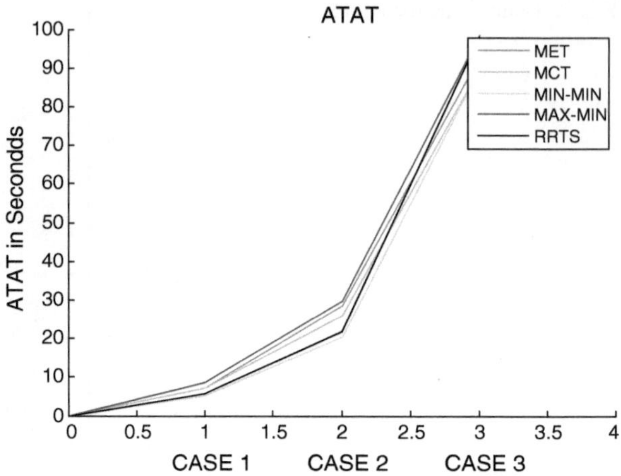

Fig. 9 Comparison of ATAT for case 1, case 2, and case 3

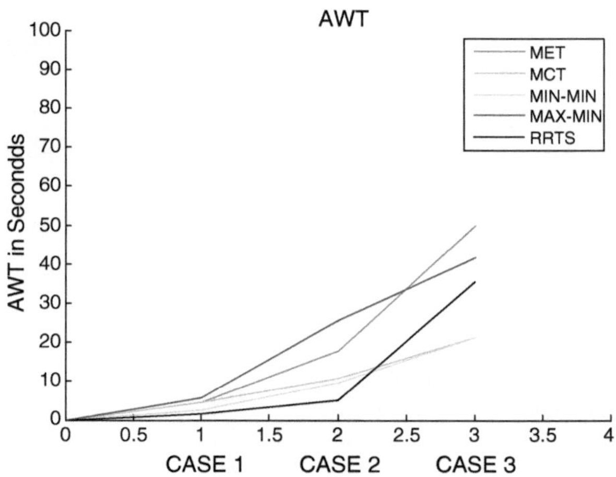

Fig. 10 Comparison of AWT for case 1, case 2, and case 3

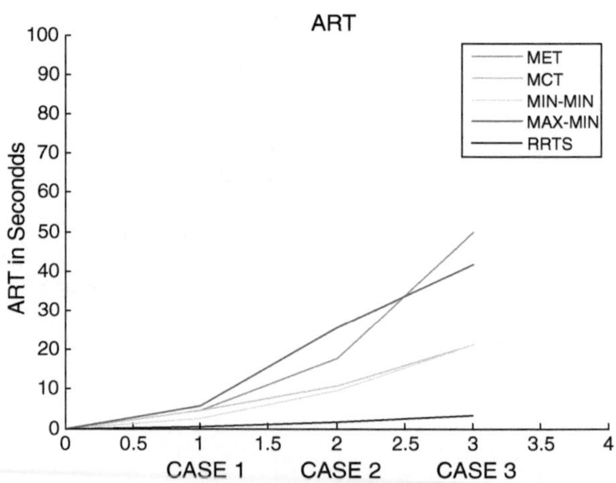

Fig. 11 Comparison of ART for case 1, case 2, and case 3

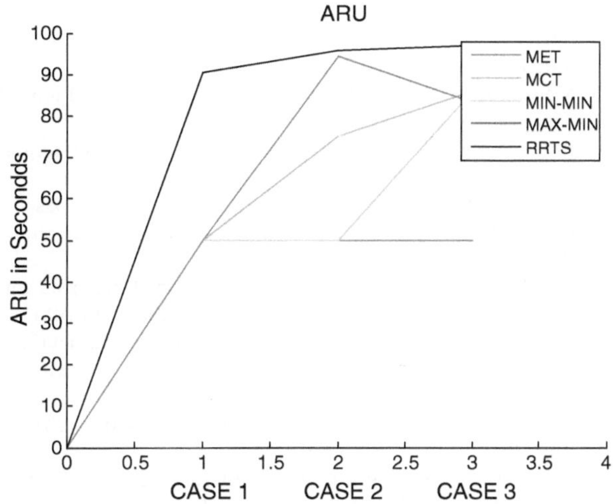

Fig. 12 Comparison of ARU for case 1, case 2, and case 3

Conclusion

From the above experimental results, RRTS shows better performance than the other existing algorithms. In this paper, the completion time and starvation of the tasks are minimized. As there is no priority to the tasks and by using the concept of RR, starvation is reduced to a greater extent. The idea of using dynamic time slice leads to the reduction of response time and Makespan. Resource utilization is fully achieved in our algorithm by which none of the resource is completely idle. In comparison, resources are more utilized in our approach. The fastest resource in grid environment is 100 % busy, it means complete resource utilization is accomplished.

In the future, we can implement this approach in real-time task scheduling. We will develop this algorithm by using the concepts of deadline; priority and effective dynamic time slice to make the algorithm more robust. Applying the algorithm in a grid environment is a great challenge in task scheduling.

References

1. Freund, R.F., Gherrity, M., Ambrosius, S., Campbell, M., Hensgen, D., Keith, E., Kidd, T., Kussow, M., Lima, J.D., Mirabile, F., Moore, L., Rust, B., Siegel, H.J.: Scheduling resources in multi-user, heterogeneous, computing environments with smartnet. IEEE, pp. 184–199 (1998)
2. Xiaoshan, H., Sun, X.H., Laszewski, G.V.: Qos guided min–min heuristic for grid task scheduling, NSF Grant Nos. EIA-0224377, ANI-0123930, EIA-0130673, ARO Grant No. DAAD19-01-1-0432, May 2003
3. Hemamalini, M.: Review on grid task scheduling in distributed heterogeneous environment. IJCA **40**(2), 24–30 (2012)

4. Liu, K., Chen, J., Jin, H., Yang, Y.: A min–min average algorithm for scheduling transaction-intensive grid workflows. In: 7th Australasian Symposium on Grid Computing and e-Research (2009)
5. Mansouri, N., Dastghaibyfard, G., Horri, A.: A novel job scheduling algorithm for improving data grid's performance. IEEE (2011)
6. Etminani, K., Naghibzadeh, M.: A min–min max–min selective algorithm for grid task scheduling, In: IEEE (2007)
7. Sun, W., Zhu, Y., Su, Z., Jiao, D., Li, M.: A priority-based task scheduling algorithm in grid. In: 3rd International Symposium on Parallel Architectures, Algorithms and Programming, IEEE, pp. 311–315 (2010)
8. Kokilavani, T., Amalarethinam, D.I.G.: Load balanced min–min algorithm for static meta-task scheduling in grid computing. IJCA **20**(2), 43–49 (2011)
9. Kamalam, G.K., Bhaskaran, V.M.: New enhanced heuristic min-mean scheduling algorithm for scheduling meta-tasks on heterogeneous grid environment. Eur. J. Sci. Res. **70**(3), 423–430 (2012)
10. Baca, D.F.: Allocating modules to processors in a distributed system. IEEE Trans. Softw. Eng. **15**(11), 1427–1436 (1989)
11. Munir, E.U., Li, J., Shi, S.: QoS sufferage heuristic for independent task scheduling in grid. Inf. Technol. J **6**(8), 1166–1170 (2007)
12. Liang, Y., Jiliu, Z.: The improvement of A task scheduling algorithm in grid computing. In: First International Symposium on Data, Privacy and E-Commerce, IEEE (2007)
13. Parsa, S., Maleki, R.E.: RASA: A new grid task scheduling algorithm. JDCTA **3**(4), 91–99 (2009)
14. Rasooli, A., Aghatabar, M.M., Khorsandi, S.: Introduction of novel rule based algorithms for scheduling in grid computing systems. In: Second Asia international conference on modelling & simulation, IEEE, pp. 138–143 (2008)
15. Buyya, R.: High Performance Cluster Computing. Pearson Education, ISBN 81-317-1693-7 (2008)
16. Xiao, Y.: Security in Distributed, Grid, Mobile, and Pervasive Computing. Auerbach Publications, ISBN-10 0-8493-7921-0, ISBN-13 978-0-8493-7921-5 (2007)
17. Yaashuwanth, C., Ramesh, R.: A new scheduling algorithm for real time system. Int. J. Comput. Electr. Eng. **2**(6), 1104–1106 (2010)
18. Yaashuwanth, C., Ramesh, R.: A new scheduling algorithms for real time tasks. IJCSIS **6**(2), 61–66 (2009)
19. Zhang, Y., Koelbel, C., Kennedy, K.: Relative performance of scheduling algorithms in grid environments. In: 7th International Symposium on Cluster Computing and the Grid, IEEE (2007)
20. Navimipour, N.J., Khanli, L.M.: The LGR method for task scheduling in computational grid. In: International Conference on Advanced Computer Theory and Engineering, IEEE, pp. 1062–1066 (2008)
21. Panda, S.K., Bhoi, S.K.: An effective round Robin algorithm using min-max dispersion measure. IJCSE **4**(1), 45–53 (2012)
22. Tang, M., Lee, B.S., Tang, X., Yeo, C.: The impact of data replication on job scheduling performance in the data grid. Future Gener. Comput Syst. 22, 254–268 (Elsevier)
23. Abdi, S., Pedram, H., Mohamadi, S.: The impact of data replication on job scheduling performance in hierarchical data grid. Int. J. Appl. Graph Theory Wireless Ad hoc Netw. Sens. Netw. 2(3), (2010)

A Network Survivability Approach to Resist Access Point Failure in IEEE 802.11 WLAN

Sourav Kumar Bhoi, Sanjaya Kumar Panda and Pabitra Mohan Khilar

Abstract IEEE 802.11 WLAN is a most promising and demandable technology for communication. But it is affected by faults in the Access Points (APs), which degrade the network performance. So the network should tolerate the faults to preserve its performance and efficiency. In our approach, we have designed a fault tolerance technique to resist the faults in APs. This consists of three phases: Design of Minimum Cost Spanning Tree, creating Node Priority Table and establishing route to connect the nodes by Network Survivability Algorithm. By this method, we get the network coverage area, priorities of the nodes according to the degree and a route to connect the nodes after AP failure. We have considered it for both Single-Point Failure and Multi-Point Failure, which show better results in tolerating the faults by utilising maximum number of nodes and making it a cost-effective model.

Keywords Minimum cost spanning tree · Network survivability · Node priority · Access points · Distributed system · Basic service set · Extended service set · Wireless LAN

Introduction

Nowadays, Wireless LAN is a popular means of technology for robust communication. It is growing day by day with a high demand. This wireless communication is used everywhere like in office, organisation, public areas etc. There are

S. K. Bhoi (✉) · S. K. Panda · P. M. Khilar
Department of CSE, National Institute of Technology, Rourkela, India
e-mail: souravbhoi@gmail.com

S. K. Panda
e-mail: sanjayauce@gmail.com

P. M. Khilar
e-mail: pmkhilar@nitrkl.ac.in

S. Sathiakumar et al. (eds.), *Proceedings of International Conference on Internet Computing and Information Communications*, Advances in Intelligent Systems and Computing 216,
DOI: 10.1007/978-81-322-1299-7_28,

many wireless technologies which are most promising for LANs like IEEE 802.11 Wireless LAN and Bluetooth [1]. IEEE 802.11 describes the features of a Wireless LAN by covering the scope of data link layer and physical layer.

The architecture of IEEE 802.11 consists of two types of services. First is Basic Service Set (BSS) and second is Extended service set (ESS). BSS consists of stations and Access Point (AP). AP is a central base station. The BSS having AP is called Infrastructure Network and the BSS having no AP is called Ad hoc network [2, 3]. Every BSS has an ID, which represents MAC ID of the AP. ESS consists of a Distribution System, which connects the APs of the BSSs [4, 5].

We know that, nowadays, mobile network is struggling a lot against the AP failure in case of natural disasters. Faults in the stations (APs) have a great impact on the network performance. Critical applications are the applications which are required to be function even if the network has faults [6]. So many fault tolerant techniques have been proposed to tolerate the faults in the APs.

Network ability to resist the faults is measured by Reliability, Availability and Survivability [7]. Our research is mainly concentrated on survivability, which describes about the network recovery from failure. In our approach, we have reduced the faults by designing an algorithm which consists of three phases: Design of Minimum Cost Spanning Tree (MCST), Creating Node Priority Table and a route to connect the nodes by Network Survivability Algorithm. Our method is a cost-effective method and the loads are balanced for better communication.

The organisation of the paper is as follows: Section “Related Work” presents the related work done in the field of network survivability. Section “Preliminaries” presents about the preliminaries, where we have discuss about the notations, assumptions, network model and network fault model. In Sect. “Proposed Algorithm”, the algorithm is proposed with the three phases. Section “Modelling and Evaluation” presents the modelling and evaluation of single-point failure and multi-point failure. Section “Conclusion” presents the conclusion of our proposed algorithm.

Related Work

In the recent years, many researches have been done on survivability of the network to tolerate faults. Sahoo et al. [1, 7] describe about the survivability against AP failure in IEEE 802.11. They have taken two main phases: *Design phase* describes how to quantify, place and set-up of APs take place according to area coverage and performance criteria and in *Fault Response phase* they consider about the reconfiguration of the active APs by connecting the nodes in order to deal with AP fault.

Zhou et al. [8] find out the minimal number of APs and their locations to achieve fault tolerance and QoS constraints satisfaction. Here, they have taken minimum APs and add the APs according to demand. By this fault tolerance is achieved. Hass et al. [9] describe a method to tolerate the failure of location database. Chen et al. [1] describe a process to enhance the connection reliability in WLANs by resisting the existence of *shadow regions* by using redundant APs.

Our algorithm is not based on the concept of redundancy and shadow access points.

Tipper et al. [10, 11] describe about a survivability analysis of Personal Communication Service (PCS) networks. The simulation model shows that user mobility can degrade the performance of the network in case of failures.

To tolerate the access-point failures in wireless networks by using additional AP as a backup [12], and that is activated when the primary previous AP fails. Another technique to tolerate AP failures is to use AP with overlapping coverage [12, 13].

In our approach, we have taken a concept of node priority by which we establishes route to connect the nodes in case of AP failure.

Preliminaries

We have considered a complete graph, where each node is connected to each other. Each node is responsible for communication. The node having high degree has high priority because of its high communication paths. The threshold value for range is assumed by considering the network coverage area. The connection range between the nodes should not be more than the threshold value. But due to the failure of AP, we have to connect the nodes to tolerate the faults which may cross the range of threshold value (these nodes act as reservation nodes in case of AP failure). So we have given flexibility in connecting the nodes if it is not coming under the range. Suppose, we have taken the threshold (weight) as 6 and the distance from one node to other is 9 (after AP failure), then we connect the nodes for fault tolerance. The degree of the node should be greater than or equal to 1. We have taken an assumption that up to 50 % of node failure of the total nodes, this algorithm works due to large number of node failures there will be generation of load on active nodes which degrades the performance.

Proposed Algorithm

Description

In our algorithm, we have proposed a technique which detects and resists the faults in the APs. In our technique, firstly, we find the MCST to reduce the cost. Second, we detect the faults by sending ping messages to check whether the APs are working or not. Third, we check the failure of the APs by connecting the nodes by a node priority algorithm. By this algorithm, we establish a route to connect the nodes in case of AP failure. Our algorithm shows better results than the algorithm proposed by Sahoo et al. [7]. In this redundant node approach [7], they have used the concept of BFS and DFS to establish the route in case of AP failure to connect the nodes. They have applied this approach to single-point failure only. But we have applied our approach to both single- and multi-point failures.

Design

Our algorithm consists of three phases. The three phases are:

(1) *Design of MCST.* This algorithm finds the MCST to reduce the cost. We have taken input as the locations of APs and range of the access points (weight). We get the output as MAC IDs for establishing network and MAC IDs for tolerating the faults by node priority algorithm.

Pseudocode for designing MCST

1. Create adjacency matrix for the ‘n’ nodes by entering the range values.
2. By taking the network within a range, we give a threshold value = ‘6’ (below ‘6’) to remove the long ranges.
3. Update the adjacency matrix after the threshold value is applied.
4. From the adjacency matrix A[i][j], find the MCST by using Kruskal’s algorithm.
5. From MCST, store the adjacent MAC IDs of each node in a separate $Adj_{ID}[]$ ($Adj_1[]$, $Adj_2[]$,..,$Adj_8[]$).

(2) *Node Priority Algorithm.* After finding the MCST, we find the priority of the APs by a node priority table. The APs having high degree have more priority than the APs having low degree. This signifies that the node having high degrees has a high importance in the network, because of its high connections. So we create the node priority table by giving priority to each node. Here, we have taken the input as MAC IDs after MST is created and output as APs priority table with priority of each node.

Pseudocode for Creating Node Priority Table:

1. Find the priority of each node by calculating the degrees of each node
2. Calculate and store the degree of each node in a separate $Deg_{ID}[]$ ($Deg_1[]$,…, $Deg_8[]$)
3. Assign priority to each node by looking for highest degree node to lowest degree node
4. According to convention, the nodes [1,2,3,4,….,n] are assigned with priority
5. The node having highest degree has high priority and the node having lowest degree has low priority
6. Node table is created by node index (N_i), Degree, Priority (P_i).

(3) *Network Survivability Algorithm*: After getting the AP priority table, we check the faults by pinging to all APs to know whether the APs are active or not

and then establish the route to connect the APs in case of failure. This algorithm uses two techniques to tolerate the faults: Single-point failure function and Multi-point failure function.

Pseudocode for Single-Point Failure function:

```
Single-Point-Failure()
    {
        if (node degree of the failed AP = 1)
        {
      fault in the network is tolerated by simply
      removing the node (failed AP)
        }
        else
        {
         establish a route from low priority node to high priority node
         (adjacent nodes of the failed AP) and connect these APs
            }
    {
```

Pseudocode for Multi-Point Failure function:

```
            Multi-Point-Failure()
    {
        if (node degree of the failed APs = 1)
           {
    fault in the network is tolerated by simply
    removing the nodes (failed APs)
           }
        else if(node degree of failed APs = 1 and other failed APs having
        degree > 1)
           {
    remove the nodes having degree 1 and establish route from low priority to
    high priority node (remaining adjacent nodes of the failed APs from
    Adj_ID[])
           }
           else if (node degree of the failed APs > 1)
           {
    remove the nodes and establish route from low priority to high priority
    node (remaining adjacent nodes of the failed APs from Adj_ID[])
           }
           else
           {
    all APs are failed and communication fails
           }
    }
```

Pseudocode for Survivability algorithm:

```
1. Ping to all the APs to know whether they are active or not
2. By getting the response we know which APs are active
3. if (all are active)
       {
               Communication starts
       }
   Else
           {
                           go to step 4
         }
4. if (Number of AP failed = 1)
     {
       Single-Point-Failure()
          }
       Else
       {
       Multi-Point-Failure()
         }
5. A new-reduced size network is generated by tolerating the fault.
6. Load is balanced and communication starts.
```

So these are the three phases used in our algorithm to tolerate the faults in the network.

Modelling and Evaluation

Modelling and evaluation show how the network is modelled and how the faults are detected and checked. In our experiment, we have considered an illustration and checked it for the two cases: Single-point failure and multi-point failure. The graphical representation of AP utilisation for multi-point failure is shown in Fig. 7. By these two cases, we come to know about the fault tolerance mechanism in the network.

Illustration

We have taken a complete graph of 8 nodes (APs). Each node is connected to each other with different weights. The graph is shown in Fig. 1.

Step 1: Create the adjacency matrix table and the updated adjacency matrix table after applying the threshold. Tables 1 and 2 show the adjacency matrix table and updated adjacency matrix table, respectively.

Fig. 1 Complete graph of access point network

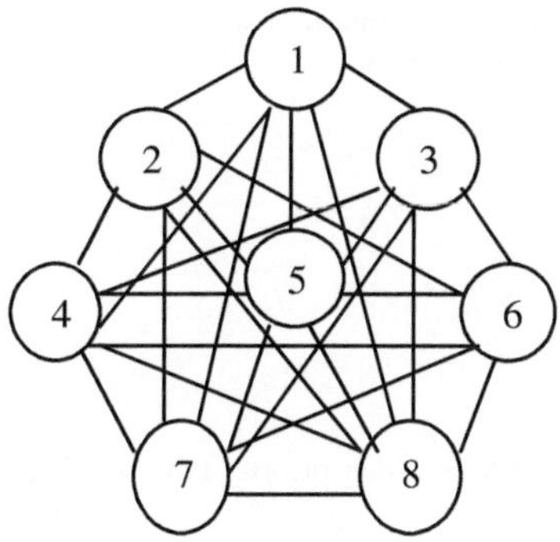

Table 1 Adjacency matrix table

	1	2	3	4	5	6	7	8
1	0	3	4	5	6	7	8	9
2	3	0	4	5	5	6	7	8
3	4	4	0	8	4	6	5	3
4	5	5	8	0	6	10	5	8
5	6	5	4	6	0	6	3	3
6	7	6	6	10	6	0	9	5
7	8	7	5	5	3	9	0	5
8	9	8	3	8	3	5	5	0

Table 2 Updated adjacency matrix table

	1	2	3	4	5	6	7	8
1	0	3	4	5	0	0	0	0
2	3	0	4	5	5	0	0	0
3	4	4	0	0	4	0	5	3
4	5	5	0	0	0	0	5	0
5	0	5	4	0	0	0	3	3
6	0	0	0	0	6	0	0	5
7	0	0	5	5	3	0	0	5
8	0	0	3	0	3	5	5	0

Step 2: Create MCST, shown in Fig. 2.

Step 3: Create Node Priority Table, shown in Table 3.

Step 4: Ping to check whether the APs are active or not. If response comes from the AP then it is active otherwise AP is failed.

Scenarios

We have taken two scenarios to represent the fault tolerance model. Two scenarios are single-point failure and multi-point failure.

Fig. 2 Minimum cost spanning tree

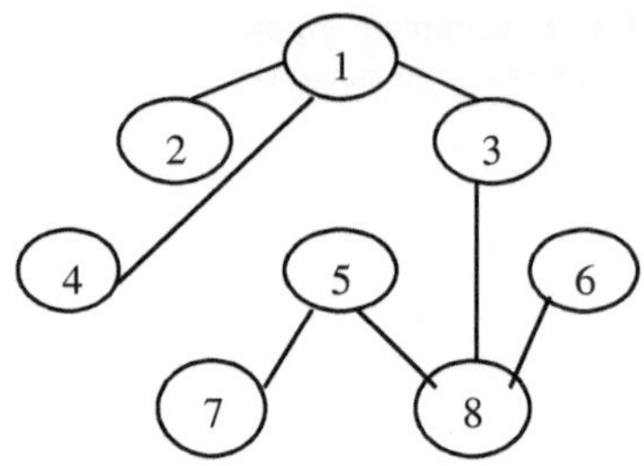

Table 3 Node priority table

Node	Degree	Priority
1	3	1
2	1	5
3	2	3
4	1	6
5	2	4
6	1	7
7	1	8
8	3	2

Fig. 3 Network after AP2 failure

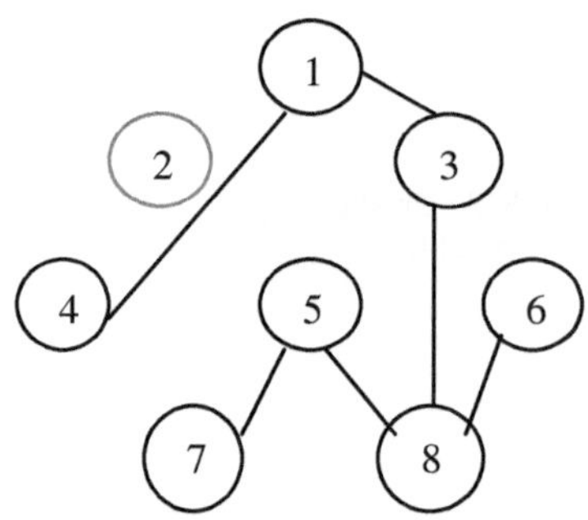

(1) *Single-Point Failure.* Single-point failure is the failure where a single AP is failed in a network. Here, we have taken the failure of AP2 and AP3.

Failure of AP2: Adjacent node of 2 is 1 and the degree of 2 is 1. So remove the node. AP2 failure is shown in Fig. 3.

Failure of AP3: Adjacent nodes of 3 are {1, 8}, priorities of 1, 8 are 1, 2, respectively. So connect the nodes from low priority to high priority as 8 → 1. AP3 failure is shown in Fig. 4.

(2) *Multi-Point Failure.* Multi-point failure is the failure where a multiple APs are failed in a network. Here, we have considered the failure of (AP1, AP5) and (AP1, AP3 and AP5) as example.

Failure of (AP1, AP5): By solving AP1 and AP5 and we get the route as 4 → 2 → 3 and 7 → 8, respectively. But 1 and 5 are failed. The remaining adjacent nodes are 4, 2, 3, 7 and 8. The priorities of 4, 2, 3, 7 and 8 are 6, 5, 3, 8 and 2, respectively.

Fig. 4 Network after AP3 failure

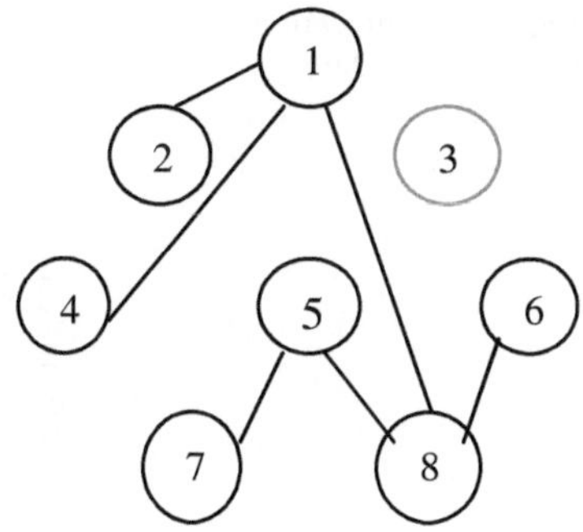

Fig. 5 Network after AP1 and AP5 failure

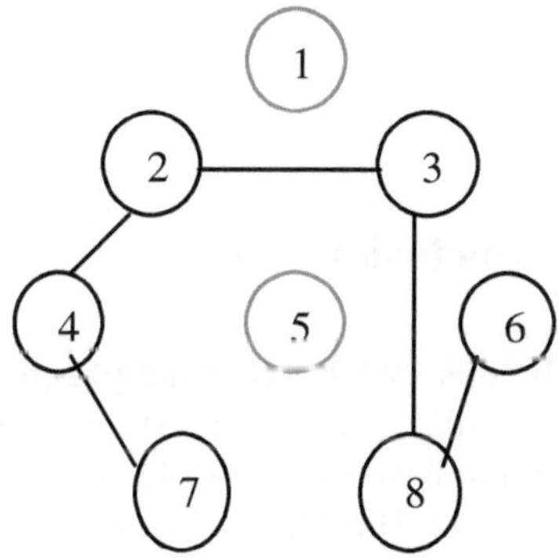

Fig. 6 AP utilisation in single-point failure

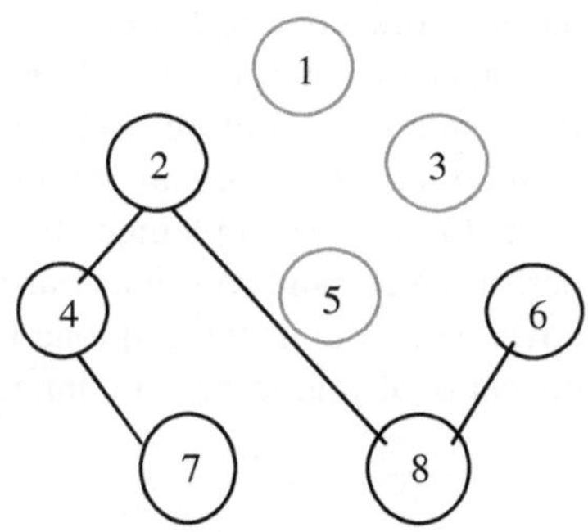

The route is established as 7 → 4 → 2 → 3 → 8. Then we connect the nodes to get the reduced fault tolerant network. AP1 and AP5 failure is shown in Fig. 5.

Failure of (AP1, AP3 and AP5): By solving AP1, AP3 and AP5, we get the routes as 4 → 2 → 3, 8 → 1 and 7 → 8, respectively. But 1, 3 and 5 are failed. So the remaining adjacent nodes 4, 2, 8 and 7. The priorities of 4, 2, 8 and 7 are 6, 2, 5 and 8, respectively. The established route is 7 → 4 → 2 → 8. Then we connect the nodes to get the reduced fault tolerant network. AP1, AP3 and AP5 failure is shown in Figs. 6, 7.

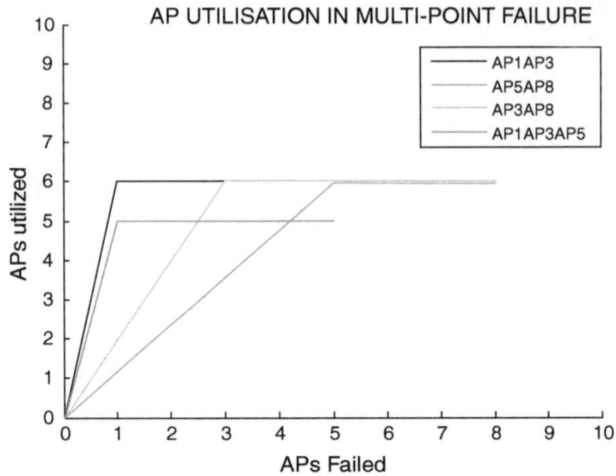

Fig. 7 AP utilisation in multi-point failure

Conclusion

In this paper, we mainly concentrated on survivability of the network which describes about the network recovery from failure. In our approach, we have minimised the faults by designing an algorithm which consists of three main phases: The phases in our algorithms are: Design of Minimum Spanning Tree, Creating Node Priority Table and Network Survivability Algorithm. By this we get the network coverage area, priorities of the nodes according to the degree and establish a route for connection of the nodes after AP failure. We have checked it for single-point failure and multi-point failure. Our method is a cost-effective method and the loads are balanced automatically for better communication.

In future, we implement this theoretical model by simulating and getting its performance and results. Further, we apply the concepts security, network coverage, redundancy and load balancing for comparisons. Further, we can use the concepts of region-based connectivity to detect the faults in a vast network.

References

1. Chen, D., Kintala, C., Garg, S., Trivedi, K.S.: Dependability enhancement for IEEE 802.11 wireless LAN with redundancy techniques. In: Proceedings of the International Conference on Dependable Systems and Networks, pp. 521–528 (2003)
2. Wireless Local Area Network (WLAN) Explained. [online]. Available: http://www.anthonycairns.com/Explained/Items_Explained_WLAN.html
3. Forouzan, B.A. Data Communication and Networking, 4th Edt., Tata Mcgraw-Hill, 2006
4. Configuring a Wireless Distribution System (WDS) with the 3Com Office Connect Wireless 11a/b/g Access Point.[online].Available:www.3com.com/other/pdfs/products/en_US/104108.pdf
5. WDS (Wireless Distribution System), ORiNOCO Technical Bulletin 046/A, February 2002

6. de Deus, F.E., Puttini, R.S., Molinaro, L.F., Kabara, J.: On survivability of IEEE 802.11 WLAN. In: Proceedings of the IEEE International Conference on Sensor Networks, Ubiquitous and Trustworthy Computing, (2006)
7. Sahoo, M.N., Khilar, P.M., Majhi, B.: A redundant neighbourhood approach to tolerate access point failure in IEEE 802.11 WLAN. In: Fourth International Conference on Industrial and Information Systems, ICIIS 2009, Sri Lanka pp. 28–31 December 2009
8. Zhou, K., Jia, X., Xie, L., Chang, Y.: Fault tolerant AP placement with QoS constraint in wireless local area networks. In: IEEE Globecom 2011 proceedings, (2011)
9. Haas, Z.J., Lin, Y.B.: Demand re-registration for PCS database restoration. Mob. Netw. Appl. **5**(3), 191–198 (2000)
10. Tipper, D., Dahlberg, T., Shin, H., Charnsripinyo, C.: Providing fault tolerance in wireless access networks. IEEE Commun. **40**, 62–68 (2002)
11. Tipper, D., Ramaswamy, S., Dahlberg, T.: PCS network survivability.In: Mobile and Wireless Communication Networks Conference Sept 1999
12. Gandhi, R.: Tolerance to access-point failures in dependable wireless local-area networks. In: Proceedings of the Ninth IEEE International Workshop on Object-Oriented Real-Time Dependable Systems, (2004)
13. Sahoo M.N., Khilar, P.M., Survivability of IEEE 802.11 wireless LAN against AP failure. Int. J. Comput. Appl. Eng. Technol. Sci. (IJCA- ETS) **174** 424–428 (2009)

Novel 2D Real-Valued Sinusoidal Signal Frequencies Estimation Based on Propagator Method

Sambit Prasad Kar and P. Palanisamy

Abstract This paper considers the problem of estimating the frequencies of multiple 2D real-valued sinusoidal signals, also known as Real X-texture mode signals, in the presence of additive white Gaussian noise. An algorithm for estimating the frequencies of real-valued 2D sine wave based on propagator method is developed. This technique is a direct method which does not require any peak search. A new data model for individual dimensions is proposed, which gives the dimension of the signal subspace is equal to the number of frequencies present in the observation. Then propagator method-based estimation technique is applied on individual dimensions using the proposed new data model. The performance of the proposed method is demonstrated and validated through computer simulation.

Keywords Array signal processing · X-texture mode signals · Signal subspace method · Two-dimensional frequency estimation

Introduction

In this paper, we consider the problem of estimating the frequencies of multiple two-dimensional (2D) real-valued sinusoids in additive white Gaussian noise. This problem is more precise case of estimating the parameters of a 2D regular and homogeneous random field from a single observed realization of it [1]. The real-valued 2D sinusoidal signal models, also known as X-texture modes. These modes

S. P. Kar (✉) · P. Palanisamy
Department of Electronics and Communication Engineering,
National Institute of Technology, Trichy, India
e-mail: sambitpk@gmail.com

P. Palanisamy
e-mail: palan@nitt.edu

S. Sathiakumar et al. (eds.), *Proceedings of International Conference on Internet Computing and Information Communications*, Advances in Intelligent Systems and Computing 216,
DOI: 10.1007/978-81-322-1299-7_29,

come into existence naturally in experimental and analytical modal and vibration analysis of circular shaped objects. X-texture modes are often used for modeling the displacements in the cross-sectional planes of isotropic, homogeneous, thick-walled cylinders [2–4], laminated composite cylindrical shells [5], and circular plates [6]. These X-texture modes have also been used to describe the radial displacements of logs of spruce subjected to continuous sinusoidal excitation [7] and standing trunks of spruce subjected to impact excitation [8–10]. The proposed signal model offers cumbersome challenges for 2D joint frequency estimation algorithms. Many algorithms for estimating complex-valued frequencies are very well-documented in the literature [11, 12] and for 1D real-valued frequencies in [13–15]. A vivid discussion on the problem of analyzing 2D homogeneous random fields with discontinuous spectral distribution functions can be found in [16]. Parameter estimation techniques of sinusoidal signals in additive white noise include the periodogram-based approximation (applicable for widely spaced sinusoids) to the maximum-likelihood (ML) solution [17–19], the Pisarenko harmonic decomposition [20], or the singular value decomposition [21]. A matrix enhancement and matrix pencil method for estimating the parameters of 2D superimposed, complex-valued exponential signals was suggested in [11]. In [22] the concept of partial forward–backward averaging is proposed for enhancing the frequency and damping estimation of X-texture modes. In [23], 2D parameter estimation of a single damped/undamped real/complex tone is proposed which is known as principal-singular-vector utilization for modal analysis (PUMA).

In this paper, we contribute to the 2D frequency estimation of multiple undamped real-valued sinusoidal signals, (X-texture mode signals), in the presence of additive white Gaussian noise and with the available finite snapshots. In our proposed method, first two new data models are formed, from the available data. Propagator method is applied in order to estimate the two-dimensional frequency individually and an efficient pairing algorithm is used to pair the estimated frequencies. The propagator method-[24] based technique is a computationally efficient method which does not require time-consuming eigendecomposition of the correlation matrix of the observation. The performance comparison between ESPRIT and propagator techniques is presented for comparison studies.

The paper is organized as follows. The signal model, together with a definition of the addressed problem is presented in Sect. "Signal Model and Problem Defination". The formation of new data model is detailed in Sect. "Proposed Method", where propagator method is summarized, that covers estimation of the two-dimensional frequency individually and the pair matching had done. The statistical performance and computer simulation is studied in Sect. "Simulation Results" and finally the conclusion is given in Sect. "Conclusion". Throughout this paper upper case, bold letters denote matrices where as lowercase bold letters are vectors. The superscript T denotes transposition of a matrix.

Signal Model and Problem Defination

The following 2D real-valued sinusoidal signal model describing $N_1 \times N_2$ samples, is considered in the presence of additive white Gaussian noise:

$$r(m,n) = x(m,n) + e(m,n) \quad \begin{aligned} m &= 1,2,\ldots,N_1 - 1 \\ n &= 1,2,\ldots,N_2 - 1 \end{aligned} \tag{1}$$

where $x(m,n) = \sum_{k=1}^{D} a_k cos(\omega_k m + \alpha_k) cos(v_k n + \beta_k)$

represents the D modes, each with amplitude a_k angular frequencies ω_k and v_k, phases α_k and β_k which are independent random variables uniformly distributed over [0, 2π] and $e(m,n)$ represents white Gaussian noise with zero mean and variance is equal to σ^2 and further assumed that α_k and β_k are independent of $e(m,n)$. The objective of this paper is to estimate ω_k and v_k, $k = 1, 2,\ldots,D$, based on the observations $r(m, n)$ for $m = 0,1, 2,\ldots, N_1-1$, $n = 0,1,2\ldots,N_2-1$.

Let us define two M $\times$ 1 snapshot vectors as given below:

$$\mathbf{y}_\omega(m,n) \triangleq \frac{1}{2}[\mathbf{y}_1(m,n) + \mathbf{y}_2(m,n)] \tag{2a}$$

$$\mathbf{y}_v(m,n) \triangleq \frac{1}{2}[\mathbf{y}_3(m,n) + \mathbf{y}_4(m,n)] \tag{2b}$$

where,

$$\mathbf{y}_1(m,n) \triangleq [r(m,n) \quad r(m+1,n)\ldots r(m+\mathrm{M}-1,n)]^{\mathrm{T}} \tag{3a}$$

$$\mathbf{y}_2(m,n) \triangleq [r(m,n) \quad r(m-1,n)\ldots r(m-\mathrm{M}+1,n)]^{\mathrm{T}} \tag{3b}$$

$$\mathbf{y}_3(m,n) \triangleq [r(m,n) \quad r(m,n-1)\ldots r(m,n-\mathrm{M}+1)]^{\mathrm{T}} \tag{3c}$$

$$\mathbf{y}_4(m,n) \triangleq [r(m,n) \quad r(m,n+1)\ldots r(m,n+\mathrm{M}-1)]^{\mathrm{T}} \tag{3d}$$

From the above set of equations we can obtain pair of expression for the two M $\times$ 1 snapshot vectors by substituting equation (3a)–(3b) in (2a) and (3c)–(3d) in (2b) as follows,

$$\mathbf{y}_\omega(\mathrm{m,n}) = \mathbf{A}(\omega)\mathbf{s}(\mathrm{m,n}) + \mathbf{g}(\mathrm{m,n}) \tag{4}$$

$$\mathbf{y}_v(\mathrm{m,n}) = \mathbf{A}(v)\mathbf{s}(\mathrm{m,n}) + \mathbf{h}(\mathrm{m,n}) \tag{5}$$

where $\mathbf{A}(\omega) \triangleq [\gamma(\omega_1)\ldots\gamma(\omega_D)]$ and $\mathbf{A}(v) \triangleq [\rho(v_1)\ldots\rho(v_D)]$ are M $\times$ D matrices, $\mathbf{s}(m, n) = [a_1\cos(\omega_1 m + \alpha_1)\cos(v_1 n + \beta_1)\ldots\ a_D\cos(\omega_D m + \alpha_D)\cos(v_D n + \beta_D)]^{\mathrm{T}}$ is the D $\times$ 1 signal matrix, $\gamma(\omega_i)$ and $\rho(v_i)$ are M $\times$ 1 vectors defined, respectively, as $\gamma(\omega_i) = [1\ \cos(\omega_i)\ldots\cos((\mathrm{M}-1)\omega_i)]^{\mathrm{T}}$ and $\rho(v_i) = [1\ \cos(v_i)\ldots \cos((\mathrm{M}-1)v_i)]^{\mathrm{T}}$. The modified M $\times$ 1 error vectors $\mathbf{g}(m, n)$ and $\mathbf{h}(m, n)$ are defined, respectively, as $\mathbf{g}(m, n) \triangleq [g_1(m, n)\ g_2(m, n)\ldots\ g_M(m, n)]^{\mathrm{T}}$ and

$\mathbf{h}(m, n) \triangleq [h_1(m, n)\ h_2(m, n)\ldots h_M(m, n)]^T$ where $g_j(m, n)= \frac{1}{2}[e(m + j-1, n) + e(m -j + 1, n)]$ and $h_j(m, n)= \frac{1}{2}[e(m, n + j-1) + e(m, n-j + 1)]$. The matrices $\mathbf{A}(\omega)$ and $\mathbf{A}(v)$ are full rank matrices because all the columns are linearly independent to each other

Enhanced Data model

Before advancing into the implementation of the algorithm, let us define another pair of new signal models based on the observations in (1) as given below:

$$\mathbf{z}_{\omega}(m, n) = \frac{1}{4}\sum_{j=1}^{4} \mathbf{z}_j(m, n) \tag{6a}$$

$$\mathbf{z}_{v}(m, n) = \frac{1}{4}\sum_{j=5}^{8} \mathbf{z}_j(m, n) \tag{6b}$$

where

$$\mathbf{z}_1(m, n) \triangleq [r(m+1, n)\, r(m, n)\, r(m-1, n)\ldots r(m-\mathrm{M}+2, n)]^T \tag{7a}$$

$$\mathbf{z}_2(m, n) \triangleq [r(m-1, n)\, r(m, n)\, r(m+1, n)\ldots r(m+\mathrm{M}-2, n)]^T \tag{7b}$$

$$\mathbf{z}_3(m, n) \triangleq [r(m-1, n)\, r(m-2, n)\, r(m-3, n)\ldots r(m-\mathrm{M}, n)]^T \tag{7c}$$

$$\mathbf{z}_4(m, n) \triangleq [r(m+1, n)\, r(m+2, n)\, r(m+3, n)\ldots r(m+\mathrm{M}, n)]^T \tag{7d}$$

$$\mathbf{z}_5(m, n) \triangleq [\mathrm{r}(m, n+1)\, r(m, n)\, r(m, n-1)\ldots r(m, n-\mathrm{M}+2)]^T \tag{8a}$$

$$\mathbf{z}_6(m, n) \triangleq [r(m, n-1)\, r(m, n)\, r(m, n+1)\ldots r(m, n+\mathrm{M}-2)]^T \tag{8b}$$

$$\mathbf{z}_7(m, n) \triangleq [r(m, n-1)\, r(m, n-2)\, r(m, n-3)\ldots r(m, n-\mathrm{M})]^T \tag{8c}$$

$$\mathbf{z}_8(m, n) \triangleq [r(m, n+1)\, r(m, n+2)\, r(m, n+3)..\, r(m, n+\mathrm{M})]^T \tag{8d}$$

By substituting (7a) to (7d) in (6a), it can be shown that the new enhanced data model $z_{\omega}(m, n)$ is simplified as,

$$\mathbf{z}_{\omega}(m, n) = \mathbf{A}(\omega)\Phi_{\omega}\mathbf{s}(m, n) + \mathbf{q_r}(\mathbf{m}, \mathbf{n}) \tag{9}$$

Similarly by substituting (8a) to (8d) in (6b) another enhanced data model $z_v(m, n)$ can be obtained as

$$\mathbf{z}_{v}(m, n) = \mathbf{A}(v)\Phi_{v}\mathbf{s}(m, n) + \mathbf{q_e}(\mathbf{m}, \mathbf{n}) \tag{10}$$

where $\mathbf{\Phi}_\omega$ and $\mathbf{\Phi}_v$ are two D × D diagonal matrices defined, respectively, as $\mathbf{\Phi}_\omega = \text{diag}\{\cos\omega_1\ \cos\omega_2 \ldots \cos\omega_D\}$ and $\mathbf{\Phi}_v = \text{diag}\{\cos v_1 \ldots \cos v_D\}$. The two M × 1 modified noise vectors $\mathbf{q}_r(m, n)$ and $\mathbf{q}_e(m, n)$ are defined, respectively, as $\mathbf{q}_r(m, n) = [q_{r1}(m, n)\ q_{r2}(m, n) \ldots q_{rM}(m,n)]^T$, $\mathbf{q}_e(m, n) = [q_{e1}(m,n)\ q_{e1}(m, n) \ldots q_{eM}(m, n)]^T$ where $q_{ri}(m, n) = \frac{1}{4}[e(m - i + 2, n) + e(m - i - 2, n) + e(m + i, n) + e(m - i, n)]$ and $q_{ei}(m, n) = \frac{1}{4}[e(m, n - i + 2) + e(m, n + i - 2) + e(m, n + i) + e(m, n - i)]$ for $i = 1, 2, \ldots, M$

Proposed Method

Let $\mathbf{w}_\omega(m, n)$ denote a new 2 M × 1 snapshot vector formed by concatenating the snapshot vectors in (4) and (9) as,

$$\mathbf{w}_\omega(m,n) = \left[\mathbf{y}_\omega^T(m,n)\mathbf{z}_\omega^T(m,n)\right]^T = \mathbf{B}(\omega)\mathbf{s}(m,n) + \varepsilon_\omega(m,n) \tag{11}$$

where $\mathbf{B}(\omega) = [\mathbf{A}^T(\omega)(\mathbf{A}(\omega)\Phi_\omega)^T]^T$ is a 2 M × D matrix and $\varepsilon_\omega(m,n) = [\mathbf{g}^T(m,n)\ \mathbf{q}_e^T(m,n)]^T$ is 2 M × 1 noise vector. Similarly, we can define another 2 M × 1 snapshot vector by concatenating the snapshot vectors in (5) and (10) as,

$$\mathbf{w}_v(m,n) = \left[\mathbf{y}_v^T(m,n)\mathbf{z}_v^T(m,n)\right]^T = \mathbf{B}(v)\mathbf{s}(m,n) + \varepsilon_v(m,n) \tag{12}$$

where $\mathbf{B}(v) = [\mathbf{A}^T(v)(\mathbf{A}(v)\Phi_v)^T]^T$ is a 2 M × D matrix and $\varepsilon_v(m, n) = [\mathbf{h}^T(m, n)\ \mathbf{q}_r^T(m, n)]^T$ is a 2 M × 1 noise vector. The problem is thus by using the new snapshot vectors to estimate the frequencies in each dimension.

Estimation of First Dimension Frequencies

The estimate of frequencies in the first dimension, that is ω_i for $i = 1, 2, \ldots, D$, is done by using the new snapshot vector $\mathbf{w}_\omega(m, n)$. The method used here will find an estimate of Φ_ω and that estimate will be used to determine the estimate of ω_i by taking operation across the diagonal elements of the estimate of Φ_ω .

The propagator method is applied to (11). We first give the definition of the propagator. The matrix $\mathbf{A}(\omega)$ partitioned into two sub matrices as $\mathbf{A}(\omega) = [(\mathbf{A_1}(\omega))^T\ (\mathbf{A_2}(\omega))^T]^T$ where $\mathbf{A_1}(\omega)$ and $\mathbf{A_2}(\omega)$ are the D × D and (M−D) × D sub matrices consisting of the first D rows and last (M−D) rows of the matrix $\mathbf{A}(\omega)$, respectively, since $\mathbf{A}(\omega)$ is a full rank matrix, clearly $\mathbf{A_1}(\omega)$ is also a full rank matrix and hence the rows of the matrix $\mathbf{A_2}(\omega)$ can be expressed as a linear combination of the rows of the sub matrix $\mathbf{A_1}(\omega)$ so equivalently there is a D × M−D linear operator $\mathbf{P}$ between $\mathbf{A_1}(\omega)$ and $\mathbf{A_2}(\omega)$ [24] such that $\mathbf{A_2}(\omega) = \mathbf{P}^T\ \mathbf{A_1}(\omega)$. Using the above fact, we can partition $\mathbf{B}(\omega)$ as,

$$\mathbf{B}(\omega) = \left[\mathbf{A}_1^T(\omega)\mathbf{B}_2^T(\omega)\right]^T$$

such that

$$\mathbf{B}_2(\omega) = \mathbf{P}_{\omega}^{\mathrm{T}}\mathbf{A}_1(\omega) = [\mathbf{A}_2^{\mathrm{T}}(\omega)(\mathbf{A}_1(\omega)\Phi_{\omega})^{\mathrm{T}}(\mathbf{A}_2(\omega)\Phi_{\omega})^{\mathrm{T}}]^{\mathrm{T}}$$

comprising the last (2 M−D) rows of $\mathbf{B}(\omega)$ and $\mathbf{P}_{\omega}$ is a (2 M−D) × D matrix whose entries are some linear combinations of the rows of $\mathbf{A}_1(\omega)$. The transpose of $\mathbf{P}_{\omega}$ is called propagator matrix of dimension D × (2 M−D). We can divide the propagator matrix as,

$$\mathbf{P}_{\omega}^{\mathrm{T}} = \left[\mathbf{P}_{\omega 1}^{\mathrm{T}}\,\mathbf{P}_{\omega 2}^{\mathrm{T}}\,\mathbf{P}_{\omega 3}^{\mathrm{T}}\right]^{\mathrm{T}}$$

Since $\mathbf{A}_1(\omega)$ is a non singular matrix, each sub matrix of the propagator matrix can be expressed as,

$$\mathbf{P}_{\omega 1}^{T} = \mathbf{A}_2(\omega)\mathbf{A}_1^{-1}(\omega) \tag{13}$$

$$\mathbf{P}_{\omega 2}^{T} = \mathbf{A}_1(\omega)\Phi_{\omega}\mathbf{A}_1^{-1}(\omega) \tag{14}$$

$$\mathbf{P}_{\omega 3}^{T} = \mathbf{A}_2(\omega)\Phi_{\omega}\mathbf{A}_1^{-1}(\omega) \tag{15}$$

The equation (14) implies that the diagonal elements of the matrix Φ_{ω} can be obtained by the D eigenvalues of $\mathbf{P}_{\omega 2}^{\mathrm{T}}$. The propagator matrix $\mathbf{P}_{\omega}^{\mathrm{T}}$ can be obtained from the autocorrelation matrix $\mathbf{R}_{\omega} = E\left[w_{\omega}(m,n)w_{\omega}^{T}(m,n)\right]$ of $\mathbf{w}_{\omega}(m,n)$. Under noise free condition, we can partition R_{ω} as $\mathbf{R}_{\omega} = [\mathbf{R}_{\omega 1}\mathbf{R}_{\omega 2}]$ so that $\mathbf{R}_{\omega 2} = \mathbf{P}_{\omega}^{T}\mathbf{R}_{\omega 1}$. But the said relation may not hold true in practice since the actual autocorrelation matrix $\mathbf{R}_{\omega}$ has to be estimated by the corresponding time autocorrelation matrix $\hat{R}_{\omega}$ based on the finite number of snapshot vector $\mathbf{w}_{\omega}(m,n)$. The estimate $\hat{R}_{\omega}$ of the autocorrelation matrix $\mathbf{R}_{\omega}$ can be computed from

$$\hat{R}_{\omega} = \frac{1}{(N_1N_2)}\sum_{n=1}^{N_2}\sum_{m=1}^{N_1} w_{\omega}(m,n)w_{\omega}^{T}(m,n) \tag{16}$$

The least-squares solution for the entries of the propagator matrix $\mathbf{P}_{\omega}$ satisfying the relation $\mathbf{R}_{\omega 2} = \mathbf{P}_{\omega}^{T}\mathbf{R}_{\omega 1}$ may be obtained by minimizing the cost function described as $\xi(P_{\omega}) = \|\mathbf{R}_{\omega 2} - R_{\omega 1}P_{\omega}\|_F^{\,2}$ where $\|.\|_{\mathbf{F}}$ denotes the Frobenius norm. The cost function $\xi(\mathbf{P}_{\omega})$ is a quadratic (convex) function of $\mathbf{P}_{\omega}$, can be minimized to give the unique least-square solution for $\mathbf{P}_{\omega}$, which can be evidently solved to found that

$$\hat{\mathbf{P}}_{\omega} = (\hat{\mathbf{R}}_{\omega 1}^{\mathrm{T}}\hat{\mathbf{R}}_{\omega 1})^{-1}\hat{\mathbf{R}}_{\omega 1}^{\mathrm{T}}\hat{\mathbf{R}}_{\omega 2} \tag{17}$$

Steps for estimating ω_i

step1: Calculate the estimate $\hat{R}_{\omega}$ of the autocorrelation matrix $\mathbf{R}_{\omega}$ using (16).
step2: Partition $\hat{R}_{\omega}$ and determine $\hat{\mathbf{R}}_{\omega 1}$ and $\hat{\mathbf{R}}_{\omega 2}$.
step3: Determine the estimate of the propagator matrix $\mathbf{P}_{\omega}$ using (17).

step4: Partition the estimate of the propagator matrix and find out $\hat{\mathbf{P}}^{T}_{\omega 2}$.

step5: Find the eigenvalues λ_i for i = 1, 2,…, D, of $\hat{\mathbf{P}}^{T}_{\omega 2}$ and then find the estimate of the first dimension frequencies ω_i for i = 1, 2,…, D, as

$$\hat{\omega}_i = \cos^{-1}(\lambda_i)$$

Estimation of Second Dimension Frequencies

The method adopted for estimating the frequencies ω_i for i = 1, 2,…,D, in the first dimension case can be used for estimating the frequencies v_i for i = 1, 2,…, D, in the second dimension. The procedure used in sub-section A of this Section can also be applied to estimate second dimension frequencies from equation (12).

Pairing Algorithm

We have now developed the propagator method to retrieve the 2D real-valued sinusoidal signal frequencies, now it is necessary to pair the separately obtained frequencies from both the dimensions. Both frequencies $(\hat{\omega}, \hat{v})$ can be paired by following the method describes as in [25] and [26]. The perfect pairing of estimated frequencies is one of the most desirable requirement of any multidimensional parameter estimation algorithm.

Simulation Results

Computer simulations have been conducted to evaluate the frequency estimation performance of the proposed algorithm. The average root-mean-square-error (RMSE) is employed for the performance measure, as well as some other simulations had done in order to show the effectiveness of the proposed algorithm. Proposed algorithm is compared to 2D-ESPRIT algorithms and CRLB.

The number of undamped sinusoids are D = 2 while $N_1 = N_2 = 50$ corresponds to the observed data and dimension of finite snapshot vector (M) is considered as M = 10. All results provided are average of 500 independent trials. The signal parameters are amplitudes $(a_1\ a_2) = (1\ 1)$, and frequencies as per the following specifications $\omega_1 = 0.28\pi, \omega_2 = 0.38\pi$ and $v_1 = 0.32\pi v_2 = 0.4\pi$.

In the first test, joint RMSE for both ESPRIT and Propagator method-based techniques are computed for various SNRs. In the second analysis, the RMSE(dB) for estimation of frequencies of first dimensions($\omega_1\ \omega_2$) versus different SNR conditions are plotted. Similar analysis also is done in case of second dimension frequencies, considering similar signal parameter as that of first test. It is observed that performance of proposed method is almost similar to that of time-consuming ESPRIT method. In fourth analysis, the average run time for both the techniques are depicted and it can be easily verified that propagator-based technique gives a faster estimate compare to the ESPRIT method (Figs. 1, 2, 3 and 4).

From all the analysis, it is clear that the proposed method provides considerable amount of efficiency compared to ESPRIT without any eigenvalue decomposition (EVD) or singular value decomposition (SVD) and is having very less computational complexity (Fig. 5).

Fig. 1 Joint RMSE versus SNR (dB)

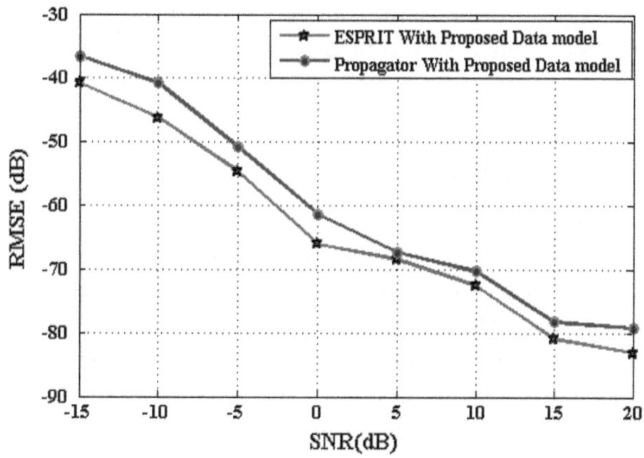

Fig. 2 RMSE for $\{\omega_k\}$ versus SNR (dB)

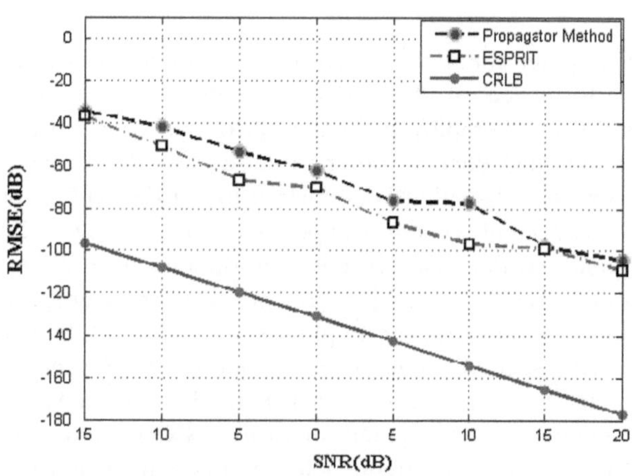

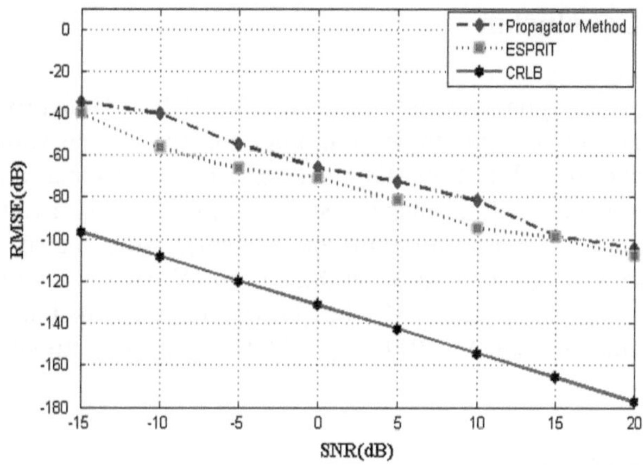

Fig. 3 RMSE for $\{\nu_k\}$ versus SNR (dB)

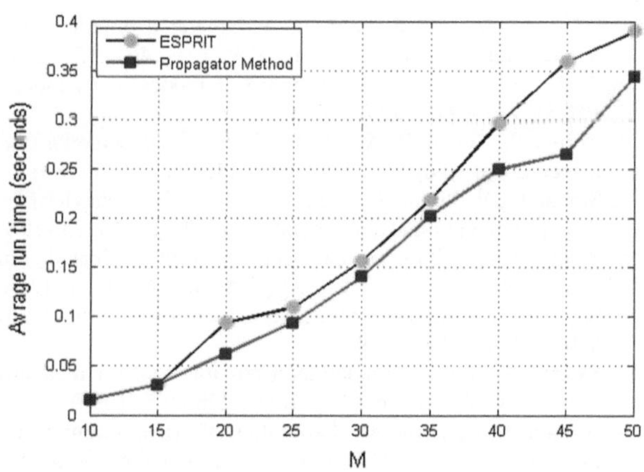

Fig. 4 Average time versus M at SNR = 10 dB

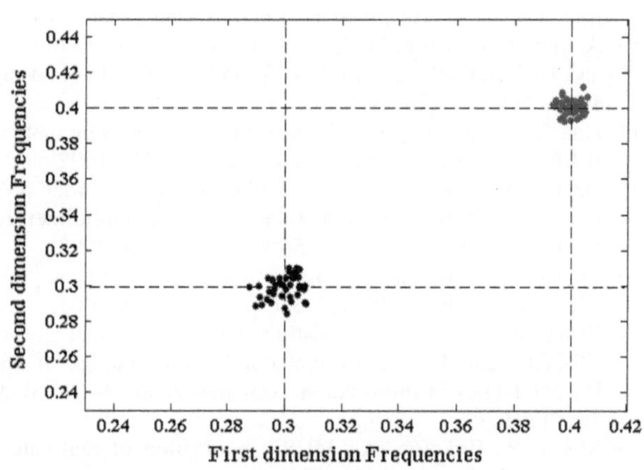

Fig. 5 2D plot of 500 independent estimates for two 2-D frequencies

Conclusion

A new approach has been developed based on propagator method for 2D frequency estimation of multiple undamped real-valued sinusoids embedded in additive white Gaussian noise. The propagator based method is a computationally efficient method since it does not require any eigenvalue decomposition or singular value decomposition of the autocorrelation matrix unlike ESPRIT. When the number samples for each dimension (N_1 and N_2) is much larger than the snapshot vector dimension (M) the propagator-based technique is computationally more efficient than ESPRIT at a cost of negligible estimation error.

References

1. Francos, J.M., Friedlander, B.: Parameter estimation of two-dimensional moving average random fields. IEEE Trans. Signal Process. **46**, 2157–2165 (1998)
2. Wang, H., Williams, K.: Vibrational modes of cylinders of finite length. J. Sound Vibr. **191**(5), 955–971 (1996)
3. Verma, S., Singal, R., Williams, K.: Vibration behavior of stators of electrical machines, Part I: theoretical study. J. Sound Vibr. **115**(1), 1–12 (1987)
4. Singal, R., Williams, K., Verma, S.: Vibration behavior of stators of electrical machines, Part II: experimental study. J. Sound Vibr. **115**(1), 13–23 (1987)
5. Zhang, X.M.: Vibration analysis of cross-ply laminated composite cylindrical shells using the wave propagation approach. Appl. Acoust. **62**, 1221–1228 (2001)
6. So, J., Leissa, A.: Three-dimensional vibrations of thick circular and annular plates. J. Sound Vibr. **209**(1), 15–41 (1998)
7. Skatter, S.: TV holography as a possible tool for mea-suring transversevibration of logs: a pilot study. Wood Fiber Sci. **28**(3), 278–285 (1996)
8. Axmon, J., Hansson, M.: Estimation of circumferential mode parameters of living trees. In: Proceedings of IASTED International Conference on Signal Image Processing (SIP'99), 1999, pp. 188–192
9. Axmon, J.: On detection of decay in growing Norway Spruce via natural frequencies. Lic. Eng. Thesis, Lund University, Lund, Sweden, October (2000)
10. Axmon, J., Hansson, M., Sörnmo, L.: Modal analysis of living spruceusing a combined Prony and DFT multichannel method for detection of internal decay. Mech. Syst. Signal Process **16**(4), 561–584 (2002)
11. Hua, Y.: Estimating two-dimensional frequencies by matrix enhancement and matrix pencil. IEEE Trans. Signal Process. **40**(9), 2267–2280 (1992)
12. Haardt, M., Roemer, F., DelGaldo, G.: Higher-order SVD-based subspace estimation to improve the parameter estimation accuracy in multidimensional harmonic retrieval problems. IEEE Trans. Signal Process. **56**(7), 3198–3213 (2008)
13. Mahata, K.: Subspace fitting approaches for frequency estimation using real-valued data. IEEE Trans. Signal Process. **53**(8), 3099–3110 (2005)
14. Palanisamy, P., Kar, S.P.: Estimation of real-valued sinusoidal signal frequencies based on ES-PRIT and Propagator method. In: Proceedings of IEEE-International Conference on Recent Trends in Information Technology, pp. 69–73, Madras Inst. of Technology, Chennai June (2011)
15. Stoica, P., Eriksson, A.: MUSIC estimation of real-valued sine-wave frequencies. Signal Process. **42**(2), 139–146 (1995)

16. Priestley, M.B.: Spectral Analysis and Time Series. Academic Press, New York (1981)
17. Walker, M.: On the estimation of a harmonic compo-nent in a time series with stationary independent resi-duals. Biometrika **58**, 21–36 (1971)
18. Rao, C.R., Zhao, L., Zhou, B.: Maximum likelihood estimation of 2-D superimposed exponential signals. IEEE Trans. Signal Process. **42**, 1795–1802 (1994)
19. Kundu, D., Mitra, A.: Asymptotic properties of the least squares estimates of 2-D exponential signals. Multidim. Syst. Signal Process. **7**, 135–150 (1996)
20. Lang, S.W., McClellan, J.H. The extension of Pisarenko's method to multiple dimensions. In: Proceedings of International Conference on Acoustics, Speech, and Signal Processing, pp. 125–128 ICASSP, paris, France (1982)
21. Kumaresan, R., Tufts, D.W.: A two-dimensional technique for frequency wave number estimation. Proc. IEEE **69**, 1515–1517 (1981)
22. Axmon, J., Hansson, M., Sörnmo, L.: Partial forward–backward averaging for enhanced frequency estimation of real X-texture modes. IEEE Trans. Signal Process. **53**(7), 2550–2562 (2005)
23. So, H.C., Chan, F.K.W., Chan, C.F.: Utilizing principal singular vectors for two-dimensional single frequency estimation. In Proceedings of IEEE International Conference on Acoustics, Speech, and Signal Processing, pp-3882-3886, Sheraton Dallas Hotel, Dallas, Texas, U.S.A. March 2010
24. Marcos, S., Marsal, A., Benidir, M.: The propagator method for source bearing estimation. Signal Process. **42**(2), 121–138 (1995)
25. Challa, R., Shamsunder, S.: High-order subspace based algorithms for passive localization of near-field sources. In Proceedings of 29th IEEE Asilomar Conference, pp. 777–781, October 1995
26. Liu, T.H., Mendel, J.M.: Azimuth and elevation direction finding using arbitrary array geometries. IEEE Trans. Signal Process. **46**(7), 2061–2065 (1998)

Creating Network Test Setup Using Virtualization and Simulation

Abhay Shriramwar

Abstract How the virtualization and simulation can be used for creating network test setup? In research and development environment, there are multiple instances when the user has to test end-to-end network setup. During the development phase, the user may not have an access of end-to-end setup as other modules may be under development phase. In this scenario, he can create end-to-end network setup using simulation and virtualization to test his real module using hardware/software in loop concept. Hence tester integrates all three—simulation, virtualization, and his actual module to get close to real-world results. In this paper, we will explore how to create such a test setup specifically for networking and telecom applications. We will also explore mathematically how close the results of these test setups are to real world. To create this end-to-end network test setup, one needs to combine VM networks, simulators or simulated network, and actual applications as shown in Fig. 1. This test setup can be configurable as per the requirements.

Keywords Virtualization · Network test setup · Simulation

Introduction

Virtual, simulated, and real network ultimately deal with packets at the fundamental level. Hence, it finally comes down to the equation for packets. If user integrates all the three networks, then

---

Paper is by product of actual research done at consortium and was converted to product used by corporates

---

A. Shriramwar (✉)
EMC, Bangalore, India
e-mail: abhay.shriramwar@emc.com

S. Sathiakumar et al. (eds.), *Proceedings of International Conference on Internet Computing and Information Communications*, Advances in Intelligent Systems and Computing 216,
DOI: 10.1007/978-81-322-1299-7_30,

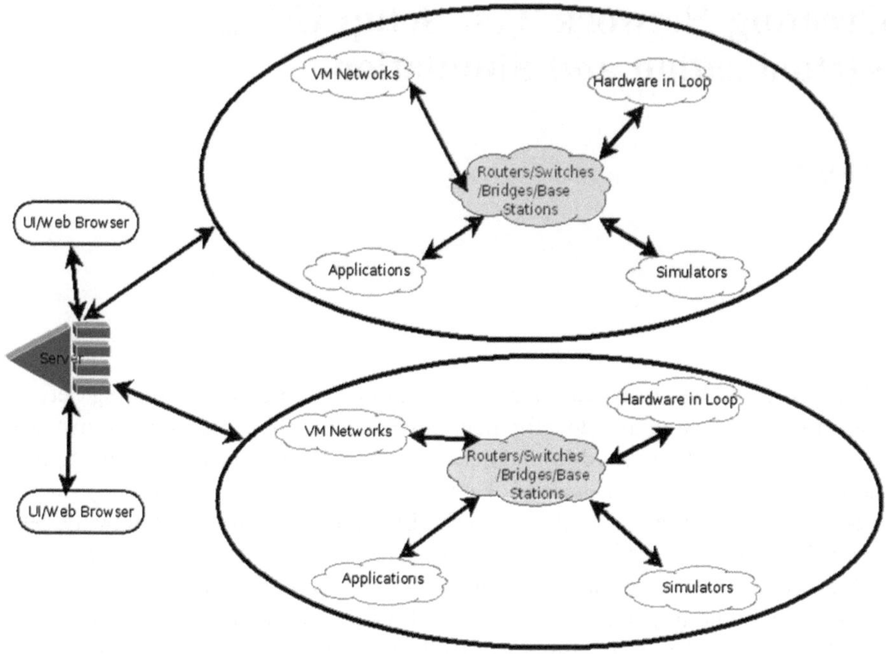

Fig. 1 Possible conceptual network test setup derived from real complex network

```
function packet_updating (packet p) {
If (packet p is at a junction of virtual simulated or real network) {
Apply junction-movement model;
} else {
if (packet p has to move from one network type to another network type)
apply network-change model
} else if (packet p has not changed the network {
apply specific-network model
}
update network statistics;
}
```

In the network test setup scenario specific network model, junction movement model, or network change model, we need to model it based on packet movement from node to node. Packets can either be in simulated network, virtual network, or real network. Packets can also be at the junction of these networks. Packets can be moving from one of these types of network to another type.

Network test setup consists of user interface for the user to interact, physical systems and virtual systems, various applications—both software and hardware, simulated nodes, or simulated networks etc. User needs to create the test setup using configuration UI and run the network test setup. During the run up, user may have to again interact with test setup using UI.

For the packet P at time T, if the user interaction time with UI and configuration delta is t, then

$$A_p(T) = C * \left(A_p(T + t) - A_p(T)\right)$$

Where C depends on network parameters like bandwidth, protocol, devices, topology etc., used in the network and network configuration.

With times, network requirement gets extremely complex as disruptive technologies get introduced in the market. What is needed is network setup that is flexible and adaptable to the changes. In this paper, we discuss how to create and use the network setup that will be useful for research and development of networks. Virtual machines integrated with applications are required for emulating various resources. Network simulators are required for simulating communication technologies in development. User interface framework for dynamic configuration and management of resources.

Architecture

Network test setup can consists of three types of networks—real networks connected with network devices like routers, bridges, base stations etc., virtual networks connected with the help of virtual bridges and simulated networks, connected with the help of simulated engines. Figure 2 shows the proposed architecture. CVMDP is a single instance process that collects the required data at run time from the complete network irrespective of its type—weather—real, virtual, or simulated. The other three processes are required to support CVMDP in the collection. System data collection process (SDCP) runs on every physical system and is responsible for collecting attributes from the physical system. Daemon process runs on every virtual machine and is responsible for the attributes of that virtual machine. Simulator controller process is again a single instance and is responsible for control of simulation weather single of distributed simulation. Database keeps all the data of the network for physical machines, virtual machines, etc. These data can range from IP addresses. CPU, memory, network interfaces etc., on the network as required for the test setup.

Figure 3 shows the integration of physical real network, virtual network, and simulated network with hardware in loop. Many such virtual machines may be running on every physical machines. Many such physical and virtual machines will form real and virtual networks. Distributed simulators may be running on multiple systems along with the real application. Every physical machine (ID) for the purpose of communication (say physical address, IP address, etc). Similarly every virtual machine (VM), simulated nodes, and hardware in loop devices or applications will have its own ID to communicate. Based on the network under test, the protocols for communication will be different.

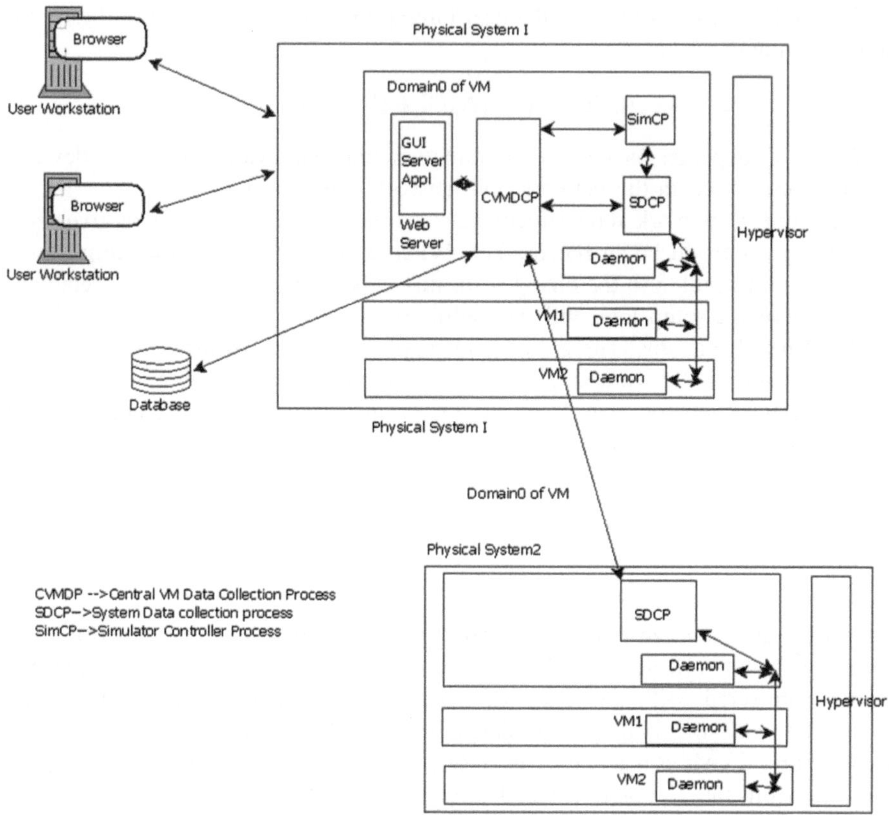

Fig. 2 Proposed architecture for network test setup using virtualization and simulation

Configuring the Network Test Setup

- Configuring the hardware to be used in the test setup, so that results stay close to actual. In this step administrator has to define the hardware, which needs to be included for the test setup. Hardware can vary from network cards, hard disks, CPU, etc. Also note that idea is to make each resource sharable so as to utilize it effectively and virtualization makes them sharable. Web GUI can be used to provision the hardware in the test setup. Database can be updated each time the new hardware is added by the tester.
- It may be required to create and run some services on each of this hardware so that real, virtual, and simulated communication channels are created.
- Configuration data can be stored in files, needed by engineers to create the flexible research test setup. This is analogous to creating VPN over the existing infrastructure. Administrator may have to specify the directory location and the

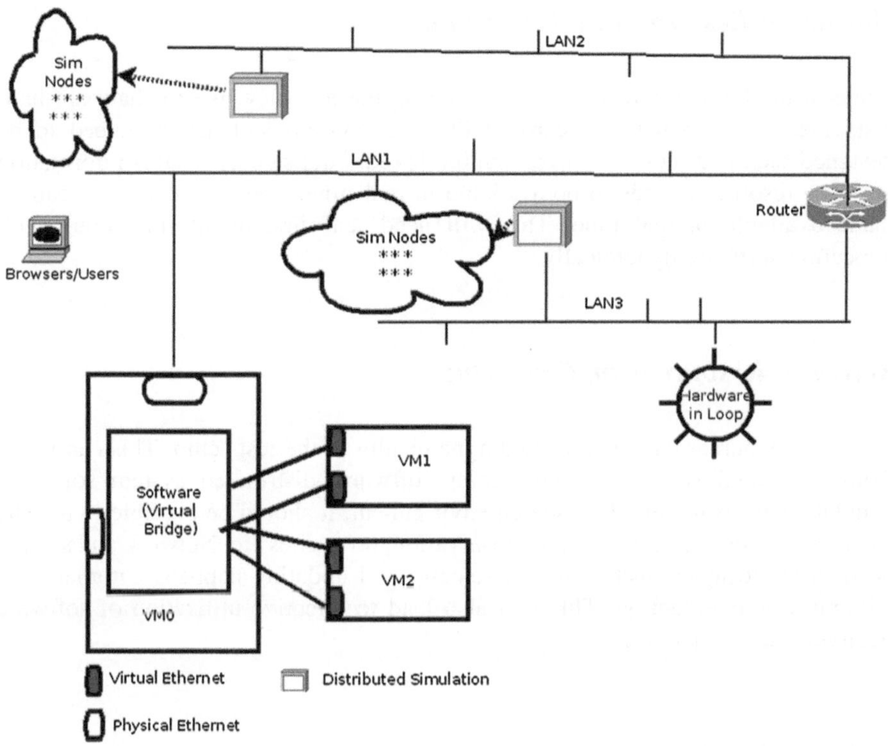

Fig. 3 Integration of virtual, simulated, and real networks

hardware where configuration file and database server and other credentials (username and password, if any) are specified.
- Any update mechanism may have to be implemented for hardware and software resources so as to monitor and track the usage.

Using the Test Setup

Users in research and development department can start using the test setup after specifying their requirements. Requirements can be specified in common configurations files, which the users can modify after having proper access permissions. This configuration file may be dynamically loaded so as to allot proper test resources. Provision can be made so as to change this requirement dynamically. Users can develop their modules independently and then integrate them.

Hardware Resources in Test Setup

Some of the hardware resources users during the test may like to share are hard disk, memory, and network cards, CPU, etc. System and network need to be designed such that these resources can be shared. First step as usual is information on these resources needs to be available in real time. Then the resources can be made available in real time. This will need real-time monitoring along with allocation software dynamically.

Software Resources in Test Setup

Software resources can be allocated dynamically in the test setup. This can range from individual software, client–server software, distributed system software, simulation software, etc. In research environment, it should be possible to use the software dynamically, meaning no reservation, no queues etc. Network and system need to be designed such that use, release, and updating happens automatically without user intervention. This will also lead to effective utilization of software licenses, usage track, etc.

Relationship Between Resources

Underlying test setup needs to understand the relationship between the resources, as this will be a basic for allotting resources. For example, physical machines contain CPU, memory, hard disks, network cards etc. Operation needs to be allowed on all these collection of resources like adding, deleting of physical machines, network cards, processors, etc. They will also have different attributes. Most of these are implementation details and can be implemented using different technologies. Figure 4 shows one such example of relationship between resources or programming languages though object oriented will suite.

Utilization in Time and Space

It should be possible to partition hardware and software resources in time and space dynamically. Multiple applications on the network will be using hard disk, ram, CPU, network cards, etc. For different applications, requirements of time and space will be different. For example, x application may need m1 memory at time t1; y application needs m2 memory at time t2 etc. Network and system should be able to allocate and deallocate this dynamically. It will be true with hardware

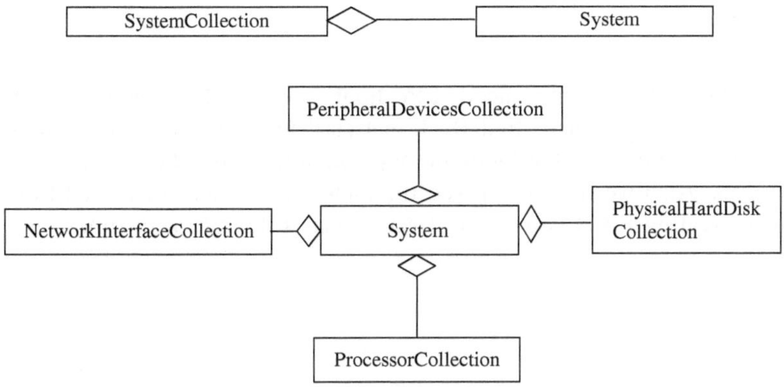

Fig. 4 Relationship between resources

resources like CPU, network cards etc too. Most virtualization middleware are able to do this.

Results and Usage Example

Let us take example of development of Base Station Router and the corresponding network where the above architecture was used for testing during the development and test results were found to be close to actual results. This development was spread across different R&D departments across the geographies. One of the departments took up the development of radio network between mobile terminal and base station. The second department in different geography took up the development of resource management—radio, OAM, etc. The third department can did the network component implementation of base station and service provider gateway and still other did the base station cluster. Since they did not have access to each other's module during development phase they used simulation and hardware in loop concept to design and develop the module. Testers defined the availability of hardware and software resources in the database, such that underlying layer understood the relationship among them and was able to view them as the collection of relevant objects. Monitoring software monitored the usage at the given time. Testers were able to access these resources by specifying them in config files. First department team simulated many mobile terminals to develop and test radio part with hardware in loop concept along with their module. Other‘s did the same for there portions. As the modules got developed, virtualization was used for integration and testing. Another advantage is one can easily isolate bugs and faults in different modules using virtualization, simulation, and hardware in loop concept.

Conclusion

In changing technology times, need of the hour is flexible R&D network test setup. We have seen in this paper that R&D network test setup can be created using virtualization, simulation, hardware in loop, and integrating them with the existing technologies. Also we have discussed the factors and issues that need to be considered in creating R&D network test setup that is flexible with times.

Automated Graphical User Interface Regression Testing

Madhumita Panda and Durga Prasad Mohapatra

Abstract Regression testing is performed after software modification to find out the correctness and quality of any software product. The modification is either corrective or adaptive. Normally in regression testing, we have to rerun the existing test cases or test scripts on the modified product. In this paper we are generating test scripts, test cases automatically for performing regression testing of graphical user interfaces. We have taken three example applications and used the capture and replay feature of the automated tool to capture the event sequences and generate test scripts automatically.

Keywords Graphical user interface · Automated regression testing · Test scripts

Introduction

The sole purpose of testing is to improve software reliability and detect bugs or errors. Regression testing comes under software quality and maintenance phase. Nowadays, most of the applications comprise Graphical User Interface (GUI). GUI is very user friendly. GUI make up a large portion of the code comprising many modern software applications. The way the test cases are implemented, relatively minor changes to the construction of the GUI can cause a large number of test cases executions to malfunction, often because GUI elements referred to by test cases have been renamed, moved, or otherwise altered [1]. Testing of GUI differs

M. Panda (✉)
Department of MCA, Seemanta Engineering College, Mayurbhanj,
Odisha 757086, India
e-mail: madhumita.panda3@gmail.com

D. P. Mohapatra
Department of Computer Science and Engineering, National Institute of Technology,
Rourkela, Odisha 769008, India
e-mail: durga@nitrkl.ac.in

S. Sathiakumar et al. (eds.), *Proceedings of International Conference on Internet Computing and Information Communications*, Advances in Intelligent Systems and Computing 216,
DOI: 10.1007/978-81-322-1299-7_31,

significantly from the testing of traditional software. Therefore, a number of specialized techniques have been adapted for GUI test automations. Automated tools are used for effective GUI testing. Although several approaches have been proposed for both GUI and nonGUI application testings, it is a little bit difficult to implement any new approach. We see that GUI regression tests have redundant steps which are critical to the functioning of the GUI and deserve sufficient testing. This would be labor intensive if performed by manual testing. However, if we can record those test sequences, then we will be able to capture the event sequences and also identify the existing critical paths. Finally, we can set up unit tests manually, for these critical states. Once the scripts are recorded then we can execute those scripts to get the desired results. For failed test sequences, we can modify our scripts manually. First we have to identify which parts of the script we want to modify. Next, performing the unit tests for these changed scripts will be much easier. The available automated tools have a capture/replay feature using which novice testers can easily generate test cases automatically. GUI is designed in such a manner that minor changes in one screen would result in the malfunction of large no of test cases related to it. Thus, automated tools are normally used for testing the GUI-related components of any software. We have selected Test-Complete, it is a perfect tool for GUI testing having many features available for GUI testing and it generates test cases and test scripts automatically which are very helpful for novice testers.

The rest of the paper is organized as follows: Section "Basic Concepts" includes basic concepts, Sect. "Related Work"includes related work performed in the same field, Sect. "Steps for Regression Testing of GUI" includes steps for performing regression testing of GUI, and Sect. "Implementations and Results" includes our implementations and results. Finally, we conclude in Sect. "Conclusion" with a discussion of future work.

Basic Concepts

This section describes the basic concepts needed for understanding GUI and the methodologies applied for their testing.

GUI

A GUI consists of a set of widgets W = $\{w_1, w_2, w_3, \ldots, w_l\}$ including buttons, panels, and test fields having a set of properties P = $\{p_1, p_2, p_3, \ldots, p_m\}$ of those widgets like background, color, size, font, and a set of values V = $\{v_1, v_2, \ldots, v_n\}$ like {red, bold, 16pt} associated with the properties [2].

Capture/Replay

All automated tools used for graphical user testing have an in build test recording feature known as capture and replay, which has the capability of capturing and replaying each and every event interactions performed by the tester manually. It also captures all the mouse movements, keystrokes including screen shots.

GUI Test Case

A GUI test case T is a pair $< S_0, e_1; e_2;\ldots; e_n >$ consisting of a state $S_0 \in S_1$, called the initial state for T, and a legal event sequence $e_1; e_2;\ldots; e_n$ [2].

TestComplete

TestComplete is a very useful tool for any software developer who needs to apply a rigorous testing methodology for testing his products. It has the amazing capability of working in five languages, including Visual C++, Visual Basic, C++ Builder, Delphi, Java, and .NET applications. It also supports a wide variety of scripting languages means we can work in the scripting language we are comfortable [3]. We strongly agree with those statements of Dave, as the tool is much more effective than the above described words and it gives the real flavor of graphical user testing. We have used it for testing our GUI applications developed in VB and Java.

Automated tool. Automated tools are capable of covering every aspect of testing like organizing manual tests, performing unit testing, functional testing, performance, and load testing. They support number of functionality, and some of them support more than one programming languages and platforms.

Language support feature. It offers systematic, automated, and structured testing, with the capability of supporting .NET, Java, Visual C++, Visual Basic, WPF (XAML), Delphi, C++Builder, and Web applications. It is equally oriented for testing 32-bit and 64-bit applications. With TestComplete we can also test Power Builder, FoxPro, Access, and other applications.

Regression Testing using TestComplete. Regression testing is the process of executing tests in a repeatable manner and comparing the latest results with previous test executions to ensure that the same outcome is achieved. Regression testing is extremely important and is the means of realizing the value of test automation. Repeatedly, executing tests over time allow us to verify whether the application is still performing in the manner in which it was intended. Ultimately, the goal of test automation is to save time, money, and improve quality. Test automation can relieve us of the task of manually testing existing functionality thus allowing testers to focus attention on areas not already covered by automation. However, the above described statements are only true as long as the existing test automation feature is working, being executed consistently and the results are clearly communicated to the people who need them.

Projects and Project Suites

Project is a test suite containing test scripts and other project items used by the tester that make up the project. A project suite is a combination of projects. To create a new project we have to select File—New—Project... and we will get the Create New Project dialog. When a new project is initially created the tester gets the chance to specify the items he needs to include. Project items can be added and removed at any point. In TestComplete, projects function as containers for different pieces of functionality related to testing and are organized on disk in a similar file/folder structure. Removing a node from Project Explorer only removes it from the logical view and does not actually delete the item from disk, so we can easily add removed items back to a project. While removing nodes from the project explorer we are only removing them from the logical view. The actual files and folders are not deleted from disk and may be added back to the project using the right-click context menu and selecting Add—Existing Item.

Features associated with New Project Wizard. The New Project Wizard is used to setup a new project along with a predefined set of testing functionality. Items that a project will require as project items can be added or removed directly from the Project Explorer panel, at any time after the project has been created [4].

Test Log. Test logs are used to save feedback from executed test scripts. These include errors, warning messages, and events and images. Logs are stored in a directory as XML files.

Log Tree. This area will have a node for each test item level used to run the test. When a project suite is run, it is shown as the parent or root node and each project is represented as child node of the project suite. The executed status of child items (the red X indicates the error status) reflects back in the parent items.

Test Log. This area contains messages coming from both TestComplete and the script code. There are six different types of messages:

Message. Can come from TestComplete or from the script code (Log.Message), will not cause the test to fail.
Warning. Can come from TestComplete or from the script code (Log.Warning), not necessary a failure for the test but is an indicator for why a test failed.
Error. Can come from TestComplete or from the script code (Log.Error), this indicates a failure for the test.
Events. Usually comes from TestComplete but can come from the script as well (Log.Event). It do not cause the tests to fail. TestComplete generates an event for every mouse click or keyboard entry in the test.

Image. Usually comes from script code (Log.Picture) but TestComplete can generate as well (Region Checkpoints).
File. There are two types of File log items. Log.File, will copy the file into the directory holding the log XML file and creates a hyperlink to the file. Log.Link just creates a hyperlink to the file without copying the file.
LockEvents. With any long running test, the test result log will be filled with hundreds if not thousands of events.

Basic Record and Playback

TestComplete has rich record and playback functionality allowing us to quickly record new test scripts. TestComplete has several options that allow us to control how recording is performed. To view or change these options select Tools—Options… and click the Engines folder and select Recording. Before starting the recording process first we have to load the specific project which needs to be tested. We have to simply select and add the project to the tested applications list. Then we have to open that project. Once we have a project loaded there are two ways to begin a new recording,

- Select Test—Record from the TestComplete main menu.
- Press the Record toolbar button on the Test Engine toolbar.

Functionalities of TestRecoder

Automated tools are normally associated with a recording tool which is associated with recording all the testers actions performed on the executable files and user screens for performing testing. Once the recorder finishes recording all the user actions are generated in the form of a test script, in the specific scripting language specified by the tester and supported by the tool. We have used TestComplete and Fig. 1 visualizes its recording toolbar. TestComplete has rich record and playback functionality allowing us to quickly record new test scripts. It has several options that allow us to control how recording is performed. To view or change these options we have to select Tools—Options… and click the Engines folder and select

Fig. 1 Recording toolbar of TestComplete

Recording, we will see the following options in the form of buttons. They are

- Start Recording,
- Stop Recording,
- Pause Recording,
- Record Script,
- Record Low Level Procedure (screen coordinates),
- Add Comment to Test,
- Run Tested Application,
- Add Checkpoint From the List,
- Run a Script Extension.

Region Checkpoints

Region checkpoints allow for the comparison of an area of the screen with a baseline. The baseline is stored in the Project Item called "Regions" under the Project Item called "Stores". A region checkpoint can be created either while recording or from the toolbar at test design time. Figure 2 shows create region checkpoint screen.

An example of creating a region checkpoint: (1) Open any GUI screen. (2) Open a project in TestComplete. (3) Click Create Region Checkpoint from the Editor's Tools toolbar. (4) Add the Stores project item if not already in the TestComplete project.

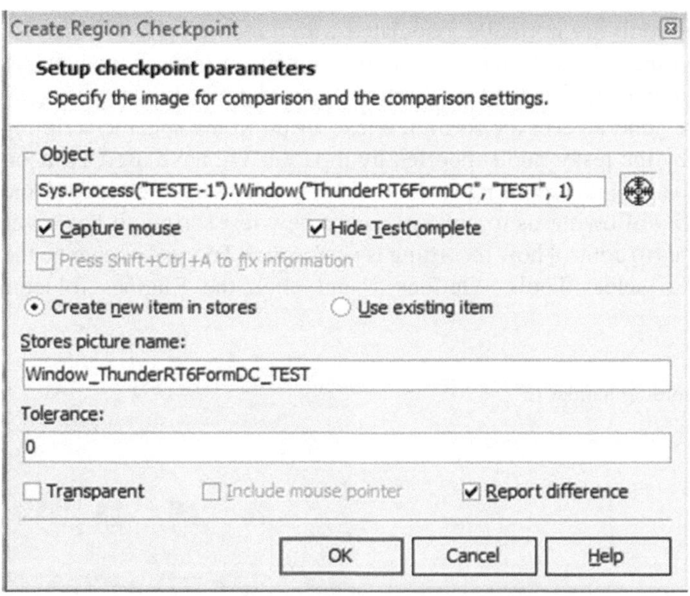

Fig. 2 Test1

Related Work

GUI testing is an area of growing importance, facing a number of severe challenges. A few methods have been proposed for GUI testing. However, it is still not clear how to define GUI test cases and how many actions should be comprised as a GUI test case. The reason GUI testing is different and difficult is that the input is interactive, whereas the output may be graphical or may be an event. An especially serious problem occurs in maintenance where changes are made to the GUI interface, and the maintenance engineer does not have a sufficient understanding of how the GUI software was designed [1].

White [5] identified the static and dynamic event interactions in a GUT, and showed how automated regression tests can be generated to test these interactions. He showed that Latin squares can be used for automated test design, Latin squares techniques are borrowed from statistical experimental design. He has compared three methods, brute force test generation, random test generation, and the use of Latin squares. The main problem associated with GUI testing is an exponentially large number of test cases. For example, for just three windows in a screen with five modes, there are potentially 1,050 possible combinations to test. GUI testing can be categorized into two categories based upon test screens involved. The problem is static when testing involves a single screen, and it is dynamic when we have to move from one screen to another to select events from GUI objects. Dynamic GUI interaction testing involves making a selection from one or more menu, buttons, etc., on one screen, which then brings up a new screen, where more selections are made, and the process is repeated.

Memon [2, 6] presented a GUI smoke regression testing process called Daily Automated Regression Tester (DART). He combined smoke tests with different types of test oracles and presented guidelines for practitioners thus helping them generate and execute the test oracles for generating test cases. Then used four GUI-based applications and generated 5000–8000 smoke tests for each application. They findout that short GUI smoke tests with certain test oracles are capable of detecting a large number of faults, some classes of faults are there which cannot be detected with smoke test. He findout that short smoke tests execute a large percentage of code.

Zunliang [7] proposed a novel Actionable Knowledge Model (AKM) for representing GUI test cases. The proposed model is adaptive to the change of GUI. They developed a multi-agent based framework using AKM for GUI application regression testing and used it to enhance the process of test case creation, execution, and repair.

Chen [8] in his paper proposed a framework of component-based library (CBL) for GUI regression. Test components, encapsulating associated properties, operations, and meta-information are defined as reusable object. They introduced three key processes of CBL in details, the first process includes CBL organization structure based on message mechanism, the second process includes the test components regeneration guidelines for GUI testing and the third process divides test procedure into several phases in order to improve the testing efficiency. The phases include message segment capture phase, component abstract phase, regenerate test component phase, and component-driven testing phase. They have experimentally proved that CBL model performances well and the proposed reusable component for GUI regression testing can regenerate a large number of test cases automatically and efficiently.

Udgata in his paper has proposed an automated test tool which will test a Web portal for all the missing links, file type mismatches, unreachable files, etc. It also reports errors in Java Script and PHP modules embedded in the HTML documents. The tool has a dry run to identify all features of html and generates a complete report regarding the Web portal thus helping in its maintenance. The tool generates test cases for running over the Web portal using regression testing techniques. They claimed that the tool can work on both static and dynamic Web pages. Dynamic content include filling up the forms, entering login information etc. It also provides GUI for easy interaction with the portal.

Grechanik [9] in his paper has described a case study taking a team of 45 professional programmers and test engineers who experimentally assessed the tool-based approach versus the manual approach for maintaining GUI directed test scripts. They recommend organizations to supply programmers with testing tools as these tools fix test scripts faster and also can unit test software. They have also recommended that productiveness of experienced test engineers is same for the manual as well as tool-based approach, so they do not need a tool for fixing test scripts.

Zhao [3] in his paper studied the modeling of the test profiles in GUI testing. Then they presented a methodology for studying the relationship between the test profiles and the fault detection in GUI testing. They also proposed a control scheme based on the above described relationship which can improve the efficiency of GUI testing.

Hui proposed a method called function-diagram-based regression testing. They described GUI by functional diagrams. Then they selected regression test cases by comparing the already existing function diagrams and the newly created function diagrams. Their paper proves both in theory and practice that this kind of regression testing strategy of test case selection can promote the efficiency of regression testing and reduce the testing costs.

Steps for Regression Testing of GUI

Normally, the GUI regression tests have redundant steps which are critical to the functioning of the GUI and require repeated testing. So it is too expensive and time consuming to perform regression testing manually. For performing regression testing we have used the following steps:

(1) Identify the executable application under test. Then open and add this application to the new project suite.
(2) For the first time tester perform the testing manually and provide all the test data out of own intuition. Adds different checkpoints to identify the GUI screens properly.

(3) Use record and playback mechanism to capture all test sequences performed during manual testing of the application under test.
(4) Records each and every mouse and keyboard interactions and generate the test script automatically in the specified scripting language.
(5) Run the generated test script to generate test cases automatically. Test cases will be generated automatically showing pass and failed event sequences, comparing all the checkpoints.
(6) Checks the generated test execution report and fix bugs detected by either changing the test script manually or re recording test scripts.

Implementations and Results

We are now presenting the details of our experiments. First we developed three applications and their modified versions for regression testing using Visual Basic and Java. Then we followed the above described steps of regression testing manually to generate automatic test scripts. Then we run those test scripts to study important characteristics of GUI regression test cases. We were interested in knowing following factors,

(1) What is the ability of TestComplete to perform GUI testing?
(2) Is TestComplete capable of identifying and automatically generate test scripts for regression testing and if scripts are generated are they capable of detecting any fault or change in the original and modified application?
(3) Can we apply the script generated for one application on the modified version of the same application and what is the fault detection capability of this old script on the new application developed?
(4) When the script is run repeatedly on the modified versions of the application is there any difficulty encountered in the rerunning of the script and if yes then what is the effect on generated test cases?
(5) Are we able to capture and record all the GUI events in the proper order and if the order gets disrupted inbetween then what is the effect?
(6) If the new version of the software is totally different from the parent version or if we are changing the logic all together then is TestComplete is capable of detecting this and what is the result shown after rerunning the existing script?

We used graphical user testing tool TestComplete to perform the regression testing of our applications. First we performed regression testing of two notepad applications developed using VB using the above described steps of testing. Then we performed regression testing of two notepad applications and two calculator applications, developed in Java. We generated automated test scripts for VB project in VB script and generated test scripts for Java projects using Java script. Then we run those scripts to find out the test cases. Then we modified the test scripts manually and tried to find out error detection capability of the scripts. Given below figures show the graphical user screens and the generated test cases.

Example-1

Figure 3 shows workspace of TestComplete, Figs. 4 and 5 show the two example notepad applications developed in VB, Figs. 6, 7 and 8 show generated test scripts and test cases for the VB applications.

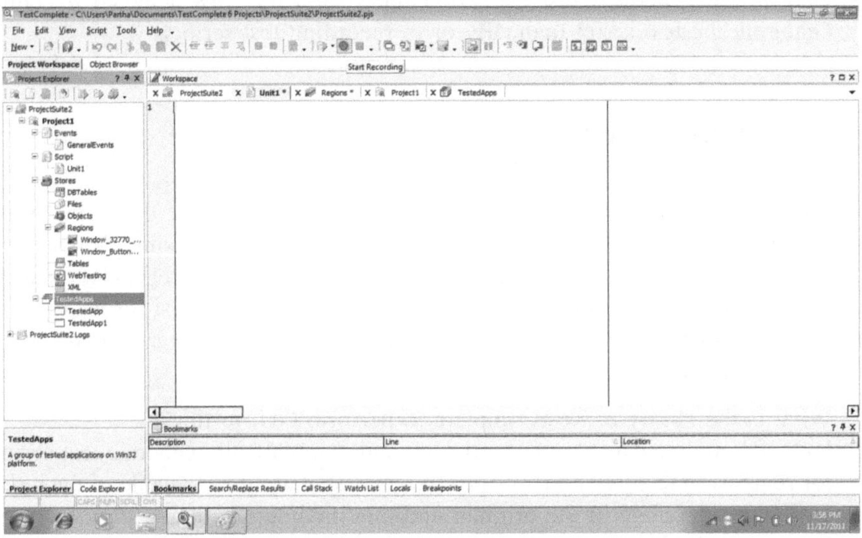

Fig. 3 Workspace of TestComplete

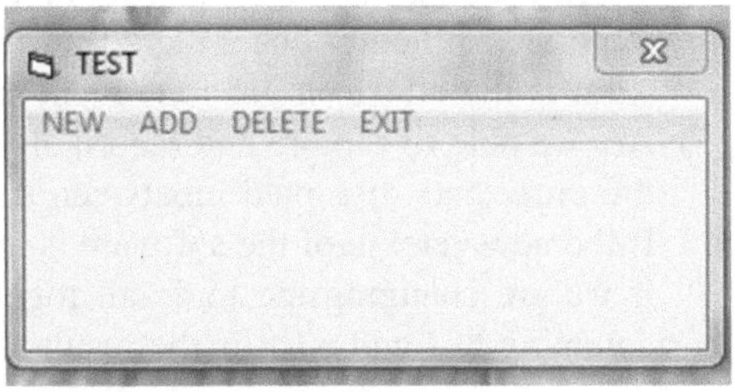

Fig. 4 Notepad application1in VB

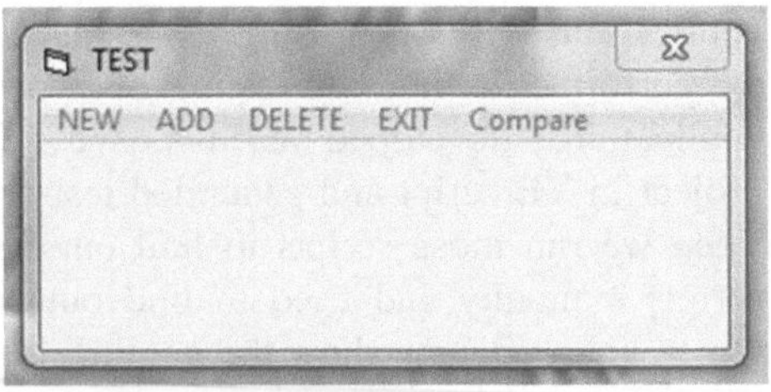

Fig. 5 Notepad2 in VB

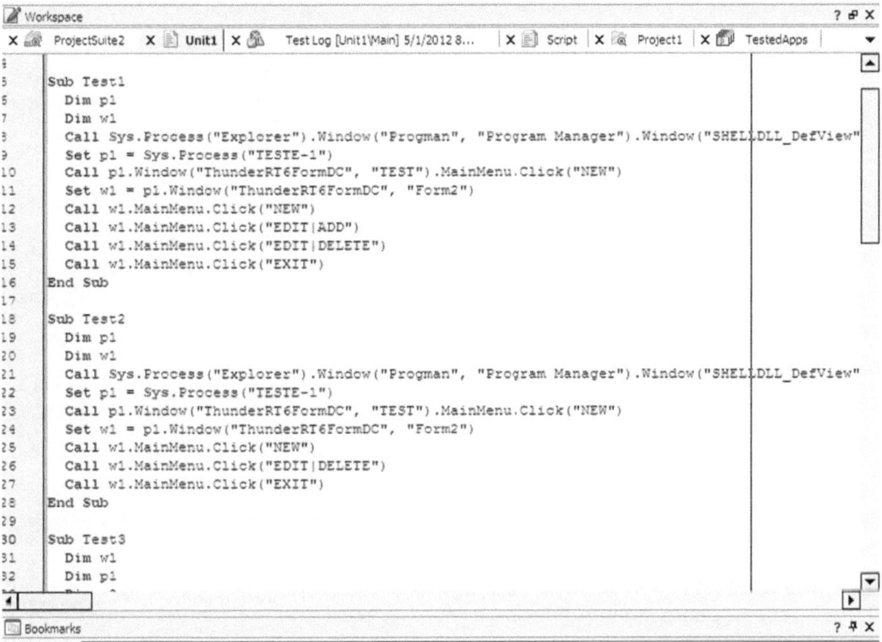

Fig. 6 Automatically generated VB script

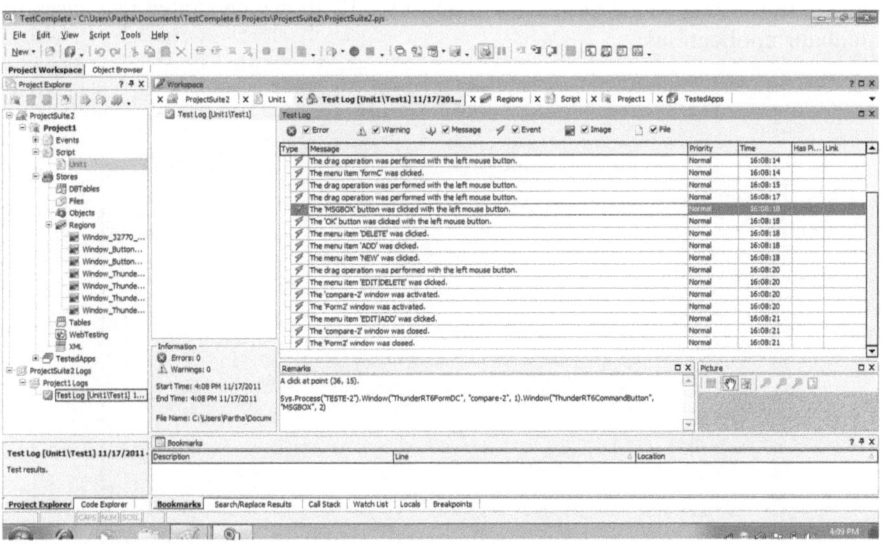

Fig. 7 Generated test cases

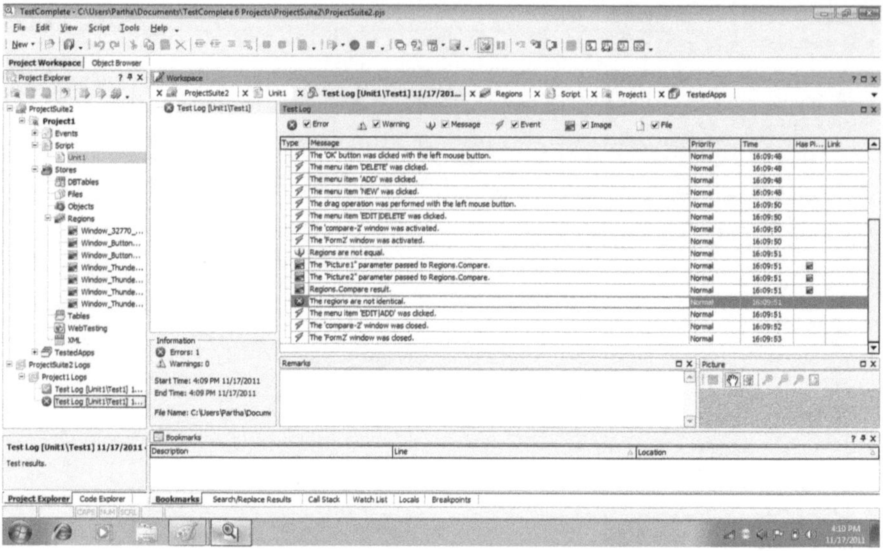

Fig. 8 Regression test cases

Example-2

Figures 9 and 10 show calculator and modified calculator applications in Java, Fig. 11 shows the generated test script and Fig. 12 shows generated test cases for calculator applications.

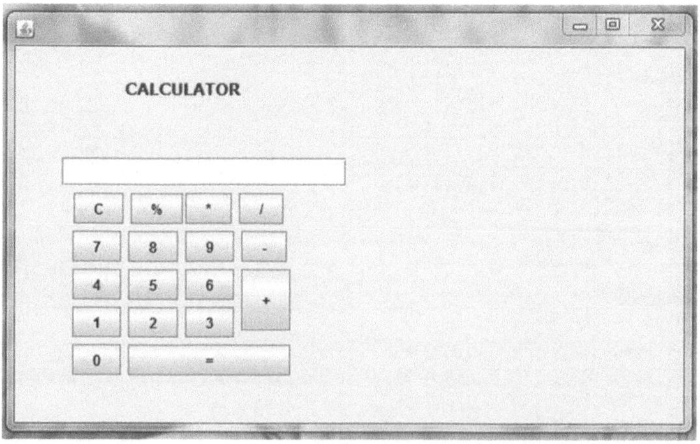

Fig. 9 Calculator in Java

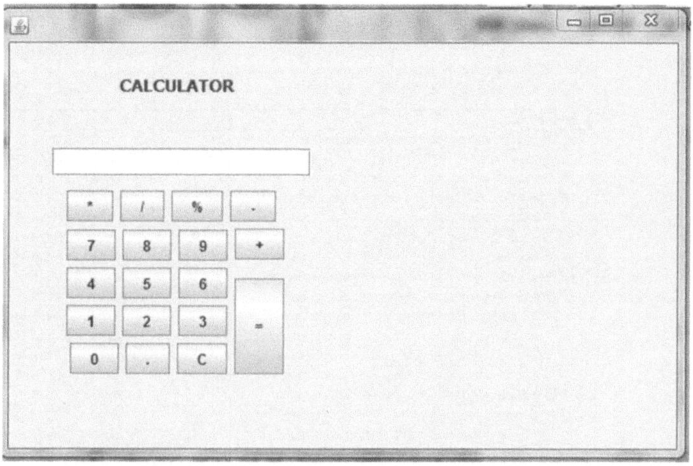

Fig. 10 Modified calculator

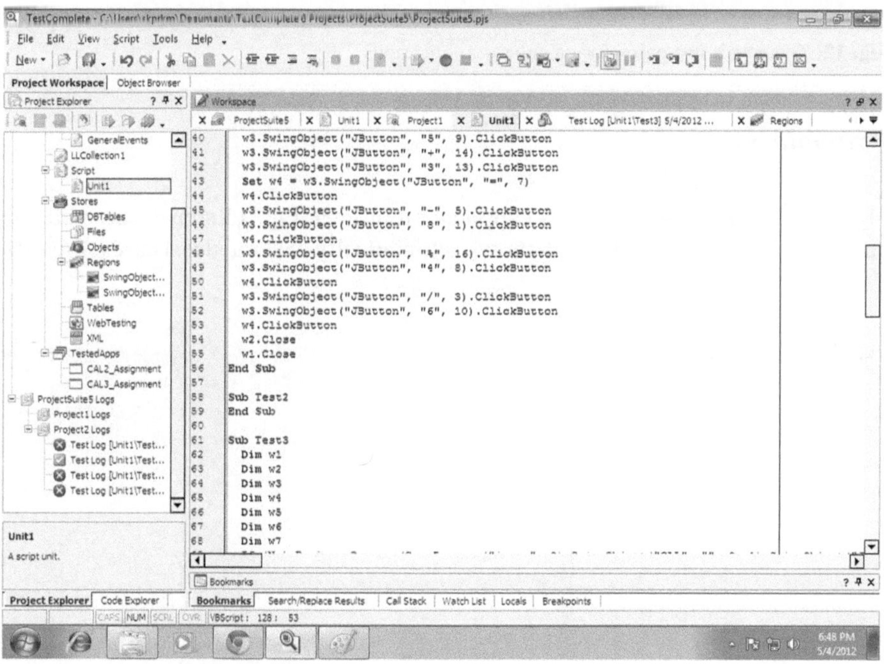

Fig. 11 Generated script for calculator applications

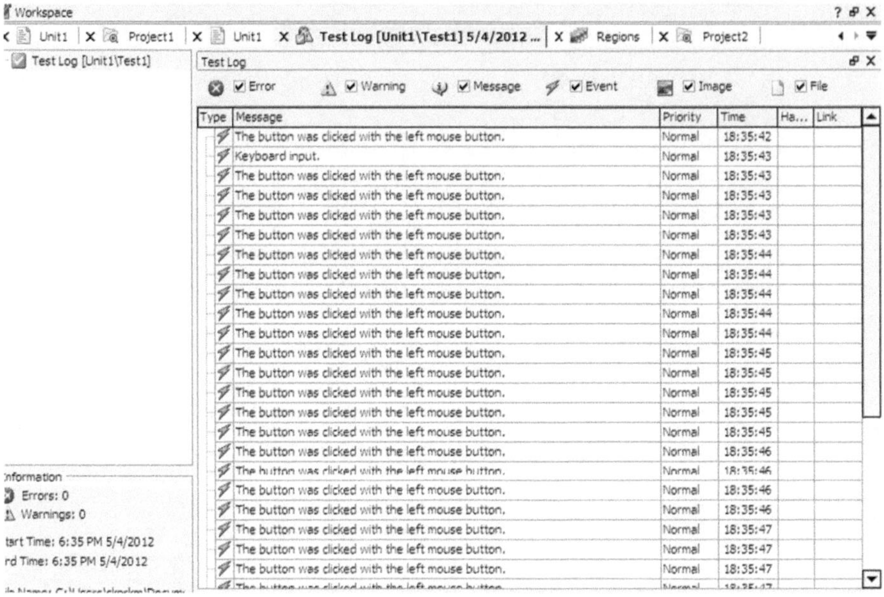

Fig. 12 Generated test cases for calculator applications

Example-3

Figures 13 and 14 shows notepad application in Java, Fig. 15 shows automatically generated Java script and Fig. 16 shows automatically generated test cases (Fig. 10).

Fig. 13 Notepad in Java

Fig. 14 Example notepad application

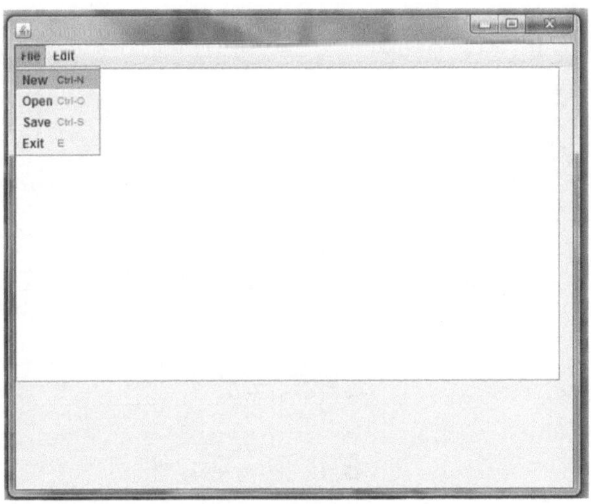

Fig. 15 Example modified notepad application

Results

First we recorded the script for one application and executed the script. Then we rerun the same script on the modified version of the application, we find that test cases fail at the mismatching sequences. We also discovered that if the application is completely modified or changed then all the test cases fail. We find out that the tool used by us, TestComplete could test the GUI applications effectively as it has the capability of capturing and replaying all the mouse and keyboard sequences and also it is capturing all the event sequences (Fig. 17).

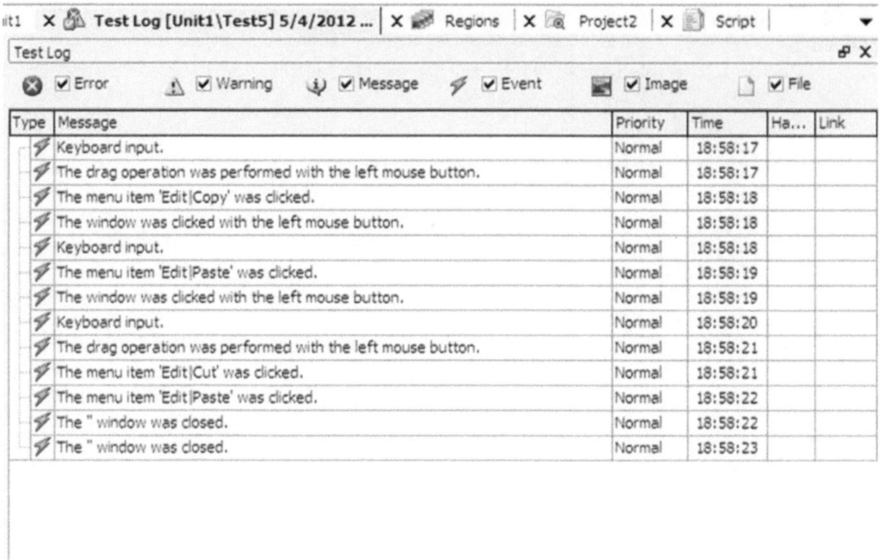

Fig. 16 Script notepad

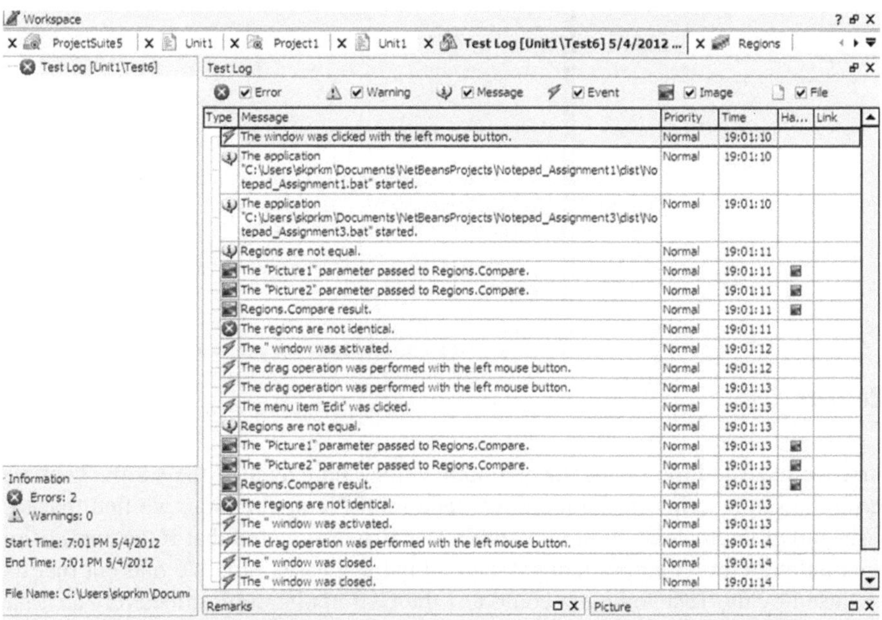

Fig. 17 Generated test cases for notepad

Conclusion

The exploration of testing techniques for GUI is still a very young field within the research field for software testing in general. In this paper, we proposed a systematic approach for regression testing of GUI. We can perform GUI testing of any applications using any automatic testing tool by following these steps. We tested three applications and their modified versions and generated scripts and test cases for those applications automatically. With TestComplete we captured the mouse movements, input sequences, and also the widgets and event sequences in their proper order of generation. Using checkpoints, we compared the widgets and identified difference between previously existing screens and modified screens. This automated tool gave the real flavor and scenario of GUI testing. Its a very challenging and newly evolving field and we can explore it further by proposing and implementing new approaches to this area.

References

1. Memon, A.M.: A comprehensive framework for testing graphical user interfaces. University of Pittsburgh, Pittsburgh (2001)
2. Memon, A.M., Xie, Q.: Empirical evaluation of the fault-detection effectiveness of smoke regression test cases for GUI-based software. Department of Computer Science and Institute for Advanced Computer Studies, University of Maryland, College Park, Maryland, ICSM, IEEE (2004)
3. Zhao, L., Cai, K.Y.: On modeling of GUI test profile. Accenture Technology Labs, Chicago, 978-1-4244-4828-9, ICSM 2009, Edmonton, Canada, IEEE (2009)
4. TestComplete 7 Made Easy. www.microway.co.nz/catalog/automatedqa/TestComplete7-comp.pdf. If applicable
5. Lee, J.: White regression testing of GUI event interactions. Department of Computer Engineering and Science, Case Western Reserve University, Cleveland (1996)
6. Memon, A.M., Xie, Q.: Studying the fault-detection effectiveness of GUI test cases for rapidly evolving software. Department of Computer Science and Institute for Advanced Computer Studies, University of Maryland, College Park, IEEE. Trans. Softw. Eng. **31**(10) (2005)
7. Yin, Z., Miao, C.: Aactionable knowledge model for GUI regression testing. Nanyang Technological University, Singapore (2005). IEEE/WIC/ACM
8. Chen, H., Zou, B.: A reusable component-based library for GUI regression testing. School of Information Science and Engineering, Central South University, China, Software School, Hunan University, Changsha, doi:10.1109-0-7695-3090-7/08, IEEE (2008)
9. Grechanik, M., Xie, Q., Fu, C.: Experimental assessment of manual versus tool-based maintenance of GUI-directed test scripts. In: 3rd International Conference on Software Testing, Verification, and Validation Workshops, Department of Automatic Control Beijing University of Aeronautics and Astronautics, Beijing. doi:10.1109/ICSTW.2010.19, IEEE (2010)

Video Search Using Map Reduce Framework in an Education Cloud

E. Iniya Nehru, P. Seethalakshmi, S. Sujatha and M. Nandhini

Abstract Video search systems have become popular in recent years. The system prompts users to give a string query and retrieves the matched video quickly from the user for playing. There is still no established search method available for scalable fast search in large distributed video databases. In video search systems, when the number of online users reaches a certain scale, it will greatly reduce the response from the server. There are many video resources stored in distributed databases which should be accessible to all the users. If many users access the videos at the same time, it may lead to increase in load on the server. To solve this problem, cloud computing technology is used. A distributed database is used for storing and indexing videos. This system uses Map Reduce paradigm for retrieving the videos in distributed fashion. Map Reduce approach allows splitting the tasks into sub-tasks and then assigning it to various virtual nodes present in the cloud, which are then processed and consolidated to give the final output. Thus, the processing speed is increased while the processing time is greatly reduced.

Keywords Cloud computing · Distributed database · Map reduce

E. Iniya Nehru (✉)
National Informatics Centre, Chennai, India
e-mail: nehru@nic.in

P. Seethalakshmi
Department of Computer Science, Anna University of Technology, Tiruchirapalli, India
e-mail: auropansee@yahoo.co.in

S. Sujatha
Department of Computer Applications, Anna University of Technology, Tiruchirapalli, India
e-mail: sujathaaut@gmail.com

M. Nandhini
Department of Information Science and Technology, Anna University, Chennai, India
e-mail: mnandhini2009@gmail.com

S. Sathiakumar et al. (eds.), *Proceedings of International Conference on Internet Computing and Information Communications*, Advances in Intelligent Systems and Computing 216,
DOI: 10.1007/978-81-322-1299-7_32, © Springer India 2014

Introduction

Cloud computing overlaps some of the concepts of distributed, grid, and utility computing, however, it does have its own meaning if contextually used correctly. Some of its characteristics are scalability, reliability, and multi-tenancy. It supports various deployment models, such as, public cloud, hybrid cloud, and private cloud, and also supports many virtualization technologies to implement the cloud abstractions.

Many platforms are available to experiment on and to deploy a prototype of an application in the cloud. Of these, Eucalyptus used for deploying video search system, is presented in this paper. Eucalyptus [1] is a software platform for the implementation of private cloud computing on computer clusters.

MySQL Cluster is one of the prominent distributed database management systems. It is used to provide high availability, high performance, and scalability. MySQL Cluster is implemented through an additional storage engine available within MySQL called NDB or NDBCLUSTER. It provides Replication, Horizontal data partitioning, Hybrid Storage, and shared nothing. The shared nothing architecture enables the system to work with inexpensive hardware and with a minimum of hardware and software. It is designed not to have any single point of failure. In a shared nothing system, each component is expected to have its own memory and disk, and use of shared storage mechanisms such as network shares and network file systems.

The above technologies are utilized to their full extent by using parallel programming paradigm. The Map Reduce framework provides a simple programming interface that is specifically designed to make it easy for a programmer to design a parallel program that can efficiently perform a data-intensive computation. Map Reduce [2] a distributed computing model has two functions, map and reduce. Map Reduce framework divides the input data files into a number of splits; each split is passed to a map task and they are processed in parallel by each node in the cluster without communicating with each other. The output of map tasks is repartitioned across all nodes of the cluster. Finally, a set of Reduce tasks are executed in parallel by each node on the partition it receives. Thus, the Map Reduce framework enables high parallel computing and reliability of data.

Hadoop [3], the Java open source implementation of Map Reduce has been applied by many vendors such as Amazon, Facebook and Yahoo, and so on. It contains the Map Reduce runtime system and distributed file system (HDFS), which provides data redundancy and ensures that the cluster data transmission between nodes is transparent. It is used to run applications in the framework for a large cluster on low-cost hardware devices.

The Apache Hadoop [3] software library allows for the distributed processing of large data sets across clusters of computers using Map Reduce programming model. Rather than rely on hardware to deliver high-availability, the library itself is designed to detect and handle failures at the application layer and delivers a highly available service on top of a cluster of computers, each of which may be prone to failures. Using the above technologies a novel, scalable, and efficient video search system has been implemented and is described in this paper.

The remainder of this paper is organized as follows. Section “Literature Review” discusses the literature survey emphasizing the research activities and related works in video search system, pattern matching, Map Reduce which has emerged in the past few years. Section “System Design” presents the overall system architecture and description of the video search on education cloud. It provides the infrastructure model and the process is carried out by the entire system and also provides control flow between each function and components. Section “Implementation” explains the details of project implementation module-wise. Section“Results and Conclusion” draws the results and conclusion.

Literature Review

Indexing helps to answer queries that can be resolved efficiently. Hash-based indexing is a technology for retrieving text-based information. It is applicable for grouping, similarity search, and classification. In paper [4] two principles for hash-based indexing are introduced, fuzzy-fingerprinting and locality-sensitive hashing.

Bitmap Indices [5] play an important role in improving query performance in large video database systems. An efficient database is used for handling large volumes of data which also supports a large number of concurrent users. Indexing scheme is generally required to achieve scalability and high throughput. Another advantage of using a bitmap indexing method is that a low-cost Boolean operation such as AND, OR, and XOR can be utilized in the analysis. The basic bitmap index [2] uses each distinct value of the indexed attribute as a key and generates one bitmap containing as many bits as the number of records in the data set for each key.

By using histogram feature [6], a fast and robust video search algorithm for videos from large video databases, based on the adjacent pixel intensity difference quantization (APIDQ) algorithm is presented. An efficient VSS (Video Similarity Search) approach is proposed in paper [7]. Based on Spatial–temporal features of video sequences, a compact video signature was computed. For the scalable computing requirement, a new search method via Signature Index Table was presented by index clustering.

Pattern matching algorithm is used in the detection engine of the intrusion detection engine of the intrusion prevention system [8]. This algorithm is used for calculating the efficiency of the intrusion prevention system. The multi-pattern matching algorithm (wu-Manber) is explained in paper [8] to improve the efficiency. By dividing the pattern group into two subgroups and dealing with the two subgroups in different ways, the QuickWu-Manber algorithm improved the efficiency of pattern matching.

In Compressed Pattern Matching algorithm [9], given a text with compressed form and a pattern the algorithm finds the occurrences in the pattern without data compression. In the Internet environment, to utilize bandwidth more effectively, it is highly desirable that data be kept and sent over the Internet in compressed form.

Pattern matching is a string searching algorithm that compares the pattern with text from distributed database. If the characters do not match, searching jumps to the next possible match. The Boyer Moore BM [10] algorithm checks whether the pattern is positioned over the text at a place, and then compares the characters of the pattern with the corresponding characters in the text one by one from the right-hand side of the pattern to the left; if the matching fails, based on the comparison information about current iteration, it right shifts the pattern for one to several characters.

Distributed Image Retrieval System [11], in which images are retrieved, and the retrieval is done from massive image data storage is speeded up by utilizing Map Reduce distributed computing model. Moreover, fault tolerance, ability to run in a heterogeneous environment, and scalability are supported in the system.

Map Reduce for Geographical Information Retrieval (MRGIR) is presented and Augmented User-based Collaborative Filtering (AUCF) is implemented within MRGIR. The MRGIR [12] is implemented in Hadoop. Experimental results show that with moderate number of map tasks, the execution time of GIR algorithms (i.e., AUCF) can be reduced remarkably.

In paper [12] Hadoop applications are developed to crunch several terabytes of data, using four to 4,000 computers. The paper describes several frameworks and utilities developed using Hadoop that increase programmer-productivity and application-performance.

System Design

The architecture of the system is described in terms of its components and their functionalities. Figure 1 shows the overall architecture of the video search on education cloud. The overall system architecture shows two major processes namely, video database creation and video retrieval. An educational database based on NPTEL video lectures is created. The video database is indexed and stored into the distributed database.

Video Retrieval involves query pre-processing; searching video, and retrieving video from distributed database using Map Reduce technique and finally producing the matched video output.

Video Indexing:

To speed-up the process of finding relevant video for an input query, indexing techniques are necessarily adopted to organize videos. Bitmap indexing is used to store the column values in bits. Each bit represents a single value. Figure 2 shows the bitmap index creation.

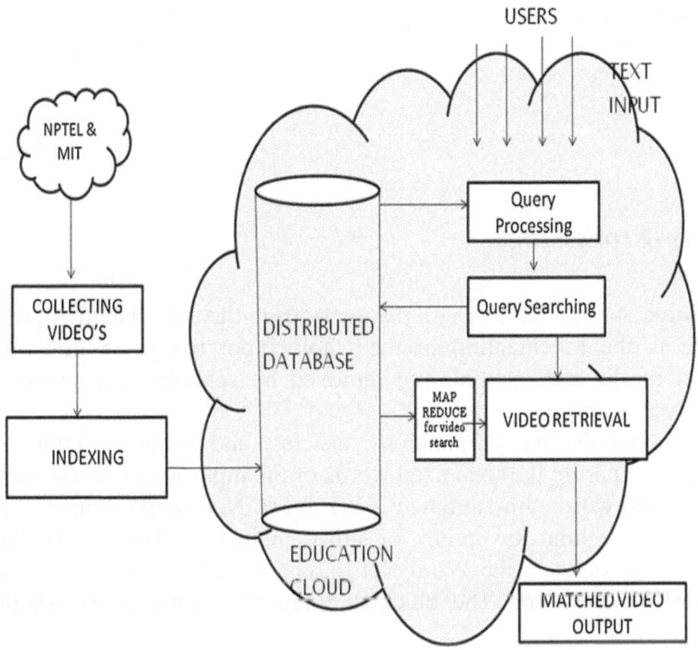

Fig. 1 System architecture of education cloud

Fig. 2 Block diagram for bit map index creation

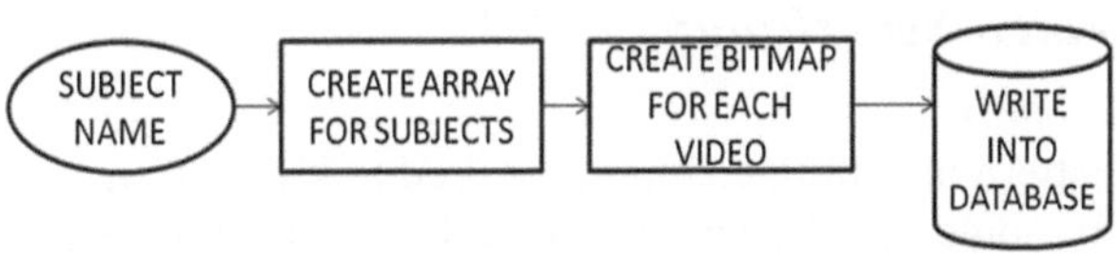

Table 1 Video table

Video id	Subject name	Video name
V1	S1	Video name 1.3 gp
V2	S2	Video name 2.3 gp
V3	S3	Video name 3.3 gp

Table 2 Bitmap index table

Video Id	S1	S2	S3
V1	1	0	0
V2	0	1	0
V3	0	0	1

In the bit map index creation, based on the number of subjects, an array is created. Then, for each video id it sets the bitmap value true according to the subject name. The created Bit Map values are stored into distributed database in tabular form. Table 1 shows the video table.

The created bit map index is shown in Table 2.

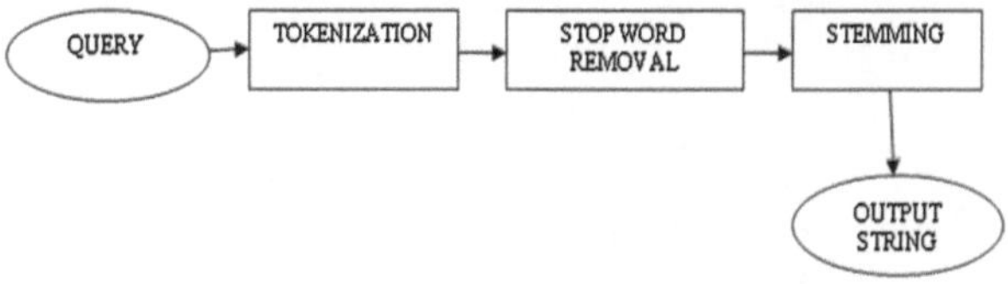

Fig. 3 Block diagram for query pre-processing

Query Pre-Processing:

Pre-Processing starts with getting input query from the user. The first process to be carried out is the tokenization of the whole input query. In this step all the punctuation marks and symbols are removed by scanning each word which is separated by a 'space'.

Next, stop words like prepositions, articles, and some common words are removed by comparing the tokenized words of the input with a standard stop word file. Finally, the words are stemmed using Word Net, which converts them into their root form without any prefix or suffix characters. The words that are not matching or not found in the Word Net like names of persons, places, and numbers, are left as they are. The block diagram shows the query pre-processing (Fig. 3).

Video Searching:

Pattern matching technique is used to search a string containing text of binary data for some set of characters based on a specific search pattern. This is achieved by Boyer Moore String Searching Algorithm. The formal definition of boyer moore string matching algorithm is: given a N length text string T[1..N] and an M length pattern string P[i + 1..i + M], if there exist an i (1 < i < N) and T[i + 1..i + M], pattern P appearances in text T at the position i, then the matching is successful. String matching is used to determine pattern P whether it is in text T.

The basic idea for this algorithm is to compare the characters of the pattern and the characters of the text from left to right; if the matching fails during comparison, then the pattern moves forward (right) one character, and the algorithm starts a new iteration of comparison.

Video Retrieval:

Map Reduce is used for huge data set processing on clusters of computers. Figure 4 shows the processing approach for Map Reduce.

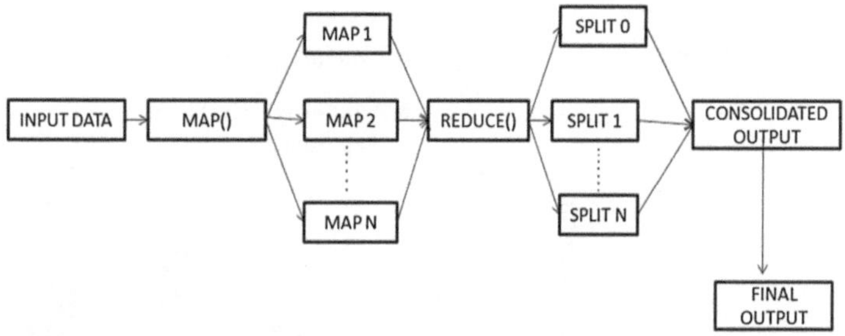

Fig. 4 Block diagram for Map reduce technique

Video retrieval is done using Map Reduce paradigm. It is designed to be capable of processing more number of videos irrespective of the size of each of them. In this paper, Map Reduce is done after getting the subject name from bitmap index table and then, the map function is called. The map function maps the entire video_ id in each task with key/value pair. After mapping, in the reduce function, all the tasks are split into subtasks and each task processes the video files in parallel. The output from all the tasks is combined to produce the final output. Each map job and reduce job of various nodes requires to communicate with the distributed database at all times.

Implementation

Distributed Database Creation

There are three nodes in MySQL Cluster; data node, management node, and sql node. Data node is used for storing the data. Management node is used for configuration and monitoring of cluster. This node is required only for node startup. MySQL node connects to all the data nodes in order to perform data storage and retrieval. This node is optional. It is possible to query data nodes directly via the NDB API. Using bitmap indexing technique, videos are stored into the distributed database for effective video search.

Video Retrieval System

After getting input query from the user, the given query is tokenized. Next, it will get the correct word from the dictionary file. Then, stop words like a, is, etc., are removed from the input. Then, stemming is done using Word Net tool. If there are words that contain 'ing,' they should be removed from the input query. Stemmed words are searched from the description of the subject table in the distributed

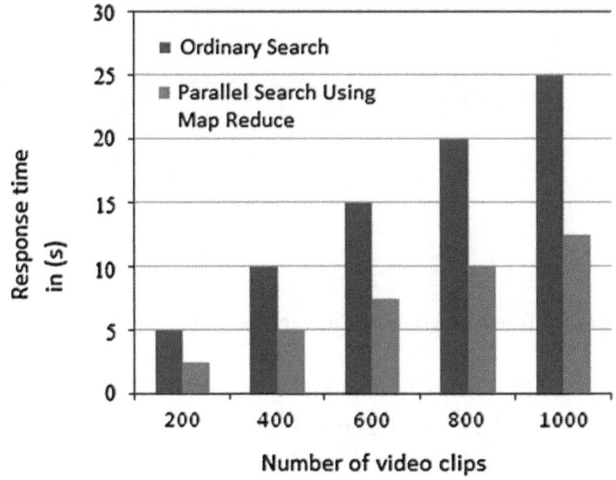

Fig. 5 Ordinary search versus parallel search using Map Reduce

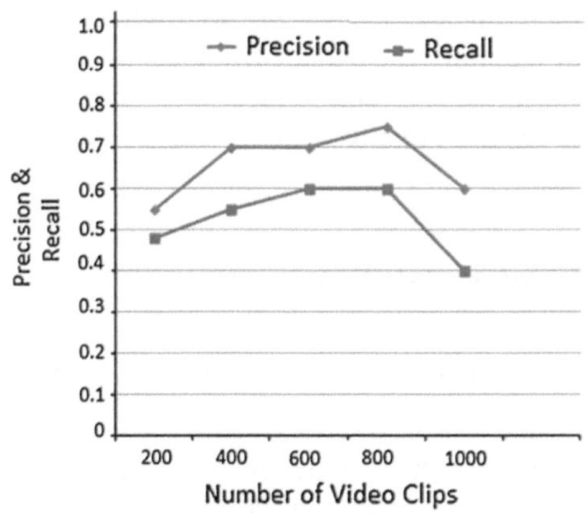

Fig. 6 Precision and recall graph for queries

database using Boyer Moore String Searching Algorithm. Retrieved subject name of the user query from the database is passed to Map Reduce program. In Map Reduce Technique, using key/value pair, each task is processed for retrieving the video in parallel. Then the video is retrieved from the distributed database and given to the user.

Results and Conclusion

Figure 5 shows the implemented video on demand using Map Reduce framework.

The average response time calculated by implementing Map Reduce technique in video search is less compared to ordinary video search. The processing speed is found to be two times faster for retrieving videos. Figure 6 shows the Precision and Recall for Retrieved videos.

$$\text{Recall} = \frac{\text{Number of relevant video clips retrieved}}{\text{Number of relevant video clips in collection}}$$

$$\text{Precision} = \frac{\text{Number of relevant video clips retrieved}}{\text{Total number of video clips retrieved}}$$

The average precision value range index varying from 0.6 to 1.0 proves that the results of the query are generally found relevant. This is because the evaluation occurs only by matching the exact word and not others; this gives the resultant video clips in more or less the same context.

The conclusion for the video search system presented in this paper solves most of the pitfalls of today's video search systems, such as, scalability, reliability, and fault tolerance. The parallel processing of video Retrieval in Map Reduce fashion ensures scalability and robustness. The use of distributed database meets the need for huge space for storage of videos and makes the system fault tolerant.

References

1. Nurmi, D., Wolski, R., Grzegorczyk, C., Obertelli, G., Soman, S., Youseff, L., Zagorodnov, D.: The eucalyptus open source cloud computing system. In: Proceedings of Cluster. Computing, and, the, Grid (2009)
2. Yang, G.: The application of Map Reduce in the cloud computing. In: Proceedings of Intelligence Information Processing and Trusted Computing (IPTC) (2011)
3. White, T.: Hadoop: The Definitive Guide. O'Reilly (2010)
4. Stein, B., Potthast, M.: Applying hash-based indexing in text-based information retrieval. In: Proceedings of DIR (2007)
5. Goyal, N., Zaveri, S.K., Sharma,Y.: Improved bit map indexing strategy for data warehouses. In: Proceedings of Information Technology (2006)
6. Lee, F., Kotani, K., Chen, Q., Ohmi, T.: Fast search algorithm for short video clips from large video database using a novel histogram feature. In: Proceedings of Computational Intelligence for Modeling Control and Automation (2008)
7. Lu, B.F., Cao, H.B., Cao, Z.: An efficient method for video similarity search with video signature. In: Proceedings of Computational and Information Sciences (ICCIS) (2010)
8. Qiang, Z.: An improved multiple patterns matching algorithm for intrusion detection. In: Proceedings of Intelligent Computing and Intelligent Systems (ICIS) (2010)
9. Ho, M.H., Yen, H.C.: A dictionary-based compressed pattern matching algorithm. In: Proceedings of Computer Software and Applications Conference (2002)
10. Xiong, Z.: A composite boyer-moore algorithm for the string matching problem. In: Proceedings of International Conference on Parallel and Distributed Computing, Applications and Technologies (2010)
11. Zhang, J., Liu, X., Luo, J., Lang, B.: DIRS, Distributed, image, Retrieval, system, based, on Map Reduce. In: Proceedings of Pervasive Computing and Applications (ICPCA) (2010)
12. Wu, Z., Mao, B., Cao, J.: MRGIR: Open geographical information retrieval using Map Reduce. In: Proceedings of Geo Informatics (2011)

Multimedia Service Delivery on Next-Generation Platforms: A Survey on the Need for Control, Adaptation and Inter-Mediation

C. Balakrishna

Abstract With the increasing popularity of interactive and innovative multimedia streaming applications on mobile devices, the next-generation communication platforms are expected to be ready for such multimedia service delivery. This paper highlights the technical hurdles that arise due to end-user/device mobility and investigates an optimal multimedia service delivery mechanism to deal with mobility-related issues. The proposed solution called the "Mobility and Quality" Service enabler uses media control and mediation to offer cross-layered, application-independent solution to multimedia service delivery. The paper also presents preliminary experiments and results.

Keywords SIP · IMS · Multimedia streaming · Media control

Introduction

In the past decade, there had been a steady shift in the medium of communication from fixed to mobile communications [1]. Proliferation of mobile devices in the market and the increase in mobile phone subscribers stands proof for this transition [2]. This has led to the constant evolution of mobile access and network technologies. The latest cellular access technologies such as UMTS and HSDPA provide connection bandwidth of 300 Kbps to 10 Mbps and the emerging LTE [3] technology is expected to offer high speed broadband connectivity of up to 100 Mbps. Similarly, the communication content has also undergone a transition from being purely text based to multimedia-rich content. Particularly, streaming media over IP networks has become increasingly popular. This has triggered the

C. Balakrishna (✉)
Faculty of Advanced Technology, University of Glamorgan,
Pontypridd, CF37 1DL, UK
e-mail: balakris@edgehill.ac.uk

S. Sathiakumar et al. (eds.), *Proceedings of International Conference on Internet Computing and Information Communications*, Advances in Intelligent Systems and Computing 216,
DOI: 10.1007/978-81-322-1299-7_33, © Springer India 2014

rapid evolution of video compression techniques in the form of series of codecs such as MPEG-1, MPEG-2, MPEG-4, and H.264 [4]. The more recent codec, such as MPEG-4, can transmit DVD/VCR quality video.

With the current trends, it is evident that majority of the content on the next-generation communication network is going to be multimedia and it is also fairly evident that major hurdle for multimedia service delivery is the heterogeneous communication environment and mobility. This article aims to investigate the optimal multimedia delivery mechanisms on the next-generation communication platforms. The paper highlights the need for media control and multimedia adaptation via inter-mediation. The article begins with the discussion on how mobility and the need for service quality affect the multimedia service delivery. Section "MNQ Architecture" presents the design requirements for the optimal multimedia service delivery mechanism. Section "Experimentation and Results" discusses the proposed service-enabler based solution to the problem. And the paper concludes by presenting the experimentation and results.

Critical Issues Associated with Mobility and Multimedia Service Delivery

The essential problem associated with mobility is that the onus of preserving the user's access to the same service quality despite changes to the access network or the access device lies on the service provider. Existing IP-level mobility solutions [5, 6] are insufficient to address this issue since they are not aware of the application semantics. It is impractical to make each application to be mobility aware as it will greatly increase the development overhead. On the other hand, we cannot discount the huge delays the application-layer solutions could introduce to the time-critical real-time multimedia applications. Furthermore, the mobility we are referring to could be user-initiated or network-initiated. Hence, the optimal multimedia service delivery mechanism should have the following attributes:

- It should be cross-layered, one which re-uses the robustness of the IP-layer mobility management frame-works as well as takes into consideration, the application-layer semantics.
- The solution should be independent of the handover mechanisms, which could either be user-initiated or network-initiated. And, the handover decision, handover trigger and management are best left to protocols at the IP-layer that shall ensure that it is a make-before-break handover [7, 8].
- The solution should work on as many access networks as possible. It is likely that different access networks may have different detection and handover policies that are better suited for those networks. The solution architecture should consider both terminal-centric mechanism as well as the network-initiated mechanism individually and its functional operation should be designed to vary accordingly.

- The solution should be designed to incorporate application-neutral mobility functions that eliminate the need to make each application mobility-aware.

Issues Associated with Service Quality Adaptation

Although each service comes with a unique set of requirements for network performance, end users expect services to work seamlessly across a wide range of user devices and access networks. As opposed to the currently available best-effort service on the Internet, next-generation communication platforms aim to offer quality of service (QoS) mechanisms enabling predictable and enhanced service quality to the end-users. For the next-generation mobile operators, the goal of QoS negotiation is to determine the best service configuration and network resources allocation that would maximise user-perceived service quality. Reaching this goal involves end-to-end (E2E) application-level QoS negotiation and signalling via core network entities. With respect to the IP Multimedia Subsystem [IMS] [12], the procedures for negotiating multimedia session characteristics are specified by the 3GPP and are based on the IETF's Session Initiation Protocol (SIP) [9], Session Description Protocol (SDP) [10] and their extensions. As the demands for multimedia-rich applications that are customised to end-user preferences and capabilities increases, the networks will face complex and dynamically changing QoS requirements. Although existing 3GPP specifications describe procedures for QoS negotiation and signalling for multimedia applications, the support for more advanced services, involving interactive applications with diverse and interdependent media components and QoS based on adaptive user-perceptible service quality is not addressed yet and presents an open area for research. This creates a need for more enhanced mechanisms (beyond those currently specified) to meet the demands of future advanced IP multimedia services to be supported by IMS and other next-generation communication platforms.

Hence, the proposed solution should have the following attributes:

- It should be based on a "Quality Adaptation model" that supports the signalling, negotiation and adaptation of service quality for media-rich services, addressing issues arising due to the heterogeneous environment of NGNs.
- The framework proposed should take into account the terminal and access network constraints as well as user preferences expressed in terms of application characteristics. And, the solution should map user-preferences/application requirements to transport-level QoS parameters.
- Lastly, the solution should have "Minimal Deployment Overhead" with no introduction of new protocol headers or procedures within the IMS core, nor the requirement for new capability support on the client.

It can be inferred from the above arguments, that both service continuity and quality adaptation are functionalities that need to be implemented independent of the applications. Hence, we propose a service-enabler based solution. Service enablers are reusable components and application-independent building blocks that offer generic functionality to support a diverse range of multimedia applications. It should act as a point of convergence and consolidation in a heterogeneous environment, while offering network agnostic and device-agnostic APIs to application developers. These service enablers should be designed to be deployed once and be re-used across multiple applications [11].

MNQ Architecture

The proposed solution is specifically designed to work on IP multimedia subsystem as standardised by the 3GPP. However, the principle on which the solution is built is generic enough to function on any next-generation communication platform.

Figure 1 shows a more detailed architecture of the MnQ Service enabler that spreads across the service, session control and transport layers, towards an integrated solution for multimedia streaming applications. Various building blocks and entities within the separate layers described below fulfil specific tasks and interact with each other in the overall architecture.

Control Layer entities Session Control is the most important task of the control Layer. It consists of the Session Control Function which is responsible to control and manage all incoming multimedia session requests from different users and forward their request to the specific MnQ service-enabler-based on the subscription or service profile. The MnQ architecture reuses the 3GPP standardised components for this purpose. S- CSCF forms the heart of the Session Control Function. Session Control Interface (SCI) interacts with the service-enabler layer via the 3GPP specified ISC (IMS Service Control) interface. This interface runs normal SIP protocol as defined by RFC 3261 [38] with additional enhancement to signify "origination" or "termination" call leg towards the application server. While the Session Control Filter (SCF) utilises the initial filter criteria (iFC) to filter the incoming requests based on the user profile. iFCs are generic rules that apply to any incoming SIP message. They analyse the direction of the message as to where did it originate from or is it addressed to the user, the method as to if it is an INVITE, a SUBSCRIBE, a MESSAGE or OPTIONS, the existence of headers, and the value of these headers. On registration the S-CSCF receives user profile from HSS and the "Filter Criteria" are stored in the user profile and determine the services that are applicable to the collection of Public User Identities of the profile. S-CSCF assesses the criteria in the order of their priority. Each Filter criteria contains trigger points, which are boolean conditions. Based on the trigger point, the request goes to the corresponding AS.

Profile Manager (PM) maintains the user profile and the service profile. When the S-CSCF receives a initial multimedia SIP request, upon applying the iFCs to

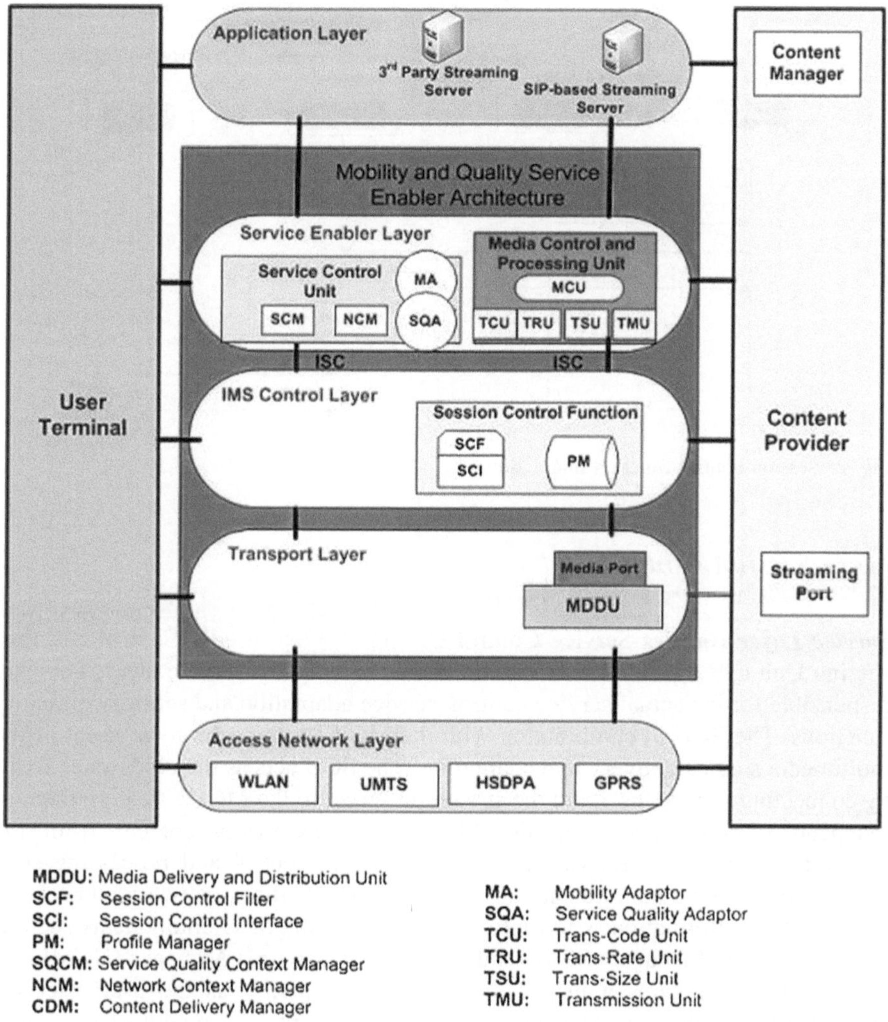

Fig. 1 MnQ architecture

it, the SIP message is forwarded to one or more application servers which hosts the MnQ service enabler. Similarly, when an application server generates a SIP message, the role of the S-CSCF will be to route it to another SIP entity (application server or end-device).

In this interaction, the S-CSCF only acts as a proxy between two SIP entities. The real service control is what happens at the SIP entity that hosts the MnQ service enabler in the service plane of the architecture. Figure 2 shows the interactions within the MnQ components in the control layer.

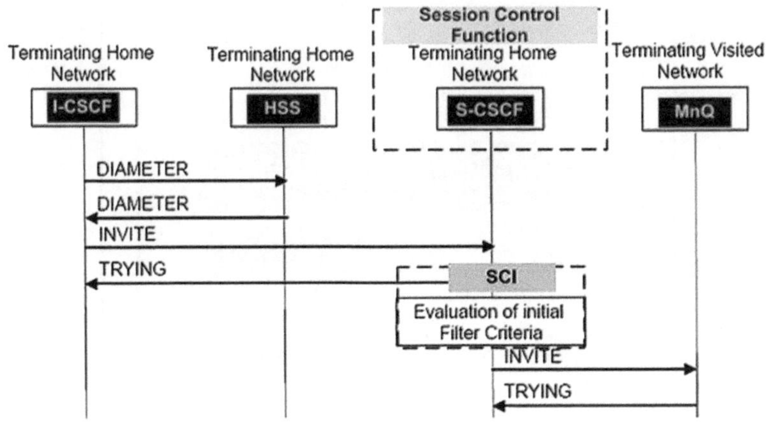

Fig. 2 Session control function interaction

MnQ Session Control

Service Layer Entities Service Control Unit (SCU) and Media Control and Processing Unit (MCPU) are the two main components of the service layer. They are responsible for the actual service control, service adaptation and service continuity functions. The SCU in combination with the MCPU acts as the focal point of the multimedia service delivery and adaptation. The SCU acts as the conference focus by connecting the end-user and the streaming server to the MCPU treating them as independent conference legs, the MCPU receives media content from the streaming server and performs the tasks of media control and processing such media mixing and media translation as per the user's current network and service context. This information is fed into the MCPU by the SCU components such as Service Context Manager (SCM) and the Network Context Manager (NCM).

The SCM maintains a list of parameters that qualify as the user's service context information derived from the Session Description Protocol of the initial INVITE request forwarded from the S-CSCF. SCM not only maintains the list, but is responsible for making Service quality adaptation decisions when the end-user's context changes. These decisions are passed on to the Service Quality Adaptor (SQA) which then initiates the appropriate Service Quality Adaptation procedures.

The NCM is responsible for storing a list of parameters that concern the end-user's network context. The parameters are derived from the 3GPP's SIP header. NCM makes service continuity and service quality adaptation decisions when the end-user's networking context changes such as when the end-user moves from one point of attachment to another or from one access device to another.

Adaptation Threshold The Service Control Manager (SCM) and the Network Control Manager (NCM) store the required quality-related and network-related parameters as explained above. They are responsible for making the adaptation decision based on "Adaptation Threshold triggers". If the adaptation threshold is set too low, the adaptation logic may lower service quality a little too early and if the threshold is set too high, then the adaptation logic may kick off a little too late. As a result, adaptation may bring a negative effect in providing perceived quality of experience to the end-users. Hence, experiments are being carried out to determine the optimal threshold for adaptation. The optimal threshold value is being implemented in the adaptation mechanism of the MnQ service enabler, so that adaptive behaviour kicks off when it is really needed from end-users perspective.

Mobility Adaptor (MA) is a component that is responsible for maintaining the service continuity to the end-user during terminal mobility or session mobility scenarios. It is a Third-Party-Call Control (3PCC) implementation on the SIP application server, which enables the renegotiation of new network parameters with the MCPU and the end-device whose networking context is changing. The MA ensures that media server streams the content to the new network interface still maintaining the stream to the old network interface of the end-device. Upon the completion of handover by the end-device, the duplicate stream to the old interface is disconnected by the 3PCC component in the MA. SQA is responsible for initiating the Service Quality Adaptation of the multimedia stream, when the end-user's service context changes. It is a Back-to-Back User Agent (B2BUA) implementation on the SIP application server that enables renegotiation of service-related parameters between the end-user and the media server. The MCPU performs the actual service quality adaptation.

SIP application server which forms part of the SCU is the brain of the architecture which contains the application logic, although the actual content resides elsewhere, while the media server that forms part of the MCPU provides the muscle by enabling playing and recording of media streams, translating and audio-video synchronisation. The media server is controlled by the application server using sip signalling and a markup language such as VoiceXML (VoXML).

MnQ Media Control

The MCPU performs the function of media control and media processing. It consist of a media server and a media translator. Media server acts as the middle man between the SCU and the streaming source. The SCU controls the media server using SIP and VXML, while the media server fetches the actual streaming content either using the native SIP protocol or HTTP or RTSP based on the nature of the streaming source. Based on the media control signalling information received by the SCU, media server triggers the media translation procedures. The user perceived service quality is enhanced considerably by changing the encoding rate or the sampling rate of the multimedia content. For video content, changing the image frame size when the end-user's receiver capabilities change, adds onto the quality of experience of the users.

The media translator consists of a Trans-code Unit (TCU), Trans-rate Unit(TRU), Trans-size Unit (TSU) and a Transmission Unit (TMU), each of them responsible for the trans-coding, trans-rating, trans-sizing and the transmission of the multimedia content, respectively. While the decision for the above media translation is signalled by the SCU.

We consider three major events which change end-user's context and creates a scope for service adaptation and optimisation as depicted in Fig. 3.

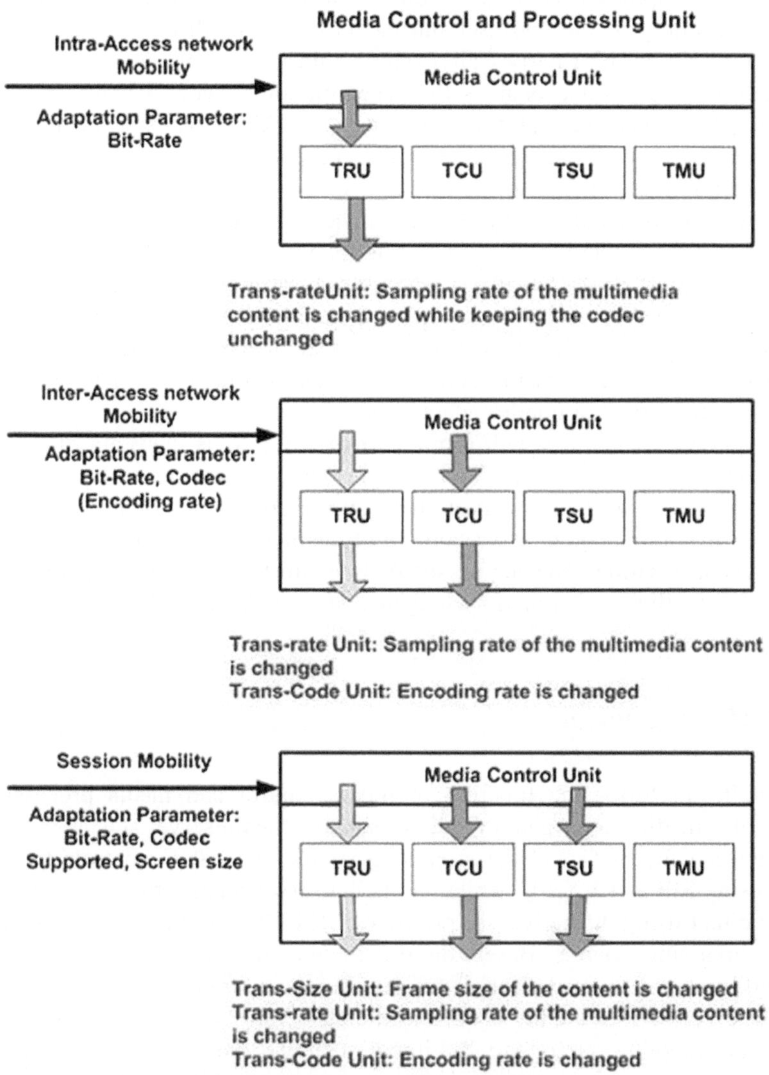

Fig. 3 MnQ media control during three different context

1. Mobility without any changes to radio access, affects the maximum available bit-rate on the end-device. The characteristics of the mobile and wireless radio access technology indicate fluctuating links, rapid changes in bit rate and bit error rates indicate that the end-user's mobile within the same access technology require service quality adaptation to suit the changing bit rates. In this situation, the Media Control Unit (MCU) activates the Trans-rate Unit (TRU) which simply changes the sampling rate of the multimedia content without changing the codec. This ensures the multimedia content adapts to the changing bit-rate without a codec-renegotiation which implies no extra SIP signalling.
2. Mobility across different access technologies creates a need for providing service continuity to the end-device as well as service quality adaptation. Each heterogeneous access technology differs in the maximum bit-rate by drastic margin, hence a mere change in sampling rate of the multimedia content would not suffice, but a change in the encoding rate would adapt the multimedia to the end-user's needs. Service continuity is ensured by the MA and the MCU, while service adaptation is performed by the Trans-code Unit (TCU) and the Trans-rate Unit (TRU).
3. Session mobility occurs when the end-user shifts from one device to another during mid-session. Each access device differs in its display capabilities and may differ in the access technology supported by each of them. This triggers the need to adapt the multimedia content to the end-user's device and network context. The MCU initiates the adaptation process by triggering Trans-size Unit (TSU) which adapts the frame size of the multimedia content to the device's display capability. Trans-rate Unit (TRU) and Trans-code Unit (TCU) are also triggered appropriately.

Experimentation and Results

Figure 4, depicts the experimental scenario that was set up to study the MnQ performance and user-perceptible quality of experience for the SIP-based streaming application. The end-device is a multi-mode terminal (in this instance a laptop with a 3G dongle and Wifi interface). However, SIP session communication takes place on one of the interfaces which will take precedence over the others as per user configuration. In our case, WLAN interface has the highest precedence followed by UMTS and finally GPRS interface. The client using the mobile device is "Kapanga" [13] IP softphone. The scenario researched in this paper is as follows:

1. The IMS user agent on the multi-mode end-device registers with the IMS network.
2. Upon successful registration, it sends a request to the IMS operator's P-CSCF requesting to be connected to the SIP-based interactive multimedia streaming

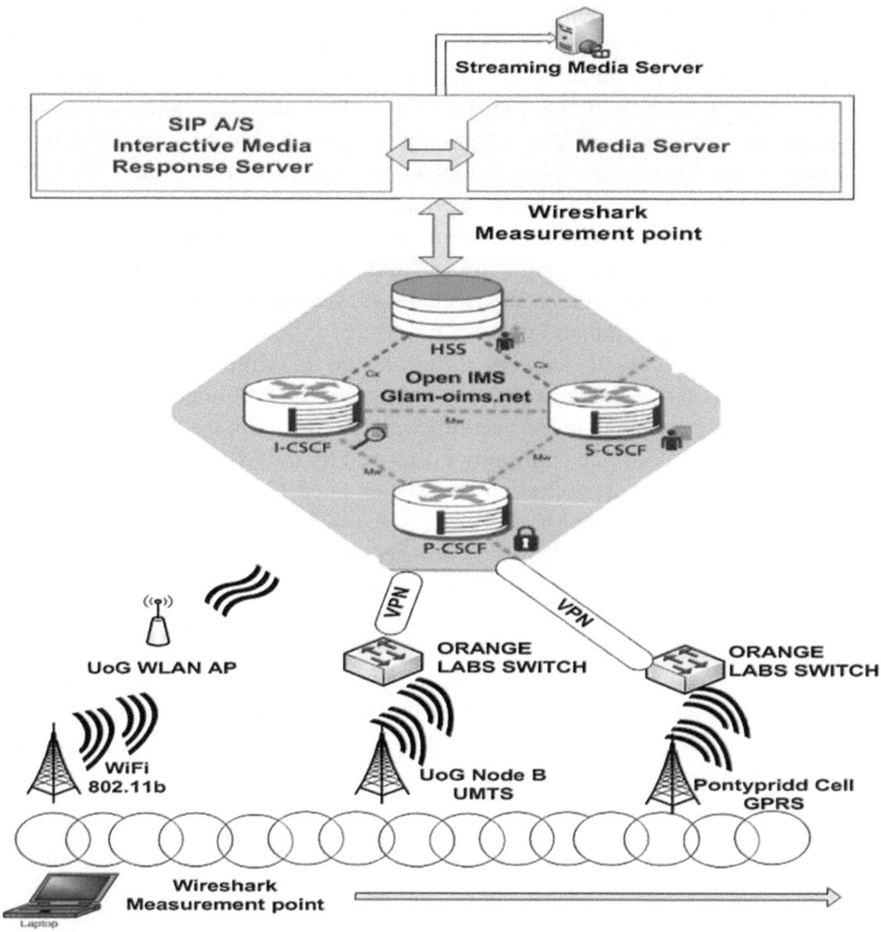

Fig. 4 Soft-handoff with MnQ

application. The chosen application is a server-to-client type served in an interactive response mode.

3. S-CSCF receives the initial invite and checks for the initial Filter Criteria (iFC) and forwards the request to the designated SIP application server.
4. SIP application server performs the initiation of the SIP multimedia session, the IMR (Interactive Media Response) Service establishes the SIP session with the end-device by interacting with the media server that provides media control and media interaction functionality.
5. Media server fetches the multimedia content from the streaming server and streams it to the end-device. In our case, the media server acts as a streaming server, hence the streaming content is locally fetched.

To study MnQ performance, the MnQ service enabler is deployed on the SIP server and the media server. Along with hosting the IMR service, the SIP A/S and media server combination perform the functions of the MnQ service enabler. The end-user is provisioned with multimedia streaming service by the concept of service chaining.

The evaluation was performed predominantly on WLAN access network (IEEE 802.11b) and UMTS network in a semi-open office environment. One must note that the practical data rates observed by the user are noticeably lower compared to the theoretical rates due to various factors degrading the nominal throughput.

Service Continuity Evaluation

Measurements are made from time when the handoff is initiated in the Orange cell from the UMTS interface to time when the handover is completed in the Wifi cell. UMTS access is provided by the Orange mobile operator and the Wifi access is provided by a local wireless access point attached to the University of Glamorgan backbone. When preparing for the soft-handoff, session control messages are exchanged between the SCU of the MnQ Service Enabler and the end-terminal which is indicated by Graphs 1 and 2 in Fig. 5. During the transient period of the hand-off, data flow happens on the UMTS interface as well as on the Wifi interface as indicated by Figs. 3 and 4. Upon the completion of the handoff, data flow to UMTS interface is cut off and the data flow over the Wifi interface continue until the sessions are terminated.

To demonstrate the effect of soft handoff, the end-device moves from UMTS to Wifi 15 sec after the streaming begins. Because of the delay associated with the mobile device/IMS User agent's address configuration, the soft handoff procedure could not be initiated before 36:8572s. The soft handoff initiation point, indicated in Graph 1 is shown in further details in Graph 2. The vertical notches in the plot indicate duplicated RTP packets received at MH, As expected, no packet loss was

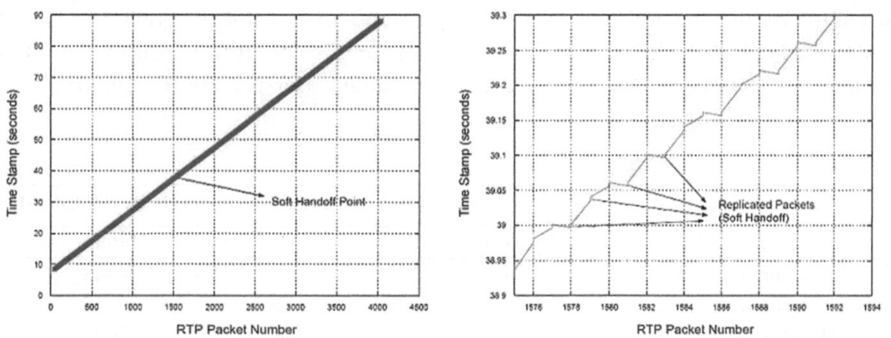

Fig. 5 Soft-handoff -RTP packet stream as measured on mobile device

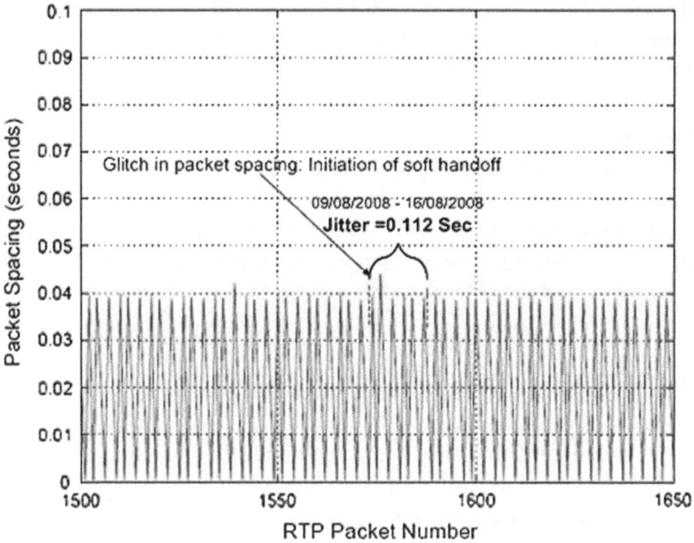

Fig. 6 Soft-handoff -RTP packet spacing/jitter

observed in the RTP stream. Figure 6 shows the spacing in seconds between the consecutive RTP packets. For the purpose of clarity, a portion of the stream (from packet 1,500 to 1,650 only) is shown with a glitch in the spacing indicating the initiation of the soft handoff procedure. The delay jitter is measured as 0.112s and it remains under control all the time and has no long-term effect on the streaming RTP traffic. The low jitter value ensures smooth streaming of RTP packets, thus maintaining the QoS of the multimedia streaming application.

Codec Performance

Various audio codecs were evaluated for user-perceptible quality at varying signal strengths of the WLAN and UMTS access networks. Table 1 shows the list of codecs tested and the bandwidth they consume. The results show the QoS parameter boundaries that are essential to maintain user-perceptible quality.

Table 1 Audio codec performance comparison

Codec	Good-to-medium quality Threshold		Low-to-poor quality Threshold	
	Loss (%)	Jitter (ms)	Loss (%)	Jitter (ms)
G.711 (64 kbps)	<4	<30	4–14	30–45
G.726 (32 kbps)	<4	<35	4–N.A	35–55
GSM (13 kbps)	<2.5	<32	2.5–15	32–50
G.729	<4	<38	4–N.A	38–N.A

Conclusion

This paper presents a novel service-enabler based solution for optimal multimedia service delivery. The proposed solution is specially designed for 3GPP standardised IMS platforms; however, the design principles are applicable to any generic next-generation platform. The MnQ service enabler adapts the service quality of multimedia streaming content based on the user's context that includes mobility, network and service parameters. The adaptation is done at the application-level through mediation by the service enabler. Preliminary results indicate that the MnQ service enabler offers better service quality to multimedia streaming content.

References

1. Keshav, S.: Why cell phones will dominate the future internet? SIGCOMM Comput. Commun. Rev. **35**(2), 83–86 (2005). ISSN 0146-4833
2. Mobile, GSM, global, handset, base station and regional cellular statistics. http://www.cellular.co.za/stats/stats-main.htm. Accessed 08 Aug 2008
3. GPP: evolved universal terrestrial radio access (E-UTRA); Long term evolution (LTE) physical layer; General description. http://www.3gpp.org/ftp/Specs/html-info/36201.htm (2007). Accessed Dec 2007
4. Iain, E., Richardson, G.: H.264 and MPEG-4 video compression: video coding for next-generation multimedia. Wiley, New York (2003)
5. Petrescu, A., Devarapalli, V., Wakikawa, R., Thubert, P.: Network Mobility (NEMO) basic support protocol. RFC 3963, internet engineering task force. http://www.rfc-editor.org/rfc/rfc3963.txt (2005). Accessed Jan 2005
6. Kong, K.-S., et al.: Mobility management for all-IP mobile networks: mobile IPv6 vs. proxy mobile IPv6. IEEE Wireless Commun. **15**(2), 36–45 (2008)
7. Polidoro, A., Salsano, S., Veltri, L., Ordine, A.: Architecture and testbed implementation of vertical handovers based on SIP session border controllers. Wireless Pers. Commun. **43**(3):1019–1034 (2007). ISSN 0929-6212
8. Kung, H., Chung W.: SIP-based user-oriented handoff control for video phone. Kumamoto, Japan (2008)
9. Camarillo, G., Johnston, A., Peterson, J., Sparks, R., Handley, M., Rosenberg, J., Schulzrinne, H., Schooler, E.: SIP: session initiation protocol. RFC 3261, internet engineering task force. http://www.rfc-editor.org/rfc/rfc3261.txt (2002). Accessed June 2002
10. Jacobson, V., Handley, M., Perkins, C.: SDP: Session description protocol. RFC 4566, internet engineering task force. http://www.rfc-editor.org/rfc/rfc4566.txt (2006). Accessed July 2006
11. Henrikson, E., Garcia-Martin, M., Mills, D.: Private header (P-Header) extensions to the session initiation protocol (SIP) for the 3rd- Generation Partnership Project (3GPP). RFC 3455, internet engineering task force. http://www.rfc-editor.org/rfc/rfc3455.txt (2003). Accessed January 2003
12. Al-Begain, K., Balakrishna, C., Galindo, L.A., Morro, D.: IMS: a development and deployment perspective, Wiley, New York (2009)
13. Kapanga IP soft-phone. http://www.kapanga.net,asseenon08.08.2008 (2008)

Generating Test Data for Path Coverage Based Testing Using Genetic Algorithms

Madhumita Panda and Durga Prasad Mohapatra

Abstract In this paper, we have developed an approach to generate test data for path coverage based testing using genetic algorithm. We have used control flow graph and cyclomatic complexity of the example program to find out the number of feasible paths present in the program and compared it with the actual number of paths covered by genetic algorithm. We have used genetic algorithm for generating test data automatically. We have shown that our algorithm is giving cent percent coverage, successfully covering all feasible paths. In our approach, we have observed that genetic algorithm is much more effective in generating test data within less time period, giving better coverage.

Keywords Control flow graph · Cyclomatic complexity · Connection matrix · Genetic algorithm · Test data · Fitness function · Path coverage

Introduction

Software testing is a necessary evil in the process of software development. It adds nothing to the functionalities of software but spends more than 50 % of labor and money. It is not possible to guarantee a product as error free without testing it thoroughly.

M. Panda (✉)
Department of MCA, Seemanta Engineering College, Mayurbhanj,
Odisha 757086, India
e-mail: madhumita.panda3@gmail.com

D. P. Mohapatra
Department Of Computer Science and Engineering, National Institute Of Technology,
Rourkela, Odisha 769008, India
e-mail: durga@nitrkl.ac.in

S. Sathiakumar et al. (eds.), *Proceedings of International Conference on Internet Computing and Information Communications*, Advances in Intelligent Systems and Computing 216,
DOI: 10.1007/978-81-322-1299-7_34,

In testing phase, the most important phase is the careful design of test data. Test data is supplied as input for getting expected output in black box testing. Test data is also needed for covering all paths in white box testing. Many testing strategies are available for white box testing like statement coverage, branch coverage, condition coverage, decision coverage, and path coverage. In white box testing, the path based or path coverage approach is considered the most crucial one. Here there are possibilities of discovering up to 65 % errors of the program source code. Earlier, many strategies have already been followed to generate test data automatically [1, 2] but till yet no approach has been able to fulfill the tester's need. White box testing is still performed manually. Evolutionary algorithms are capable of searching the entire solution space within specified time period. Therefore, the modern researchers are trying to implement evolutionary algorithms in the process of automatic test data generation (TDG). Many approaches have already been proposed by different researchers to apply genetic algorithm for generating test data for path testing.

It is normally too difficult to trace any program and find out the feasible paths existing in it. First, we have to convert application under test to any intermediate graphical forms like control flow graph, control dependency graph, or data dependency graph, so that we can trace the feasible paths either on the basis of data flow or on control flow. Genetic algorithms [3, 4] fall under dynamic TDG category and this technique is also termed as intelligent TDG technique.

We have proposed a technique for automatic TDG using genetic algorithm which is guided by the control flow graph, cyclomatic complexity of the program. The advantage of our approach is that it covers all feasible paths at least once.

The rest of the paper is organized as follows: section "Basic Concepts" includes Basic concepts, section "Genetic Algorithms" includes basic concepts of Genetic algorithms, section "Related Work" includes related work performed in the same field, section "Implementations" includes our implementations, section "Experiments and Results" includes our experiments and results. Finally, we conclude in the last section "Conclusion" with a discussion of future work.

Basic Concepts

In this section, we discussed the basic concepts and terminologies needed for understanding our approach.

Control Flow Graph

The control flow graph is a flow graph meant for the graphical representation of control structure of any program. Control flow graphs are also known as directed graphs. A directed graph (V, E) consists of a set of vertices V and a set of edges E that are ordered pairs of elements of V. A control flow graph consists of a start node, an end node, connecting edges, decision nodes, junction nodes, and bounded regions [4].

Path Testing Terminology

The terminologies included in path testing are path, length of the path, and independent path. *Path* A path through a program is a sequence of instructions or statements. In the graph it is represented as a sequence of nodes and edges beginning at the start node and terminating at the end node.

Length of path. The length of a path is measured by the number of links present. It does not include the number of instructions or statements executed along the path.

Independent Path. An independent path is any path through the program that introduces at least one new set of statements or includes any new condition or a new edge which does not exist in the previous path. Here in our example, we have considered control flow graph as from this graph we can clearly trace the minimum number of feasible paths present by considering the cyclomatic complexity of the graph.

Cyclomatic Complexity

McCabe introduced the concept of measuring the logical complexity of a program by considering its control flow graph [5]. He stated that the complexity of a program can be found out by considering the number of paths in the control flow graph of the program. He has taken into consideration the independent paths only. He proposed the following equation for computing the cyclomatic complexity.

$$V(G) = e - (n + 2) \tag{1}$$

where,

V(G) = Number of independent paths in a control flow graph
e = Number of edges present in the graph
n = Number of nodes of the graph

Graph Matrix

A graph matrix is a square matrix whose rows and columns are equal to the number of nodes in the control flow graph. Each row and column identifies a particular node. The entries in the matrix represent a connection between the nodes.

Connection Matrix

A matrix defined with link weights is known as connection matrix. This matrix is like graph matrix, the only difference is that in this matrix the link weights are added to each cell entry. When connection exists the link weight is one (1) otherwise zero (0).

Genetic Algorithm

Genetic algorithms (GAs) are mainly metaheuristic search algorithms inspired from nature [3]. These algorithms accept a possible set of solutions known as chromosomes and apply three operators' selection, crossover, and mutation for selecting best solutions out of the possible set of solutions [3, 4].

Chromosome representation and correct fitness evaluation are the keys of the success in GA applications. The beauty of GAs comes from their simplicity and elegance as robust search algorithms as well as from their power to discover good solutions rapidly for difficult high-dimensional problems covering large solution spaces.

Steps of Genetic Algorithm

The process involved in GA optimization problems is based on that of natural evolution and broadly works as follows:

(1) Randomly generates an initial population of potential solutions.
(2) Evaluate the suitability of each solution with the help of fitness function.
(3) Check whether the results obtained are satisfying stopping criterion.
(4) Select two solutions biased in favor of fitness using selection.
(5) Apply crossover to the solutions at a random point on the selected strings to produce new solutions.
(6) Mutate the new solutions based on a mutation probability.
(7) Goto step 2 (Fig. 1).

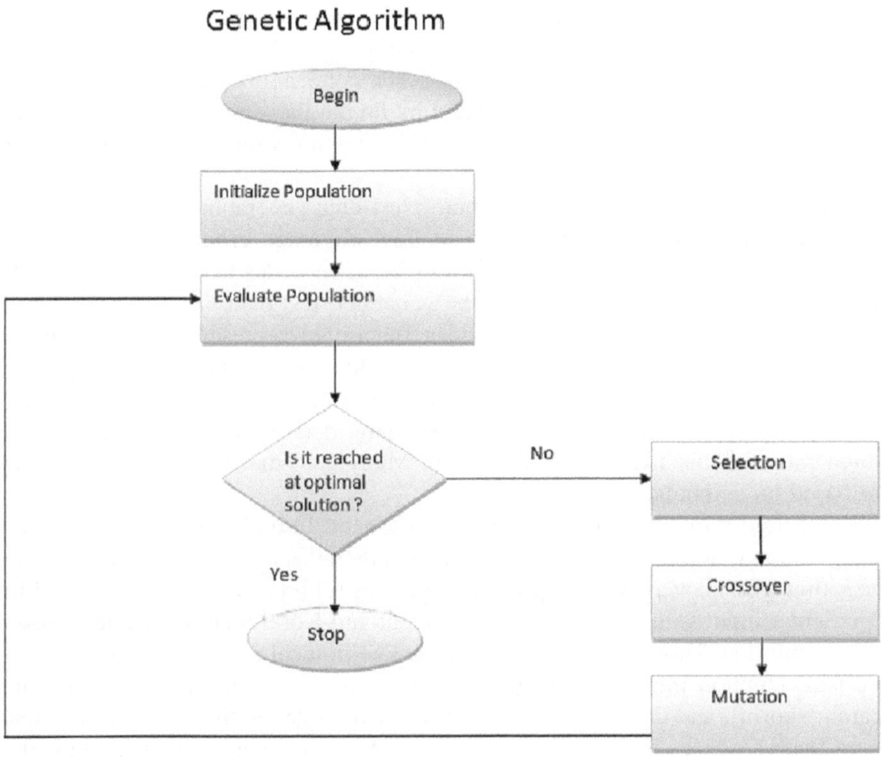

Fig. 1 Block diagram of genetic algorithm

Related Work

Harman et al. [1] suggested that the techniques to automate TDG must include a variety of functional and nonfunctional test adequacy criteria and must either implicitly or explicitly solve problems involving state propagation and constraint satisfaction. They have shown how optimization techniques associated with search-based software engineering (SBSE) have been used to automate the process of TDG.

Michael et al. [6] have extended the work of dynamic TDG by using function minimization method. They have used GAs to minimize the fitness function. They have also examined the effect of program complexity on TDG process.

Diaz et al. [7] proposed an efficient testing technique that combines Tabu search with Korel chaining approach. Their technique automatically generates test data in order to obtain branch coverage in software testing. They have stated that the metaheuristic tabu search technique is very effective in generating test data.

Hermadi et al. [8] presented a GA-based approach that tries to generate a test data that is expected to cover a given set of target paths.

Srivastava et al. [6] proposed a technique for generating test cases using 'path coverage testing criteria' and GA. This technique was based on the criticality of the path. In this technique, the edges of the control flow graph are assigned weights. The fitness function is calculated by taking the sum of the weights of all the edges of a particular path. The path with maximum value of fitness function is the most critical.

Ghiduk et al. [9] presented an automatic test-data generation technique. It uses GA to generate test data satisfying data flow coverage criteria. The technique applies the concepts of dominance relations between nodes to define a new multi-objective fitness function to evaluate the generated test data.

Symbolic execution [10, 11] provides functional representation to program paths and assigns symbolic names for input values. It evaluates a path by interpreting the statements and predicates on the path in terms of these symbolic names [11]. Symbolic execution requires the systematic derivation of these expressions which can take much computational effort. Many auto-mated test data generators are based on symbolic execution [12].

Korel used a modified form of gradient descent. Initially, the input values are changed slowly for getting the proper direction then the values are changed rapidly for getting good coverage. This process continues till no change is encountered for a particular input value. Both Gradient descent and Korel method can get stucked in local minima. This problem of getting stucked in local minima has given rise to heuristic function minimization methods like simulated annealing [Kirup], tabu search [skorin], and GA [13]. Korel [14] focused on generating software test data using 'path coverage testing criteria'. It was a dynamic path testing technique that generates test cases by executing the program with different possible test case values. Ghiduk et al. [9] proposed an approach to generate test data using 'du (definition use) paths coverage testing criteria'. In their work, they focused upon generating the dominance tree from the control flow graph of the program.

Implementation

In our approach, we have selected control flow graph as the program intermediate form. We are generating test data for path coverage based testing as path testing provides the best code coverage leading to through testing. Here the basic path set provides us the number of test cases to be covered thus ensuring the number of test cases expected for full coverage. We have used GA for generating test data. At the same time, we have also considered cyclomatic complexity along with connection matrix for analyzing the basic paths and comparing our results. We find that the GA is generating test data with 100 % coverage of all paths.

Genetic Algorithm for White Box Testing

For performing path coverage testing using GAs we have used the following steps,

(1) Write program in any programming language.
(2) Instrument the program lines of code.
(3) Generate control flow graph for this program.
(4) Prepare connection matrix of the control flow graph.
(5) Find out all path sequences present in the graph.
(6) Find out cyclomatic complexity of the graph.
(7) Apply GA to generate test data.
(8) Compare traced paths with cyclomatic number.
(9) Stop execution.

Working of the Algorithm

We have considered a program greatest number among three as our example program for generating test data using GAs, shown in Fig. 2. In our implementation, first we instrumented our program and prepared the control flow graph of example program, which is shown in Fig. 3.

Fig. 2 Example program

```
#include<stdio.h>
#include<math.h>
void main(){
int a,b,c;
printf("Enter the value of a,b and c\n");
scanf("%d %d %d", &a, &b, &c);
if(a>b)
{
If(a>c)
{
printf(" A is greatest among three\n");
}
else
{
printf("C is the greatest among three\n");
}
}
else if(b>c)
{
printf("B is greatest among three\n"); a
}
else
{
printf("C is greatest among three\n");
}
```

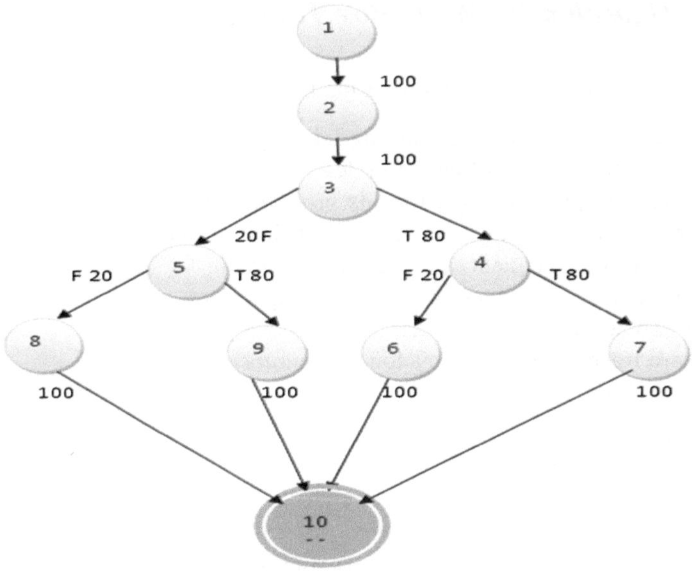

Fig. 3 Control flow graph of example

To trace the traversed paths we have used a weight-based approach i.e., we have assigned weights to each edge traversed. We have assigned 100 to each sequential statement covered. We have assigned 20 to each false edge encountered and 80 to each true edge, as shown in Fig. 3. Then we prepared the connection matrix and find out the number of paths present in our program, as shown in Table 1. We found four paths in our example program, greatest number among three. The node sequences for each path are

(1) Path 1: 1-2-3-4-7-10
(2) Path 2: 1-2-3-4-6-10
(3) Path 3: 1-2-3-5-9-10
(4) Path 4: 1-2-3-5-8-10

Table 1 Connection matrix

Node numbers	1	2	3	4	5	6	7	8	9	10
1	0	1	0	0	0	0	0	0	0	0
2	0	0	1	0	0	0	0	0	0	0
3	0	0	0	1	1	0	0	0	0	0
4	0	0	0	0	0	1	1	0	0	0
5	0	0	0	0	0	0	0	1	1	0
6	0	0	0	0	0	0	0	0	0	1
7	0	0	0	0	0	0	0	0	0	1
8	0	0	0	0	0	0	0	0	0	1
9	0	0	0	0	0	0	0	0	0	1
10	0	0	0	0	0	0	0	0	0	0

We have taken here Roulette wheel selection, the commonly used reproduction operator. We selected a string for the mating pool on the basis of probability proportional to its fitness function.

Fitness Function

$$f(x) = \sum_{i=1}^{n} (p_i) \tag{2}$$

where, $p_{(i)}$ = fittest path.

$$p_i = \sum_{i=1}^{v} (w_i) \tag{3}$$

where,

$W_{(i)}$ = weight assigned to the fittest path.

We have used (crossover) one-point (or single) crossover, we selected two input data as potential parents by selection process. Then we allowed these to exchange substring information at a random position in the data to produce two new values. Crossover happened according to a crossover probability p_c, which is an adjustable parameter. For each parent selected, a random real number r is generated in the range [0, 1], if $r < p_c$ the parent was selected for crossover.

After that, the selected data are flipped randomly. Each pair of parents generates two new paths, called offspring. The crossover technique used is one point crossover done at the midpoint of the input bit string. In this technique, right half of the bits of one parent are swapped with the corresponding right half of the other parent.

Here we have performed mutation on a bit-by-bit basis. Every bit of every chromosome in the offspring has an equal chance to mutate (change from '0' to '1' or from '1' to '0'), and the mutation occurred according to a mutation probability p_m, which is also an adjustable parameter.

Experiments and Results

We have used matlab for simulating our program. As we have used GA for generating test data targeting white box path testing, so we renamed it as (simple GA for white box path testing) SGAWP.

Parameter Settings for SGAWP

We have set our parameters as follows for our experiment.

(1) Population size: Initially 500 then 1000.
(2) No. of generations: First 10 then 100.
(3) Chromosome Length 45 bits.
(4) Cross-over probability 0.7.
(5) Random probability 0.5.
(6) Mutation probability 0.01.

Results

The given below tables and graphs show the results obtained. Table 2 shows the number of test data generated for four paths in ten generations considering population size as 500. Table 3 shows the number of test data generated for four paths. Figures 4 and 5 show the generation of test data using genetic algorithm.

Table 2 Generated test data in ten generations each with population size 500

Number of generations	Path 1	Path 2	Path 3	Path 4
1	138	84	165	113
2	199	85	137	79
3	230	72	121	77
4	256	81	106	57
5	302	71	86	41
6	321	69	76	34
7	335	67	73	25
8	346	68	65	21
9	365	51	60	24
10	368	56	58	18

Table 3 Generated test data in ten generations each with population size 1,000

Number of generations	Path 1	Path 2	Path 3	Path 4
1	264	173	327	236
2	373	163	273	191
3	478	143	229	150
4	560	121	199	120
5	621	136	162	81
6	638	131	155	76
7	672	122	141	65
8	720	108	118	54
9	714	114	132	40
10	716	111	129	44

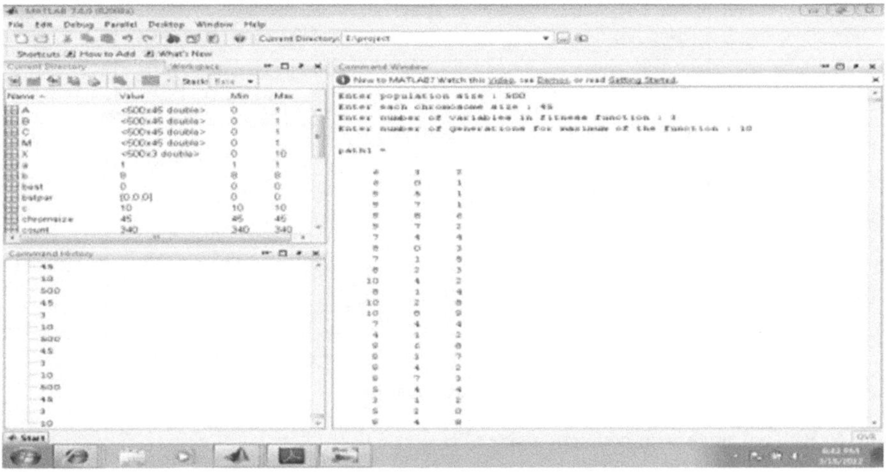

Fig. 4 Test data generation

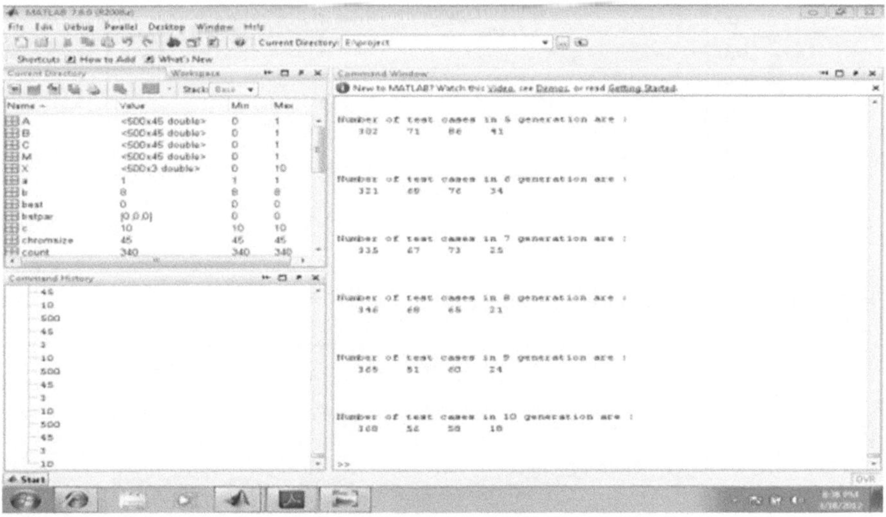

Fig. 5 Generated test data for individual path coverage

Figure 6 shows the results obtained for path coverage in ten generations when population size was 500. Figure 7 shows the results obtained for path coverage in ten generations when population size was 1,000. Figure 8 shows the results obtained for path coverage in 100 generations when population size was 1,000.

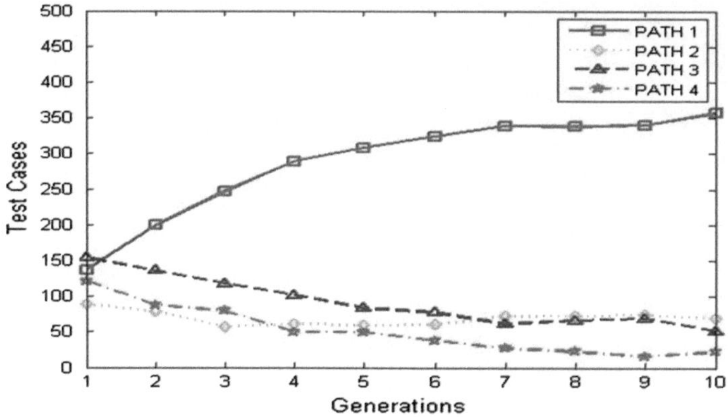

Fig. 6 Path coverage with population size 500

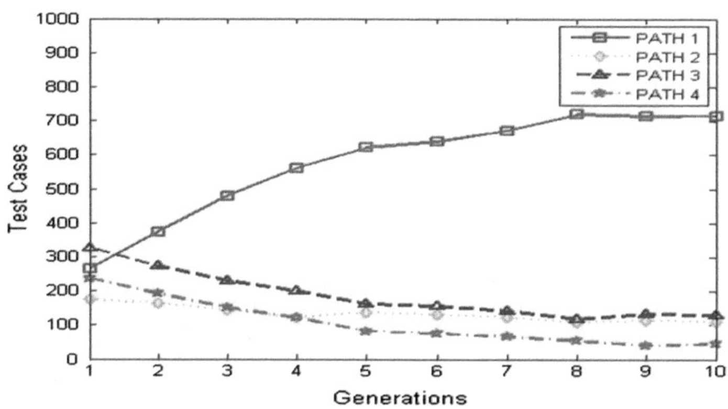

Fig. 7 Path coverage with population size 1000

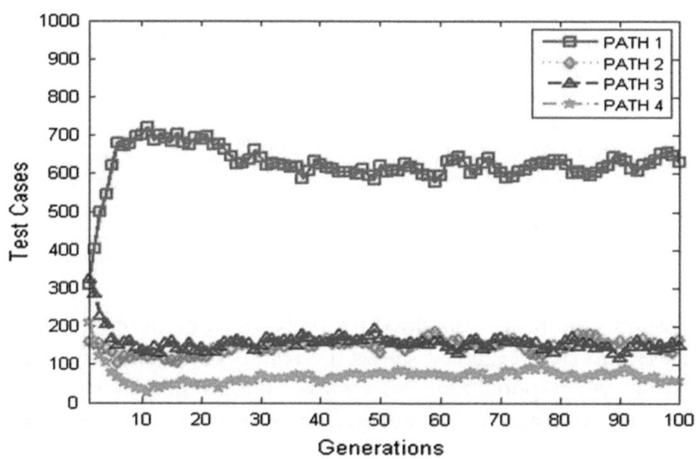

Fig. 8 Path coverage in 100 generation and population size 1000

Conclusions

Here in this paper, we have used the GA to find a better solution for path coverage based testing. Here we are generating test data automatically and we are able to cover all paths. Thus, we can use GA and any such algorithm for generating test data automatically within less time showing better coverage. In our future approach, we are planning to implement the above concept of TDG using other nature inspired algorithms like PSO and ACO.

References

1. Harman, M., McMinn, P.: A theoretical and empirical study of search-based testing, local, global, and hybrid search. IEEE Trans. Softw. Engg. **36**(2) (2010)
2. Pargas, Harrold, Peck: Test -data generation using genetic algorithms. J. Softw. Test. Verif. Realiab. Wiley (1999)
3. Goldberg: Genetic algorithms in search, optimization and machine learning. Addison-Wesley, Massachusetts (1989)
4. Mathew: Genetic algorithm. IIT, Bombay. vmtom@civil.iitb.ac.in
5. Mall, R.: Fundamentals of software engineering 3rd Edn. PHI Publications, New Delhi
6. Michael, C.C., Graw, G.M., Schatz, M.A.: Generating software test data by evolution. IEEE Trans. Softw. Engg. **27**(12) (2001)
7. Diaz, Tuya, Blanco: Automated software testing using a metaheuristic technique based on tabu search, pp. 310–313 (2003). doi:10.1109/ASE.2003.1240327
8. Hermadi, Ahmed, King: Genetic algorithm based test data generator **1**. doi:10.1109/CEC.2003.1299560
9. Ghiduk, Harrold, Girgis: Using genetic algorithms to aid test-data generation for data-flow coverage (2007). doi: 10.1109/ASPEC.2007.73
10. Ramamoorthy, Ho, Chen: On the automated generation of program test data. IEEE Trans. Softw. Engg. **SE-2**(4) (1976)
11. Harman, King: Automated test data generation using search based software engineering. ISBN:978-0-7695-2971-2, doi:10.1109/AST.2007
12. Howden: Reliability of the path analysis testing strategy. IEEE Trans. Softw. Engg. **SE-2**(5) (1976)
13. Holland: Adaptation in natural and artificial systems, 2nd edn. MIT Press, MIT, Cambridge
14. Korel: Automated software test data generation. IEEE Trans. Softw. Engg. **16** (1990)

Media Streaming Using Multiple Description Coding in Overlay Networks

Sachin Yadav, Shailendra Mishra and Ranjeeta Yadav

Abstract In this paper we examine the performance of two types of Overlay networks i.e. Peer-to-Peer (P2P) & Content Delivery Network (CDN) media streaming using Multiple Description Coding (MDC). In both the approaches many servers simultaneously serve one requesting client with complementary descriptions. This approach improves reliability and decreases the data rate a server has to provide. We have implemented both approaches in the ns-2 network simulator. The experimental results indicate that the performance of Multiple Description Coding-based media streaming in case of P2P network is better than CDN.

Keywords CDN · MDC · Video streaming · Overlay network · P2P

Introduction

Media streaming received lot of attention in the past few years. As a consequence, live and on-demand media streaming is today widely used to stream TV & radio channels, TV shows, or arbitrary audio & video media. During this time several

Please note that the LNCS Editorial assumes that all authors have used the western naming convention, with given names preceding surnames. This determines the structure of the names in the running heads and the author index.

S. Yadav (✉) · S. Mishra · R. Yadav
Research Scholar, Uttrakhand Technical University, Dehradun, India
e-mail: sac.yaduvanshi@gmail.com

S. Mishra
e-mail: skmishra1@gmail.com

R. Yadav
e-mail: ranjeeta29@gmail.com

S. Sathiakumar et al. (eds.), *Proceedings of International Conference on Internet Computing and Information Communications*, Advances in Intelligent Systems and Computing 216,
DOI: 10.1007/978-81-322-1299-7_35,

approaches have been devised to tackle the media-streaming problem. The first one is to use a client–server model, where a single server is the media provider and multiple clients are the media consumers. The second one is to use a peer-to-peer approach where the clients help the server in delivering the media content by having the roles of consumers and providers at the same time.

Both schemes have their advantages and disadvantages. The client–server approach has the advantage that the client receives the content directly from the server with the minimum delay but at the cost of overwhelming the server in particular situations (for instance at high rate hours: e.g. football/basketball games etc.). As a result, the server's bandwidth can quickly become a bottleneck in the system due to the large number of client requests. On the other hand, in the peer-to-peer approach algorithms are devised to multicast the content between clients. In this case the clients have an active role in distributing the media content to other clients and thus remove the pressure from the server node. In this way, scaling the system functionality to a large number of consumers becomes a reality. However, this solution has its drawbacks too. Specifically, these algorithms have to tackle a high dynamic system, where clients can come and leave suddenly without any prior knowledge or guarantees.

Today's video streaming systems are mostly based on the client server model of Content Delivery Networks (CDN) which leads to several problems. The most important ones are:

1. *Flash crowd.* Large numbers of streaming servers are not able to feed more than a few hundred streaming sessions simultaneously [1].
2. *Bandwidth cost.* It can be a significant problem to the content provider. In contrast, these costs are shared by every participant in the P2P streaming network.
3. *Single point of failure.* Like any client–server model, the server is the single point of failure.

P2P networks offer characteristics and possibilities which cannot be provided by CDNs as proposed in [2]. As we show in this work, the performance of media streaming can he better in a P2P network, although the probability that one stream breaks is higher [3, 4]. The reason for this is that the replication rate of the video streams in a P2P network is typically significantly higher than in a CDN, due to the large number of participating hosts. In Gnutella for example, every peer shares an average of 500 files [5] and many peers host the same file.

Using MDC in a P2P streaming scenario is illustrated in Fig. 1. Peer p1 wants to receive video file S which is available in the MDC format on p3, p5 and p6. In this example the video is encoded using two descriptions D1 & D2. Peers p3 and p6 are chosen based on the distance from server to the receiver, and they simultaneously serve the video file S, each one providing a complementary description. If both the descriptions are received at the receiving peer p1, it will experience the highest quality. If any of the descriptions are affected by packet loss or excessive delay, the receiver can still decode and display video S but at the expense of a degradation of the quality, as the descriptions are independently decodable.

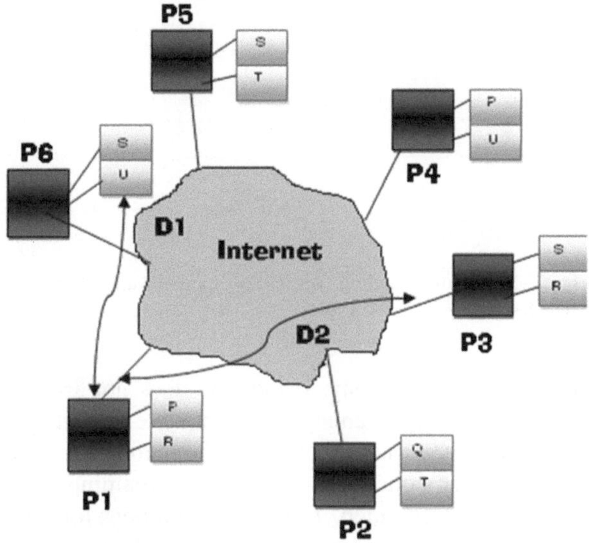

Fig. 1 Video streaming using multiple description coding in a P2P network. Peer P1 is simultaneously serve by the closest available peers P6 and P3 with descriptions D1 and D2 respectively

Multiple Description Video Coding

Multiple Description coding (MDC) is a coding technique that fragments a single media stream into n sub streams ($n \geq 2$) referred to as descriptions. The packets of each description are routed over multiple, (partially) disjoint paths. In order to decode the media stream, any description can be used; however, the quality improves with the number of descriptions received in parallel. The idea of MDC is to provide error resilience to media streams. Since an arbitrary subset of descriptions can be used to decode the original stream, network congestion or packet losses which are common in best-effort networks such as the Internet will not interrupt the stream but only cause a (temporary) loss of quality. The quality of a stream can be expected to be roughly proportional to data rate sustained by the receiver.

This property makes MDC highly suitable for lossy packet networks where there is no prioritization among the packets. The principle of MDC encoding/decoding is illustrated in Fig. 2. For a general overview on Multiple Description Coding (MDC) refer to [6].

Video Streaming Over Internet

Media streaming systems are distinct from the file sharing systems [7], in which a client has to download the entire file before using it. Real-time multimedia, as the name implies, has timing constraints. For example, audio and video data must be

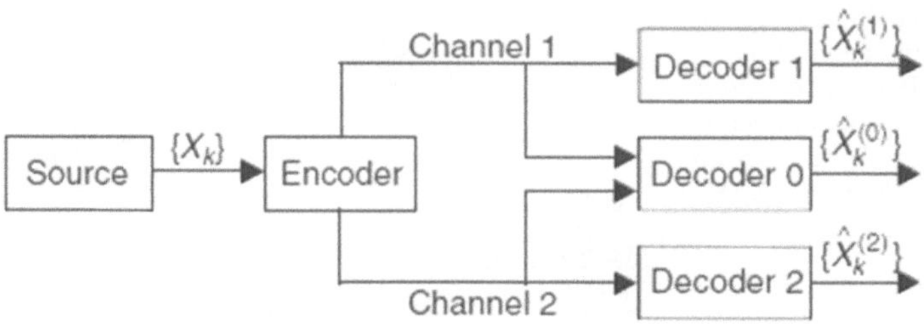

Fig. 2 MD source coding with two channels and three receivers. The general case has M channels and 2^{M-1} receivers

played out continuously. If the data does not arrive in time, the play out process will pause, which is annoying to human ears and eyes. Real-time transport of live video or stored video is the predominant part of real-time multimedia. In this paper, we are concerned with video streaming, which refers to real-time transmission of stored video. There are two modes for transmission of stored video over the Internet, namely the download mode and the streaming mode (i.e., video streaming). In streaming mode, the video content need not be downloaded in full, but is being played out while parts of the content are being received and decoded. Due to its real-time nature, video streaming typically has bandwidth, delay and loss requirements. However, the current best-effort Internet does not offer any quality of service (QoS) guarantees to streaming video over the Internet. In addition, for multicast, it is difficult to efficiently support multicast video while providing service flexibility to meet a wide range of QoS requirements from the users. Thus, designing mechanisms and protocols for Internet streaming video poses many challenges. It has been demonstrated in [2] that using MDC in combination with packet path diversity significantly improves the robustness of a real-time video application.

In Fig. 3, raw video and audio data are pre-compressed by video compression and audio compression algorithms and then saved in storage devices. Upon the client's request, a streaming server retrieves compressed video/audio data from storage devices and then the application-layer QoS control module adapts the video/audio bit-streams according to the network status and QoS requirements. After the adaptation, the transport protocols packetize the compressed bit-streams and send the video/audio packets to the Internet. Packets may be dropped or experience excessive delay inside the Internet due to congestion. To improve the quality of video/audio transmission, continuous media distribution services (e.g., caching) are deployed in the Internet. For packets that are successfully delivered to the receiver, they first pass through the transport layers and are then processed by the application layer before being decoded at the video/audio decoder. To achieve synchronization between video and audio presentations, media synchronization mechanisms are required. From Fig. 3, it can be seen that the six areas are closely related and they are coherent constituents of the video streaming architecture.

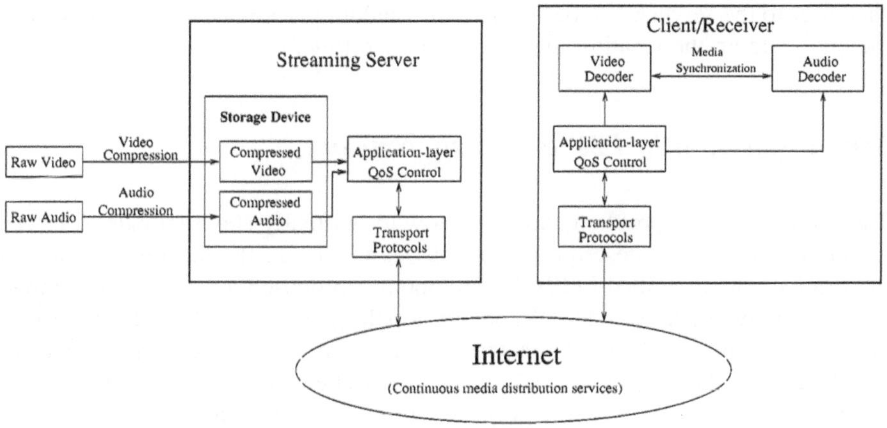

Fig. 3 Architecture for video streaming

Modeling

We use the following methodologies in our simulations to reflect the real-world network situations.

Modeling Availability in P2P Networks

In P2P networks, peer and content availability poses a challenging problem to be solved. Availability of a peer in a P2P network is quite unpredictable, depending primarily on human presence. In our experiments we model peer availability as a 2 state markov process, having the states ON and OFF. The average lifetime of a peer in a Gnutella network is found to be about 30 min [8]. For our experiments we take a Gaussian distribution of ON time, which has a mean of 30 min. To model the availability of content among the peers, we randomly choose peers having a particular media file. We vary the percentage of peers having the file from 5 % to 50 %.

Server Placement in CDN

The server placement problem addresses how to optimally place a number of servers in order to maximize the quality at the end user. In our experiments we varied the number of servers to obtain measurement of Quality of Service, such as packet loss and response time. For a particular number of servers, we placed the servers randomly in the network and measured the average round trip-time from each user to the servers. We performed this random placement 10 times and chose the one yielding the smallest average round-trip-time.

Server Selection in P2P and CDN Network

The server selection problem addresses how to optimally choose a pair of servers to get complementary descriptions in order to maximize the perceived quality at the receiver. As described in [9] Apostolopoulos proposed a path diversity model which requires the knowledge of network topology, including knowledge of joint and disjoint links, and loss characteristics for each link. In our experiments we simply choose the closest two servers for each client request. For P2P case, we choose the closest two serving peers having the required content.

Content Distribution Across Servers in CDN

This problem addresses how to optimally distribute the Multiple Description streams in an existing set of servers. In this paper we assume that all the CDN servers contain both the descriptions, which simplifies the server selection problem by merely choosing the two closest servers.

Network Load

To simulate the network load, we created random TCP connections originating from arbitrary nodes, on the average 3 new connections per second, each connection lasting for 1 min.

Result

We implemented both the P2P and CDN approaches within the Network Simulator ns-2 [10]. The topology was created using the GT-ITM topology generation tool with the transit-stub model, having 100 nodes. A video file of 1 min duration, having a data-rate of 100 Kbit/s was selected for all the simulations. Each packet contains 1000 bytes. In both the CDN and P2P based systems, there is one new request every second, originating from an arbitrary node. In P2P network, the file is streamed from two closest available peer nodes with complementary descriptions, whereas in CDN, the same is served by two closest CDN servers. It was assumed that a peer can serve only one request at one time, while a CDN server can serve a maximum of 200 streams simultaneously.

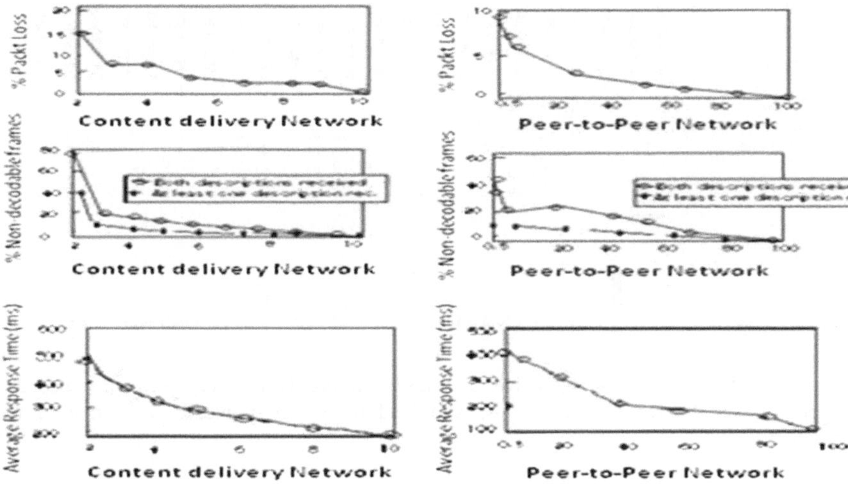

Fig. 4 Performance of P2P and CDN networks using MDC: top packet loss rate varies with varying number of CDN servers and content availability in P2P network middle. Number of decodable frames increases with increasing number of servers and availability bottom. Average response times for P2P and CDN

Figure 4 shows the results obtained through simulations. Three performance parameters, namely the rate of packet loss, number of non-decodable frames and the average response time, i.e. the time to receive the first video packet after the request has been sent, are compared for P2P and CDN networks. For the count of non decodable frames, it is assumed that the descriptions contain an Intra frame once in every second, and in case of a packet loss for the P-frames, all the subsequent frames become non-decodable, until the next I-frame is received. Because of MDC coding, the receiver can still view with a reduced frame rate, unless both the descriptions are corrupted simultaneously. This is shown in Fig. 5, where description s1 contains a packet loss, but s2 is received error-free. The receiver can view with ½ the original frame rate until the next I-frame is received in s1.

The simulation results indicate that the performance of a P2P network is comparable to that of a CDN, even at the high unavailability of peers and content in the P2P network.

Related Work

Peer-to-peer based media streaming approaches using multiple serving hosts have been proposed in [11] and [12]. In [2] MDC-based distributed video streaming has been proposed for content delivery networks. Our work is inspired by this work and we use the same multiple description encoding technique for a P2P network.

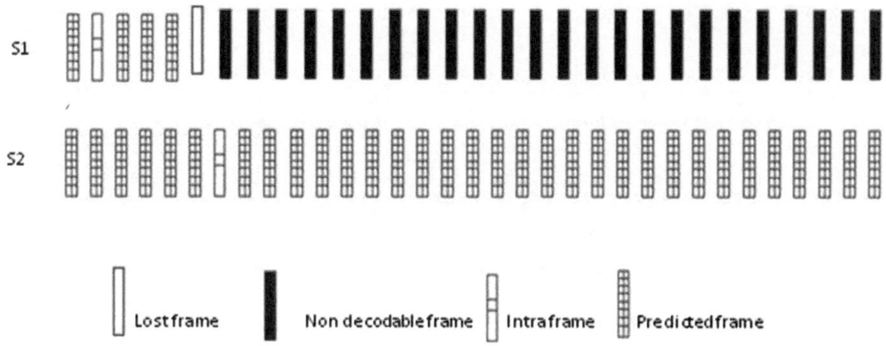

Fig. 5 Impact of packet loss in MDC-based video streaming. Only stream s2 can be decoded completely. s1 is affected by packet loss and lead to locally reduced frame rate of the reconstructed video

Conclusion

In this paper we presented a performance comparison of P2P media streaming with CDN – based media streaming, both employing MDC. The P2P approach takes advantage of multiple supplying peers to combat the inherent limitations of the P2P network and the best effort Internet. The media content is encoded using a multiple description encoder which allows realizing distributed streaming from more than one peer. In the final paper we plan to also provide experimental results on video dispersion, i.e. the time it takes to be able to satisfy a large number of streaming requests for a new video that is injected into the network, for both the P2P and CDN network.

References

1. Stading, T., Maniatis, P., Baker, M.: Peer-to-Peer Caching Schemes to Address Flash Crowds. Paper presented at: 1st International Peer To Peer Systems Workshop (IPTPS), 2002
2. Apostolopoulos, J.G.: Reliable video communication over lossy packet networks using multiple state encoding and path diversity. VCIP, 2001.
3. Baghwan, R., Savage, S., Voelker, G.M.: Understanding availability. In: Proceedings of the International Workshop on Peer-to-Peer Systems (IPTPS), February 2003
4. Saroiu, S., Gummadi, P.K., Gribble, S.D.: A measurement study of peer-to-peer file sharing systems. In: Proceedings of Multimedia Computing and Networking (MMCN), January 2002.
5. Chu, J., Labonte, K., Levine, B.N.: Availability and locality measurements of peer-to-peer file systems, in ITCom: Scalability and Traffic Control in IP Networks. In: Proceedings of SPIE. vol. 4868, July 2002.
6. Goyal, V.K.: Multiple description coding: compression meets the network. IEEE Signal Process. Mag. **18**(5), 74–94 (2001)
7. Xu, D., Hefeeda, M., Hambrusch, S., Bhargava, B.: On peer-to-peer media streaming. In: Proceedings of IEEE ICDCS, Vienna, Austria, July 2002

8. Adar, E., Huberman, B.: Free Riding on Gnutella. First Monday **5**(10), (2000)
9. Apostolopoulos, J., Wong, T., Tan, W., Wee, S.: On multiple description streaming with content delivery networks. In; Proceedings of IEEE INFOCOM, June 2002
10. The Network Simulator, http://www.isi.edu/nsnam/ns/
11. Hefeeda, M., Bhargava, B.: On-Demand Media Streaming Over the Internet. CERIAS TR 2002- 20, Purduc University, June 2002.
12. Nguyen, T., Zakhor, A.: Distributed video streaming. In: Multimedia Computing and Networking 2002, Multimedia Conference, January 2002.

DSR and DSDV Routing Protocol Analysis Using NS2 and Association Rule Mining Technique

Vinay Yadav and Divakar Singh

Abstract Ad hoc network (MANET) has no fixed networking infrastructure, and consists of mobile nodes that communicate with each other. Since nodes are mobile, routing in ad hoc network is a challenging task. Efficient routing protocols can make better performance in such networks. In this study, we are comparing the performance of two prominent reactive and proactive routing protocols for MANET, Dynamic Source Routing (DSR), and Destination Sequenced Distance Vector (DSDV). We are going to compare the DSR and DSDV mobile ad hoc network routing protocol using Network Simulator 2 (NS2). Our objective is to give performance comparison of DSR and DSDV based on CBR connection with varying speed and also various network parameter and measured performance metrics, such as packet delivery ratio for these two routing protocols. We have also use association rule mining technique–Apriori algorithm for analyzing major dropping node in NS2 simulation. We generate association rules for min_support threshold and min_confidence threshold.

Keywords DSR · DSDV · MANET · NS2 · APRIORI

Introduction

Mobile ad hoc network is a collection of wireless mobile nodes dynamically forming a temporary network of moving nodes or routers. The term "Ad-Hoc" implies "can take different forms" and can be "standalone, mobile or networked".

V. Yadav (✉) · D. Singh
BUIT, Barkatullah University, Bhopal, India
e-mail: writetoyadav@gmail.com

D. Singh
e-mail: divakar_singh@rediffmail.com

S. Sathiakumar et al. (eds.), *Proceedings of International Conference on Internet Computing and Information Communications*, Advances in Intelligent Systems and Computing 216,
DOI: 10.1007/978-81-322-1299-7_36, © Springer India 2014

The nodes are free to move randomly and organize themselves arbitrarily; thus, the network's wireless topology may change rapidly and unpredictably [1]. Mobile ad hoc network [2] is infrastructure less network due to mobile routers.

Due to these properties, MANET has wide application in industrial and commercial field involving cooperative mobile data exchange, inexpensive alternates to cellular-based mobile network infrastructures. Application of MANET in the locations where setting of infrastructured networks is difficult task and also in emergency disaster relief operations after natural hazards like earthquake. It is essential to restore communication networks in large-scale disasters by repairing the infrastructure as quickly as possible and taking appropriate measures to control congestion [1]. Communication and sharing of information in emergencies are also possible via ad hoc networks, which take full advantage of the features of wireless communication [3] including fast and temporary setup and terminal portability and mobility. Ad hoc networks can enable communication among temporarily assembled user terminals without relying on the conventional communication infrastructure.

Our goal is to carry out a systematic performance study of Destination Sequenced Distance Vector (DSDV) [4] and Dynamic Source Routing (DSR) [2]. Organization of the paper is as follows. In the Section "Routing Protocols", routing protocols of MANETs [5–7] are briefly reviewed. Section "Previous Work", describes the previous work. Section "Simulation and Results" presents the simulation and results followed by their interpretations and Conclusion.

Routing Protocols

Routing is the process of selecting path in a network along which to send data or physical traffic [6]. The performance of MANIT is related to efficiency of routing protocol in adopting to frequently changing network topology and link status [7].

Routing protocols in MANETs are primarily classified depending on:

- Routing/Network structure
- Routing strategy
- Routing information

Depending on the network structure routing protocols are classified as:

- Flat routing—no assumption for subnetting, no correlation in addressing.
- Hierarchical routing—involves subnetting, cluster formation, and hierarchical addressing.
- Geographic position assisted routing—routing based on geographic position of nodes.

According to the routing strategy the routing protocols can be categorized as:

- Table-driven (Proactive)
- On-demand or source initiated (Reactive)
- Hybrid (mix of proactive and reactive)

Depending on routing information stored in routing table and the way it is stored, routing protocols can also be classified as:

- Link state protocols—routers using a link state routing protocol maintain a full or partial copy of the network topology and costs for all known links.
- Distance-vector protocols—routers using a distance-vector protocol keep only information.

Dynamic Source Routing Protocol

The DSR protocol is a reactive routing protocol, which allows mobile sources to dynamically discover paths toward any desired destination. Every data packet includes a complete list of nodes, which the packet must pass before it reaches the destination. Hence, all nodes that forward or overhear these packets may store important routing information for future use. Even though nodes may move at any time and even continuously, DSR can support fast network topology changes. Moreover, DSR can support asymmetric links; it can successfully find paths and forward packets in unidirectional link environments. Moreover, like AODV, it has a mechanism for on-demand route maintenance, so there are no periodic topology update packets. When link failures occur, only nodes that forward packets through those links must receive proper routing advertisements.

Destination Sequenced Distance Vector (DSDV)

The DSDV routing protocol is a proactive routing protocol which is a modification of conventional Bellman-Ford routing algorithm. This protocol adds a new attribute, sequence number, to each route table entry at each node. Routing table is maintained at each node and with this table, node transmits the packets to other nodes in the network. This protocol was motivated for the use of data exchange along with changing arbitrary paths of interconnection which may not be close to any base station.

Each node in the network maintains routing table for the transmission of the packets and also for the connectivity to different stations in the network. These stations list for all the available destinations, and the number of hops required to reach each destination in the routing table. The routing entry is tagged with a sequence number which is originated by the destination station. In order to maintain the consistency, each station transmits and updates its routing table periodically. The packets being broadcasted between stations indicate which stations are accessible and how many hops are required.

Association Rules

It is the techniques for data mining and knowledge discovery in database. Items that occur often together can be associated to each other. These together occurring items form a frequent item-set. Conclusion based on the frequent item sets form association rules. support and confidence are two measures of rule. They respectively reflect the usefulness and certainty of discovered rules. Association rules are considered interesting if they satisfy both a minimum support threshold and minimum confidence threshold [8].

Previous Work

In this section, we summarize the most valuable previous studies concerning ad hoc on-demand routing performance comparisons and table-driven protocol. Many researchers have compared and evaluated the efficiency of the most famous on-demand routing and table-driven protocols for wireless mobile ad hoc networks, in the past. Node mobility is the main network characteristic that reveals the vigilance of a routing protocol to respond to network topology changes, and to perform fast route establishment and recovery. Thus, most of the previous studies compare the behavior of the protocols as a function of the node mobility. Author in [9] makes comparison between DSR and AODV protocols using maximum connection 8, environment size 840 m $\times$ 840 m, node density 30–150 nodes with constant pause time 20 s, variable speed 5–25 m/s, and Simulation time 200 s. In previous work simulations is done between two same categories of protocol. Both DSR and AODV are reactive protocol. Their performance matrix involve packet delivery ratio (PDR), average end-to-end delay, and lost packet ratio. Consider 30 nodes as low density, 90 nodes as average density, and 150 nodes as high density. Also consider 5 m/s as low speed, 15 m/s as average speed, and 25 m/s as high speed. On different density and speed calculated values for above-mentioned performance matrix parameters for DSR and AODV.

In fact, this was our motivation to make observations between two different categories of protocol. So we take DSR (reactive protocol) and DSDV (proactive protocol).

Simulation and Result

Constant Bit Rate (CBR)

Constant bit rate means consistent bits rate in traffic are supplied to the network. In CBR, data packets are sent with fixed size and fixed interval between each data packets. Establishment phase of connection between nodes is not required here, even the receiving node do not send any acknowledgement messages. Connection is one way direction like source to destination [9].

Performance Metrics and Network Parameters

For network simulation, there are several performance metrics which are used to evaluate the performance. In simulation purpose, we have used two performance metrics.

1. Packet Delivery Ratio

Packet delivery ratio is the ratio of number of packets received at the destination to the number of packets sent from the source. The performance is better when packet delivery ratio is high [9].

2. Major dropping node

The nodes which are appearing in NS2 network information. These nodes involve in dropping of packet when packet travel from source to destination.

For simulation purpose we used random waypoint mobility model. Network simulator NS 2.34 [10, 5] has been used. To measure the performance of DSDV and DSR, we used same scenario for both protocols [9].

Simulation Parameters

In our simulation, we used environment size 840 m × 840 m, node density 30–150 nodes with constant pause time 2 s and variable speed 5–25 m/s. We did the simulation for 200 s with maximum ten connections. The network parameters we have used for our simulation purpose are shown in Table 1.

Table 1 Network parameters

Parameter	Values
Protocol	DSR and DSDV
Simulation time	200 s
Number of nodes	30, 60, 90, 120, and 150
Simulation area	840 m × 840 m
Pause time	2 s
Traffic type	CBR
Maximum speed	5, 10, 15, 20, and 25
Mobility model	Random wavepoint

Simulation Results

The performance of DSDV and DSR has been analyzed with varying speed 5–25 m/s for number of nodes 30, 60, 90, 120, and 150 under CBR connection. We measure the PDR of DSDV and DSR and the simulated output has shown by using graphs.

1. Generating trace file for DSR

Now first we change number of node (30, 60, 90, 120, and 150) and speed (5, 10, 15, 20, and 25) for generating 25 traffic pattern files and 25 node movement files. These files run though the wireless. tcl file using NS2, we get the 25 trace files.

2. Generating trace file for DSDV

We again change number of node (30, 60, 90, 120, and 150) and speed (5, 10, 15, 20, and 25). For generating 25 traffic pattern files and 25 node movement files for DSDV. These files run though the wireless.tcl files using NS2, we get the 25 trace files.

3. Create network information

These 50 trace files (25 for DSR and 25 for DSDV) run on trace graph an application of microsoft window and generate network information. From network information we measure two factor:

(i) PDR on various node density and speed.

$$\text{PDR} = (\text{received packet/sent packet})*100$$

(ii) Major dropping nodes for DSR in node density 30 and speed 5, 10, 15, 20, and 25 m/s.

(i) *Graph*

We have drawn graphs for both DSR and DSDV simulation tables with varying node density and speed (Fig. 1, 2, 3, 4, and 5).

After analysis of DSDV and DSR the results have been shown in a Table 2. We define a standard for simulation results. We consider 30 nodes as low density, 90 nodes as average density, and 150 nodes as high density. We also consider 5 m/s as low speed, 15 m/s as average speed, and 25 m/s as high speed.

The standard for PDR values (approx.) defines below:

High: >=75 %
Average: 60–74 %
Low: <=59 %

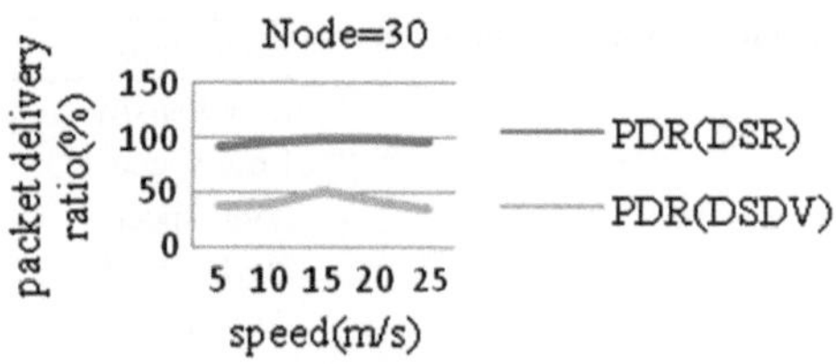

Fig. 1 Speed versus packet delivery ratio %. PDR of 30 nodes

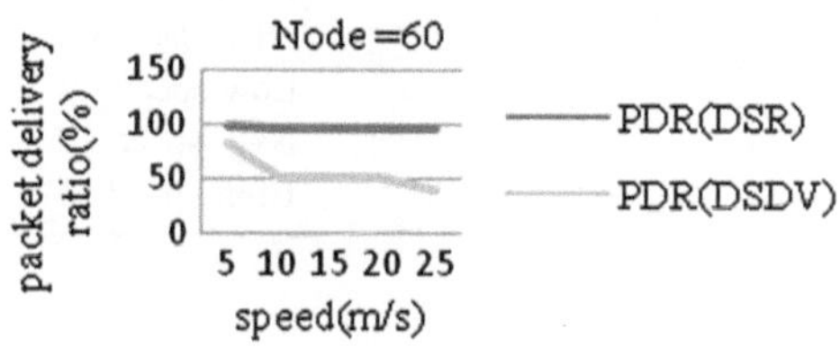

Fig. 2 Speed versus packet delivery ratio %. PDR of 60 nodes

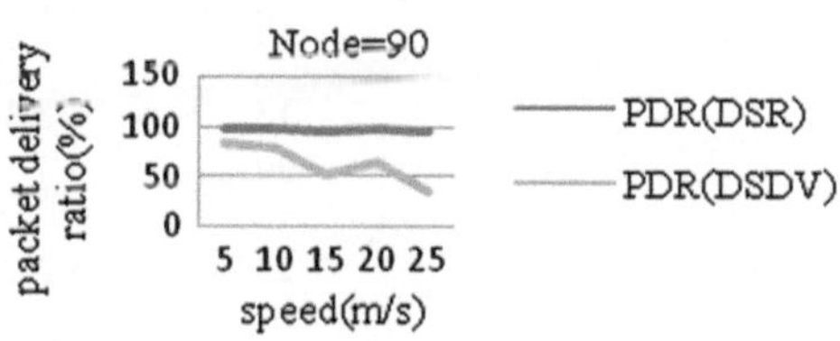

Fig. 3 Speed versus packet delivery ratio %. PDR of 90 nodes

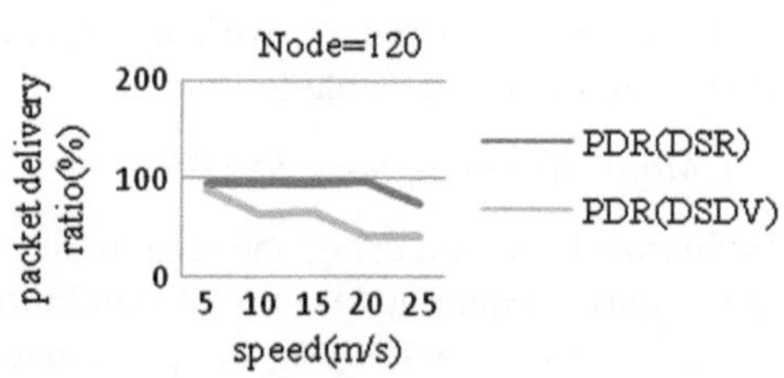

Fig. 4 Speed versus packet delivery ratio %. PDR of 120 nodes

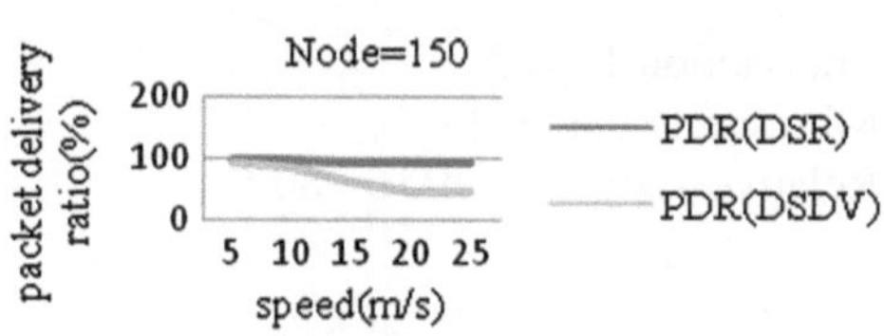

Fig. 5 Speed versus packet delivery ratio %. PDR of 150 nodes

Table 2 Analysis table

Node density	Packet delivery ratio	
Low density	*DSR*	*DSDV*
Low speed	High	Low
Avg. speed	High	Low
High speed	High	Low
Avg. density		
Low speed	High	High
Avg. speed	High	Low
High speed	High	Low
High density		
Low speed	High	High
Avg. speed	High	Average
High speed	High	Low

Table 3 Dropping node in DSR

Speed0	Dropping nodes
5	1 3 4 5 6 7 9 10 14 17 21 26
10	0 1 2 3 4 5 7 8 9 11 13 14 15 16 17 18 19 20 21 23 25 26 28
15	1 2 3 4 6 7 8 9 10 12 13 14 16 24 25 27
20	0 4 5 6 9 10 11 12 14 15 16 17 19 20 21 22 23 24 25 27 28 29
25	0 2 3 4 5 6 10 11 12 13 14 15 16 18 19 20 21 23 24 25 26 27 28 29

Based on our standard, we can summarize the following differences between DSDV and DSR in Table 2.

(ii) *Major dropping node in DSR*

In this work, we are using one data mining technique that may find major dropping node (max_freqent node) in NS2 simulation. For this purpose, we are using one of *association rules mining* technique called *apriori algorithm.* We are using node density = 30 and speed 5–25 m/s in DSR protocol and find dropping nodes shown in Table 3.

Transaction D = 5
Min_sup count = 4
Relative support = 4/5 = 80 %

Scan D for count of each candidate

→ **C1** :<0:3> <1:3> <2:2>

<3:4> <4:5> <5:4> <6:5> <7:3> <8:2> <9:4> <10:4>

<11:3> <12:3> <13:3> <14:5> <15:3> <16:4> <17:3>

<18:2> <19:3> <20:3> <21:4> <22:1> <23:3> <24:3>

<25:4> <26:2> <27:3> <28:3> <29:2>

↓ Compare candidate support count
With minimum support count

L1: <3:4> <4:5> <5:4> <6:5> <9:4> <10:4> <14:5> >
<16:4> <21:4> <25:4>

↓ Generate C2 candidate from L1

C2: <3,4> <3,5> <3,6> <3,9> <3,10> <3,14> <3,16>

<3,21> <3,25> <4,5> <4,6> <4,9> <4,10> <4,10>

<4,14> <4,16> <4,21> <4,25> <5,6> <5,9> <5,10>

<5,14> <5,16> <5,21> <5,25> <6,9> <6,10> <6,14>

<6,16> <6,21> <6,25> <9,10 <9,14> <9,16> <9,21>

<9,25> <10,14> <10,16> <10,21 > <10,25> <14,16>

<14,21> <14,25> <16,21> <16,25> <21,25>

↓ Scan D for count of each candidate

C2: <3,4:4> <3,5:3> <3,6:3> <3,9:3> <3,10:3> <3,14:4>

<3,16:3> <3,21:3> <3,25:3> <4,5:4> <4,6:5> <4,9:4>

<4,10:4> <4,10> <4,14:5> <4,16:4> <4,21:4> <4,25:4>

<5,6:3> <5,9:3> <5,10:3> <5,14:4> <5,16:4> <5,21:4>

<5,25:3> <6,9:3> <6,10:4> <6,14:4> <,16:3> <6,21:3>

<6,25:3> <9,10:3> <9,14:4> <9,16:3> <9,21:3>

<9,25:3> <10,14:4> <10,16:3> <10,21:3> <10,25:3>

<14,16:4> <14,21:4> <14,25:4> <16,21:3> <16,25:4>

<21,25:3>

↓ Compare candidate support count
With minimum support count

L2: <3,4:4> <3,14:4> <4,5:4> <4,6:5> <4,9:4> <4,10:4>

<4,14:5> <4,16:4> <4,21:4> <4,25:4> <5,14:4>

<5,16:4> <5,21:4> <6,10:4><6,14:4> <9,14:4><10,14:4>

<14,16:4> <14,21:4> <14,25:4> <16,25:4>

Generate C3 candidate from L2

C3: <3,4,14> <4,5,14> <4,5,16> <4,5,21> <4,6,10> <4,6,14> <4,9,14> <4,10,14> <4,14,16> <4,14,21> <4,14,25> <4,16,25> <5,14,16> <5,14,21> <6,10,14> <14,16,25>

Scan D for count of each candidate

C3: <3,4,14:4> <4,5,14:4> <4,5,16:3> <4,5,21:4> <4,6,10:4><4,6,14:5> <4,9,14:4> <4,10,14:4> <4,14,16:4> <4,14,21:4> <4,14,25:4> <4,16,25:4> <5,14,16:3> <5,14,21:4> <6,10,14:4> <14,16,25:4>

Compare candidate support count
With minimum support count

L3: <3,4,14:4> <4,5,14:4> <4,5,21:4> <4,6,10:4> <4,6,14:5> <4,9,14> <4,10,14:4> <4,14,16:4> <4,14,21:4> <4,14,25:4> <4,16,25:4> <5,14,21:4> <6,10,14:4> <14,16,25:4>

Generate C4 candidate from L3

C4: <4,5,14,21> <4,6,10,14>

Scan D for count of each candidate

C4: <4,5,14, 21:4> <4,6,10,14:4>

Generating rules for max freq. set L1 <4, 5, 14,21> & L2 <4, 6,10,14> .

Nonempty subset of L1 (4,5,14,21) are - (4) (5) (14,21) (4,5) (4,14) (4,21) (5,14) (5,21) (14,21) (4,5,14) (4,14,21) (5,14,21) (4,5,21)

(i) Association rules

1. 4->5^14^21 confidence = 4/5 = 80 %
2. 5->4^14^21 confidence = 4/4 = 100 %
3. 14->4^5^21 confidence = 4/5 = 80 %
4. 21->4^5^14 confidence = 4/4 = 100 %
5. 4^5->14^21 confidence = 4/4 = 100 %
6. 4^14->5^21 confidence = 4/5 = 80 %
7. 4^21->5^14 confidence = 4/4 = 100 %
8. 5^14->4^21 confidence = 4/4 = 100 %
9. 5^21->4^14 confidence = 4/4 = 100 %

10. 14^21->4^5 confidence = 4/4 = 100 %
11. 4^5^14->21 confidence = 4/4 = 100 %
12. 4^14^21->5 confidence = 4/4 = 100 %
13. 5^14^21->4 confidence = 4/4 = 100 %
14. 4^5^21-> 14 confidence = 4/4 = 100 %

At relative support = 4/5 = 80 % it means that there are 80 % chance at node density 30 and speed from 5 to 25 m/s node (4, 5, 14, and 21) including together in dropping the packet.

Set minimum confidence threshold = 100 % so output rules are:

a. 5->4^14^21 confidence = 4/4 = 100 %
b. 21->4^5^14 confidence = 4/4 = 100 %
c. 4^5->14^21 confidence = 4/4 = 100 %
d. 4^21->5^14 confidence = 4/4 = 100 %
e. 5^14->4^21 confidence = 4/4 = 100 %
f. 5^21->4^14 confidence = 4/4 = 100 %
g. 14^21->4^5 confidence = 4/4 = 100 %
h. 4^5^14->21 confidence = 4/4 = 100 %
i. 4^14^21->5 confidence = 4/4 = 100 %
j. 5^14^21->4 confidence = 4/4 = 100 %
k. 4^5^21->14 confidence = 4/4 = 100 %

Rule no. (a) Say that when a packet drops on node 5 it is 100 % chance that will also drop on node 4, 14, and 21. From rule no. (c) If packet drop on 4 and 5 it has 100 % chance to drop on 14 and 21. We can describe each association rule in similar manner.

Nonempty subset of L2 (4, 6, 10, 14) are:

(4) (6) (10) (14) (4, 6) (4, 10) (4, 14) (6, 10) (6, 14) (10, 14) (4, 6, 10) (4, 6, 14) (4, 10, 14) (6, 10, 14)

(ii) Association rules

1. 4->6^10^14 confidence = 4/5 = 80 %
2. 6->4^10^14 confidence = 4/5 = 80 %
3. 10->4^6^14 confidence = 4/4 = 100 %
4. 14->4^6^10 confidence = 4/5 = 80 %
5. 4^6->10^14 confidence = 4/5 = 80 %
6. 4^10->6^14 confidence = 4/4 = 100 %
7. 4^14->6^10 confidence = 4/5 = 80 %
8. 6^10->4^14 confidence = 4/4 = 100 %
9. 6^14->4^10 confidence = 4/4 = 100 %
10. 10^14->4^6 confidence = 4/4 = 100 %
11. 4^6^10->14 confidence = 4/4 = 100 %
12. 4^6^14->10 confidence = 4/5 = 80 %
13. 4^10^14->6 confidence = 4/4 = 100 %

14. 6^10^14->4 confidence = 4/4 = 100 %

There are 80 % chance at node density 30 and speed from 5 to 25 m/s, nodes (4, 6, 10, and 14) are including together in dropping the packet.

Set minimum confidence threshold = 100 % so output rules are:

a. 10->4^6^14 confidence = 4/4 = 100 %
b. 4^10->6^14 confidence = 4/4 = 100 %
c. 6^10->4^14 confidence = 4/4 = 100 %
d. 6^14->4^10 confidence = 4/4 = 100 %
e. 10^14->4^6 confidence = 4/4 = 100 %
f. 4^6^10->14 confidence = 4/4 = 100 %
g. 4^10^14->6 confidence = 4/4 = 100 %
h. 6^10^14->4 confidence ==4/4 = 100 %

Rule no. (a) say that if a packet drop on node 10 then it is 100 % chance it will also drop on 4, 6, and 14. From rule no. (b) if a packet drop on 4 and 10 it has 100 % chance to drop on 6 and 14. We can describe each rule in similar manner.

Conclusion

This paper illustrates:

(i) Differences between DSDV and DSR

Based on CBR connection with various network parameters. In our analytical Table 2, we have given decision based on the graph. This will definitely help to understand the performance of these two routing protocols. The performance of these two routing protocols shows some differences in low and high node density.

In low density with low speed, the PDR is high for DSR and low for DSDV. If the speed is average then PDR for both routing protocols is same as per as low speed region. If the speed is high the PDR for DSR is high and low for DSDV.

In high density with low speed PDR of both routing protocols is high. If the speed is average then PDR is high for DSR and average for DSDV. If speed is high the PDR for DSR is high but low for DSDV.

From our experimental analysis, we can conclude that if we take node density 150 and speed 15 ms DSR give good performance. When node density is 150 and speed 5 ms performance of DSDV is good.

(ii) Generate association rules for dropping node

We generate 19 association rules based on 80 % min_support threshold and 100 % min_confidence threshold. These rules are necessary because of following region:

(a) We have knowledge about which nodes are maximum involve in dropping the packet.

(b) We also predict which nodes simultaneously involve in dropping of packets.
(c) If we have knowledge about these dropping nodes we can improve performance of NS2 simulation by taking extra care and attention on these major dropping nodes.
(d) We make strategies particularly for these dropping nodes so minimize the dropping effect.

Future Work

In our future work, we plan to study the performance of these protocols under other network sceneries by varying the network size, the number of source nodes, the mobility models, and the speed of the mobile nodes. We enhance developing association rules for different node density and different routing protocols.

References

1. Kumar, D., Srivastava, A., Gupta, S.C.: Performance comparison of pro-active and reactive routing protocols for MANET computing. Communication & Application (ICCCA), 2012. doi:10.1109/ICCCA.2012.6179226
2. Jain, A., Pruthi, A., Thakur, RC., Bhatia, MPS.: TCP analysis over wireless mobile ad hoc networks, personal wireless communications. In: 2002 IEEE international conference on digital object identifier, pp. 95–99
3. Fischer, M.J., Lynch, N.A., Peterson, M.S.: Impossibility of distributed consensus with one faulty processor. J. ACM **32**(2), 374–382 (1985)
4. Kulkarni, N., Prasad, R., Comaen, H., Gupta, N.: Performance evaluation of AODV, DSDV & DSR for quasi random. Deployment of sensor nodes. In: 2011 international conference on wireless sensor networks devices and communication (ICDeCom), pp. 1–5, 2011; pp. 330–333, 2010
5. Tutorial for network simulator. http://www.isi.edu/nsnam/ns/tutorial/
6. Tuteja, A., Gujral, R., Thalia, S.: Comparative performance analysis of DSDV, AODV and DSR routing protocols in MANET using NS 2. In: international conference on advances in computer Engineering, IEEE, 2010. 978-07695-4058-0, pp 330–333
7. Tomar Geetam Singh, Sharma Trupti, Bhattcharrya Debnath, and Kim Tai-Hoon.: Performance comparison for AODV, DSR & DSDV under various network condition: A survey. In:2011 IEEE international conference on ubiquitous computing and multimedia application. doi:10.1109/UCMA.2011.9
8. Han, J., Kamber, M.: Data mining concept & techniques. Morgan Kaufmann Publishers an imprint of Elsevier. ISBN:978-1-55860-901-3
9. Paul, B., Ibrahim, Md., Abu Naser Bikas, Md.: Experimental analysis of AODV & DSR over TCP & CBR connection with varying speed and node density in VANET. Int. J. Comput. Appl. **24**(4) 0975–8887 (2011)
10. The Network Simulator - ns-2. http://www.isi.edu/nsnam/ns/

Innovative Fusion of Ear and Fingerprint in Biometrics

C. Malathy, K. Annapurani and A. K. Sadiq

Abstract Biometric-based personal identification is regarded as an effective method for identification. In multimodal systems more than one biometric sample feature is used for identification which makes duplications of feature and spoofing nearly impossible. This paper proposes an identification method for a multimodal biometric system using two traits, i.e., ear and fingerprint. A new technique is introduced here for choosing the region of interest of the ear and fingerprint image. The two local points, including the Canal Intertranguiano and the starting point of the Helix are taken as the region of interest (ROI) for ear. In case of fingerprint image also the same procedure is followed to take ROI from the fingerprint image. After feature extraction of ear and fingerprint both undergoes fusion process. The fusion of these two extracted features is done using concatenation technique. Later, matching process is carried out to identify a person.

Keywords Multimodal system · Region of interest · Fusion

Introduction

Biometric refers to identifying a person on the basis of his/her physiological or behavioral characteristic. Physiological characteristic of biometrics includes fingerprint, hand geometry, palm print, iris recognition, etc. Behavioral characteristics

C. Malathy (✉) · K. Annapurani
Department of Computer Science and Engineering, SRM University, Kattankulathur, India
e-mail: malathy.c@ktr.srmuniv.ac.in

K. Annapurani
e-mail: annapoorani.k@ktr.srmuniv.ac.in

A. K. Sadiq
Ministry of Higher Education, College of Applied Sciences, Sohar, Oman
e-mail: a_k_sadiq@yahoo.com

S. Sathiakumar et al. (eds.), *Proceedings of International Conference on Internet Computing and Information Communications*, Advances in Intelligent Systems and Computing 216,
DOI: 10.1007/978-81-322-1299-7_37, © Springer India 2014

include gait, face, voice, etc. Biometrics indicators have an advantage over traditional security identification methods, because these inherent attributes cannot be easily shared and every person has unique biometric attributes. Biometrics systems are non-deterministic and inherently dynamic because of the existence of background noise, signal noise and distortion, and environment or device variations. For example, fingerprint biometric matching results can vary because of pressing-pressure at the sensor, moisture and cut on the finger, etc. Thus use of a single biometric trait is not a solution of genuine authentication because of different variations, physical and spoofing problems [1]. To make biometric systems secure from spoofing problems researchers have proposed multimodal biometric systems. Multimodal biometric system uses more than one biometric trait in order to meet stringent performance requirements. Multimodal biometric system [2] is vital to fraudulent technologies, because it is difficult to forge multiple biometric traits than single biometric traits. In this work two biometric traits are used, fingerprint and ear. Fingerprints can be represented by a large number of features, including the overall ridge flow pattern, ridge frequency, location and position of singular points, type, direction, and location of minutiae, and location of pores. All these features contribute to fingerprint individuality [3]. In this work the minutia features ridge ending, bifurcation of the fingerprint is used [4]. Human ear has also been used as a major feature in forensic science for many years. The ear can be used as an efficient biometric trait because it does not change considerably during human life. Secondly, ear features do not change with expression as face features do [5]. Moreover, in case of face and iris the color distribution is not the same but for ear the color distribution is the same. Here, a new method of choosing the region of interest (ROI) for ear is proposed. The fusion of ear features and fingerprint features increases the robustness of identification. After extraction of each individual trait's feature, fusion strategy is applied to fuse the features.

In Sect. "Preprocessing of Ear and Its ROI" we describe the preprocessing of Ear and our new ROI. Section "Preprocessing of Fingerprint and its ROI" is about the Preprocessing of Fingerprint and its ROI. In Sect. "Feature Extraction of Fingerprint and Ear Images" we apply the feature extraction techniques for both ear and fingerprint. Section "Feature Level Fusion and Authentication" deals with fusion of the extracted features, and identification. In Sect. "Experiment Results" we give the experimental results followed by conclusions and future work.

Preprocessing of Ear and Its ROI

Ear Image Enhancement

The normal ear image taken for identification may have some kind of noises in it because of sweat, hair, or maybe sometimes the sensor device which is used to take the image may not be able to take a good quality image. Thus, before going for feature extraction of the ear image it is required to preprocess the image to get clear features. The enhancement process of the image is also called the preprocessing phase of image processing. To enhance the ear image first, the median filter is used. It is mainly used to remove the 'salt and pepper' noise. The median filter preserves useful details in an image. In median filtering, output pixel value is determined by the median of neighborhood pixels, rather than the mean. The median information is much less sensitive than the mean to extreme values or outliner [5]. Figure 1 shows the Original image and Enhanced image.

Histogram Equalization

The intensity of the pixels is not well distributed into the entire range of the intensity level. Here, grayscale images of ear are taken into account, so the intensity range is 0–255. To spread the intensity onto the entire range of intensity level, histogram equalization technique is used. It is a pixel-wise operation. The original histogram is stretched and shifted to cover the entire range. To equalize, further contrast level adaptive histogram is applied. Unlike histogram equalization it operates on small data regions (tiles) rather than the entire image. Each tile's contrast is enhanced so that the histogram of each output region approximately matches the specified histogram (uniform distribution by default). The contrast enhancement can be limited in order to avoid amplifying unwanted noises. Figure 2 shows the original image and histogram equalized image of Ear.

Binarization

The enhanced and noise-free ear image is then converted into a binary image. The process of binarization converts the grayscale image into a binary image. The output image replaces all pixels in the input image with luminance greater than

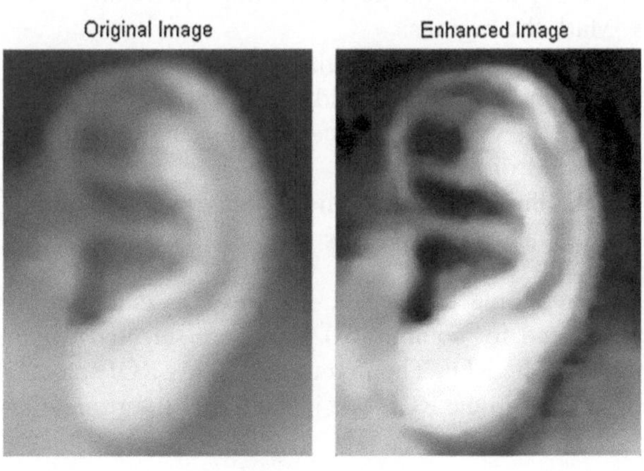

Fig. 1 The enhanced ear image

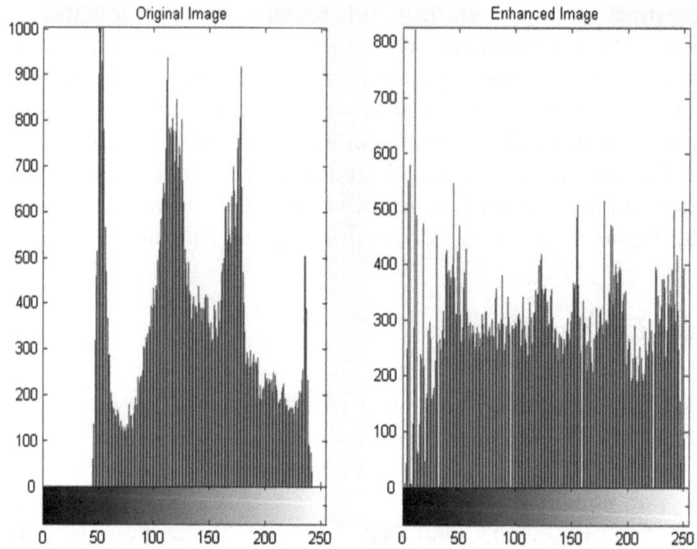

Fig. 2 Histogram equalized image

threshold level with value 1 (white) and replaces all other pixels with value 0 (black). The threshold level is in the range [0, 1], regardless of the input image. Here 0.35 is set as threshold value.

Region of Interest

The unique feature of ear is selected for feature extraction. Here, a new technique is proposed to take the ROI from the ear image.

The ROI is selected by the following steps:

- Fill the binary image with filling operation.
- Mark the regions.
- Find the centroid of each region.
- Find the nearest centroid from center of the image.
- Take the vertical distance from the centroid nearest to the center of the image to the end of the region.
- Take the midpoint of this distance which is the height of the ROI.
- Manipulate other values using the height of ROI value.

Here, some fixed values are used to make the region complete, for rows 120 pixels and for columns 70 pixels are selected which form a rectangle and this is taken as the ROI. The main goal of this ROI selection technique is to take at least two points in the ROI, one is the Canal Intertranguiano and the other is the starting

Fig. 3 ROI of ear image

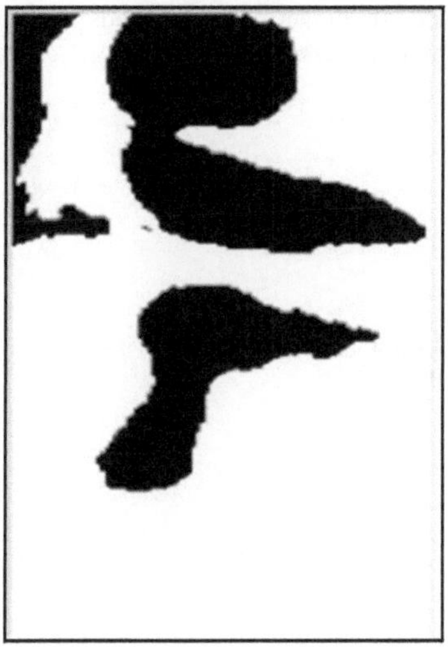

point of the Helix[6]. These two points are required to identify an ear. Figure 3 shows the ROI of Ear.

Preprocessing of Fingerprint and Its ROI

Fingerprint Image Enhancement

Every fingerprint image contains some noise; this noise is due to moisture on the finger, cut on the finger, sweat, etc. Before going to the feature extraction phase it is necessary to remove these noises. The enhancement process is more or less the same as the ear image preprocessing. Here, the median filter is again used to remove ‘salt and pepper’ noise from the fingerprint image. The fingerprint images taken here are grayscale images. After applying median filter, the operation replaces a pixel value with the median value taken from the neighborhood pixels. The image is filtered and Histogram equalization is done.

Binarization

The fingerprint image is converted into a binary image. It takes the input image, and based on the luminance level it converts the pixel value into ‘1’ or ‘0’. It has a threshold value; the luminance greater than the threshold value is converted to ‘1’ and less than that is converted to ‘0’. Figure 4 shows the Enhanced and binarized image of Fingerprint.

Fig. 4 Fingerprint enhanced and binarized image

Fig. 5 ROI of fingerprint image

Region of Interest

To extract the feature of the fingerprint the entire image is not required. The region containing the most number of ridges and furrows is selected by the ROI selection process. The process first fills the fingerprint image and then marks the regions. Once the regions get marked the centroid of those regions are calculated. Then the centroid which is nearest to the center of the image is selected. The distance from centroid to the end of the image region is calculated and the ROI top is marked at the midpoint of this distance. Other dimensional values are fixed and entered as it is done in case of ear image. Figure 5 shows the ROI of Fingerprint.

Feature Extraction of Fingerprint and Ear Images

After completion of both enhancement and ROI selection operation the images are ready for feature extraction. In this work all the feature extraction procedures are present in both earimgwork.m and finimgwork.m file. Ear image is extracted using eigenfaces. The minutia features include ridges, ridge endings, ridge bifurcations, etc. Here, the two minutia features [6] used are ridge ending and bifurcation.

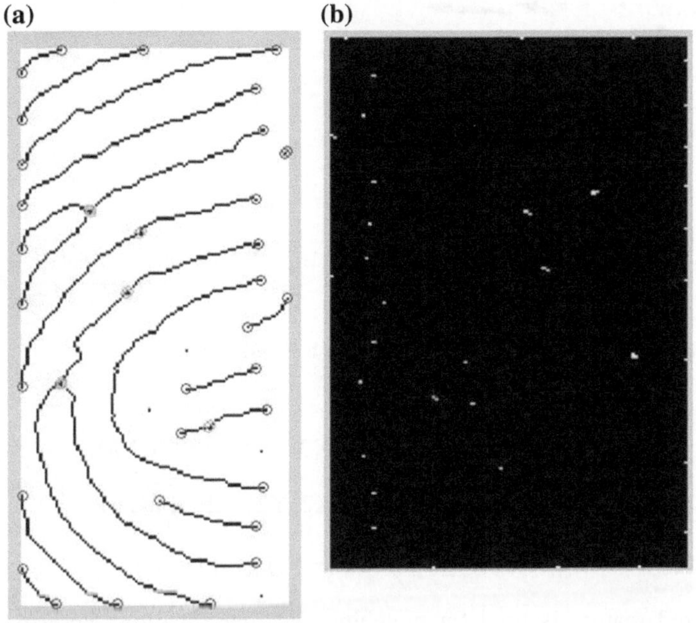

Fig. 6 **a** Marked ridge ending and bifurcations and **b** Feature points of fingerprint

The image matrix is segmented into 3 × 3 matrices and then the number of pixels is calculated which contains the value 1 around the middle pixel. If the value is 1 the center pixel is ridge ending. If the value is two 1's then there is no ridge ending or bifurcation is present inside that particular segment. If the value is three 1's then it is a bifurcation. Figure 6a. shows the marked ridge ending and bifurcations. After the entire neighborhood operation, the ridge ending and bifurcation location values are stored in zero matrixes. This matrix becomes the feature matrix for fingerprint image. Figure 6b shows the feature points of the fingerprint. The pixel position are taken and stored into a matrix F(*x*,*y*). The ear features is also extracted into the matrix E(*x*,*y*).

Feature Level Fusion and Authentication

Once both the feature matrices E(*x*,*y*) and F(*x*,*y*) are extracted, the fusion is done. To fuse these two feature matrices, feature concatenation technique is used. Each value of the E(*x*,*y*) matrix is added with the fingerprint matrix F(*x*,*y*). Each single position of the feature matrix is fused with the other. The fused matrix is stored in a different matrix [7], Fused(*x*,*y*).

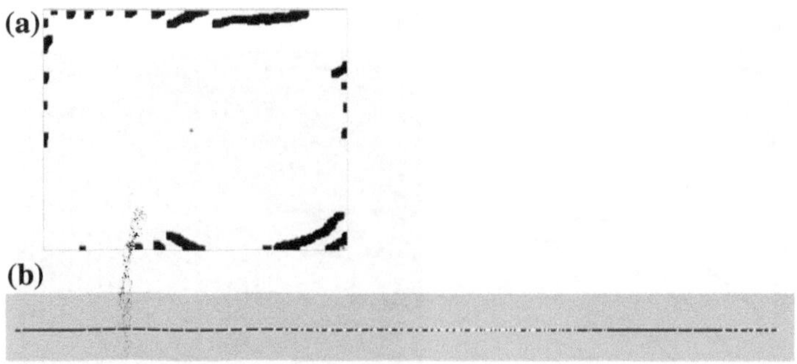

Fig. 7 **a** Marked regions of fingerprint image and **b** Fused features

$$\mathrm{E}(x,y) = \mathrm{E}(\mathrm{X1},\mathrm{Y1}),\mathrm{E}(\mathrm{X2},\mathrm{Y2})\ldots\mathrm{E}(\mathrm{Xn},\mathrm{Yn}) \tag{1}$$

$$\mathrm{F}(x,y) = \mathrm{F}(\mathrm{X1},\mathrm{Y1}),\mathrm{F}(\mathrm{X2},\mathrm{Y2})\ldots\mathrm{F}(\mathrm{Xn},\mathrm{Yn}) \tag{2}$$

$$\mathrm{Fused}(x,y) = \sum \mathrm{E}(x_\mathrm{i},y_\mathrm{i}) + \mathrm{F}(x_\mathrm{i},y_\mathrm{i}) \tag{3}$$

The fused feature is used for matching and identification. Here in this work PCA eigenvalues are used for classification and matching [8]. A total of 11 subjects are used for making the training database and for each subject ear and fingerprint images are taken.

Experiment Results

The ear image and fingerprint should be of the same number because every person has a unique fingerprint and ear image. If the fusion is not done between two respective images identification may lead to a false identification. The database is created by applying the eigenface technique. Every fused matrix represents a single column of the training database matrix. After having the entire fused matrix column in the training database the eigenvalue is calculated from that training database matrix. At the time of matching every column of the database is multiplied with the eigenvalue and the test images are also multiplied with the eigenvalue after fusion. The matching process used is eigenvalue matching technique which calculates the minimum Euclidean distance between input image eigenvalue and training database eigenvalue. Figure 7a shows the marked features of fingerprint and fig. 7b shows the fused feature of ear and fingerprint.

Conclusion

This work is focused on fusion between ear and fingerprint image. The datasets were taken from Hong Kong Polytechnic University.

http://www4.comp.polyu.edu.hk/~cslzhang/Here, 11 ear and 11 fingerprint images have been used to fuse the feature which is extracted after their enhancement process.

This gives efficient results for matching and detecting correct image pairs. In this work some consideration has been made which can be improved later on. The considerations are:

- The experiment is carried out with only 11 images.
- The fingerprint database used has good quality fingerprint images, all of which are taken at the same angle and with the same camera. Thus, orientation of ridge and other operations is not required here.

The image considered here is as if they are at right angles with the two axes. No rotation or adjustment on the images is done. There is scope to improve this technique using some more analyses on the image position with respect to both the axes. The fusion technique used here is the concatenation of both the features; it can be improved by taking features with respect to axis and other fusion techniques.

References

1. Khan, M. K., Zhang, J.: Multimodal face and fingerprint biometrics authentication on space-limited tokens. Neurocomputing **71**(13–15), 3026–3031 (2008)
2. Ross, A., Nandakumar, K., Jain, A.K.: Handbook of Multibiometrics. Springer, Heidelberg (2006)
3. Vaikole, S., Sawarkar, S.D., Hivrale, S., Sharma, T.: Minutiae feature extraction from fingerprint images. IEEE International Advance Computing Conference (IACC 2009) Patiala, India, 6–7 March (2009)
4. Kaur, R., Kamra, A.: A novel method for fingerprint feature extraction. International Conference on Networking and Information Technology, Manila, 2010
5. Choras, Michal: Ear biometrics based on geometrical feature extraction. Electron. Lett. Comp. Vis. Image Anal. **5**(3), 84–95 (2005)
6. Palmer, L.R., Al-Tarawneh, M.S., Dlay, S.S., Woo, W.L.: Efficient fingerprint feature extraction algorithm and performance evaluation. Graz, Austria, IEEE, pp. 581–584, (2008)
7. Thillaikkarasi, J.T., Duraiswamy, K.: Cryptographic key generation from multiple biometric modalities: fusing minutiae with Iris feature. Int. J. Comp. Appl. **2**(6), 0975–8887 (2010)
8. Lammi, H-K.: Ear Biometrics. Lappeenranta University of Technology, Department of Information Technology, Lappeenranta, Finland

Performance Analysis of Various Feature Extraction Techniques in Ear Biometrics

K. Annapurani, C. Malathy and A. K. Sadiq

Abstract Many feature extraction techniques are available to extract the features of ear. Here in this paper, we have concentrated on analyzing the best feature extraction method. We have analyzed linear and nonlinear feature extraction methods like Linear Discriminant Analysis (LDA), Principal Component Analysis (PCA), and Kernel Principal Component Analysis (KPCA). Also we have combined LDA and PCA methods, so that the best properties of the two methods are taken. For experimentation, we have used the ear images obtained from publicly available sources. The experimental results have showed that the combination of LDA and PCA gives good performance in both verification rate and false acceptance rate compared to the other techniques individually used.

Keywords Feature extraction · PCA · LDA · KPCA · Verification rate · False acceptance rate

Introduction

Biometrics refers to the identification of humans by their characteristics or traits. The two methods of biometric identifiers include physiological and behavioral characteristics. A physiological biometric would identify by one's face, ear, DNA, fingerprint, palm print, iris, and so on. Behavioral characteristics are related to the behavior of a person, including but not limited to: typing rhythm, gait, and voice.

K. Annapurani (✉) · C. Malathy
Department of Computer Science and Engineering, SRM University, Kattankulathur, India
e-mail: annapoorani.k@ktr.srmuniv.ac.in

C. Malathy
e-mail: malathy.c@ktr.srmuniv.ac.in

A. K. Sadiq
Ministry of Higher Education, College of Applied Sciences, Sohar, Oman
e-mail: a_k_sadiq@yahoo.com

S. Sathiakumar et al. (eds.), *Proceedings of International Conference on Internet Computing and Information Communications*, Advances in Intelligent Systems and Computing 216,
DOI: 10.1007/978-81-322-1299-7_38,

A new class of biometrics based upon ear features was introduced for use in the development of passive identification systems by Alfred Iannarelli [1]. Human ear is a perfect data for passive person identification, which can be applied to provide security in the public places [2]. The ear is a unique feature of human beings. Even the ears of "identical twins" differ in some respects. The most famous work among ear identification is made by Iannarelli [1] when he gathered up over ten thousands ears and found that all were different [3]. Even the identical twins had similar, but not identical, ear physiological features [3].Various feature extraction techniques were applied for ear biometrics by different research group.

Burger and Burger [3] have proposed a feature extraction technique of 2D voronoi diagrams. Choras et al. [4] have proposed a methodology of several geometrical methods. Force field transformation method was proposed by Hurley et al. and 2D PCA method was implemented by Victor et al. The extended version of PCA is KPCA. By the use of operator kernel functions, principal components in high-dimensional feature spaces are efficiently computed, related to input space by some nonlinear map [5].

Various Algorithms of PCA, LDA, KPCA, and PCA + LDA

The main task is to improve the performance of ear recognition. Many algorithms are available; here we have considered PCA, LDA, and combination of both PCA and LDA.

Principal Component Analysis

The goal of principal component analysis (PCA) is to reduce the dimensionality of the data, while retaining as much as possible of the variation present in the original dataset [6]. PCA allows us to compute a linear transformation that maps data from a high-dimensional space to a lower dimensional subspace. PCA projects the data along the directions where the data vary the most. These directions are determined by the eigenvectors of the covariance matrix corresponding to the largest Eigenvalues. The magnitude of the Eigenvalues corresponds to the variance of the data along the eigenvector directions. The simplest approach is to think of it as a template matching problem. When performing recognition in a high-dimensional space problems arise. Significant improvements can be achieved by first mapping the data into a lower dimensional space.

Computation of the Eigenfaces: The ear images are obtained from Hong Kong Polytechnic University. http://www4.comp.polyu.edu.hk/~cslzhang/. The images must be same size.

The sample of ear images is given in Fig. 1. Every image is represented as a vector. The average ear vector is computed and subtracted from the mean. The covariance matrix is calculated. After this largest Eigenvalues are taken. For identification of the person, the distance within the ear space is computed using Mahalanobis distance method.

Linear Discriminant Analysis

LDA performs dimensionality reduction while preserving as much of the class discriminatory information as possible. It seeks to find directions along which the classes are best separated. It takes into consideration the scatter within-classes but also the scatter between-classes. It is more capable of distinguishing image variation due to identity from variation due to other sources such as illumination and expression. So it best suits for ear recognition. LDA computes a transformation that maximizes the between-class scatter, while minimizing the within-class scatter. Such a transformation should retain class separability while reducing the variation due to sources other than identity (e.g., illumination).

Steps in LDA:

Step 1: Taking two datasets.
Step 2: Finding mean of each set.
Step 3: Finding between-class scatter (variance) and within-class scatter (variance).
Step 4: Removing linear dependency and redundancy.
Step 5: Transformation of dataset to one axis.
Step 6: Finding Euclidean distance.
Step 7: Smallest Euclidean distance.

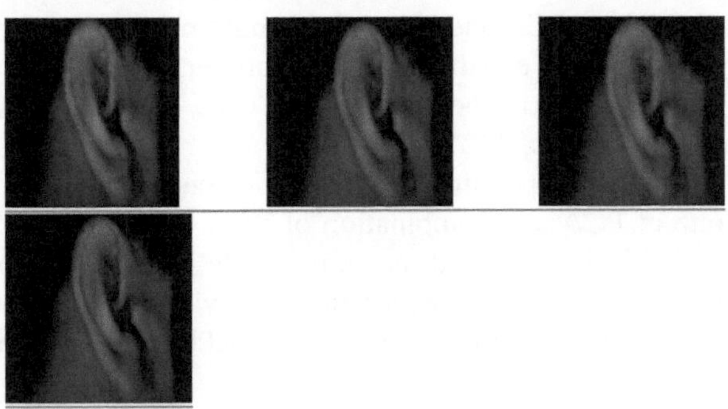

Fig. 1 Sample ear images

KPCA

It is an extension of PCA using techniques of kernel methods. Here nonlinear mapping is possible. Kernel method is transforming data or input space into high-dimensional feature space, i.e., transforming data into set of points.

Combination of PCA and LDA

PCA is based on the sample covariance which characterizes the scatter of the entire dataset; irrespective of class membership. The projection axes chosen by PCA might not provide good discrimination power. In case of LDA, it deals directly with discrimination between classes. In order to improve the performance, the best features in both the algorithms are integrated.

Steps in PCA + LDA

Step 1: Database loading.
Step 2: Partitioning of data into training and test datasets.
Step 3: Construction of PCA + LDA subspace.
Step 4: Computing similarity matrix.
Step 5: Evaluate similarity matrix and plot graph.
Step 6: Present performance metrics.

Results and Discussion

The ROC and cumulative matching curve (CMC) curves are plotted. From the graph it is observed that PCA performs well and the combination of PCA and LDA results are almost similar to PCA. KPCA does not perform well compared to linear methods of PCA and LDA. This might be due to less number of training sets. LDA also performs equally good, but can perform well if large datasets are considered.

The CMC is shown in Fig. 2. The recognition rate is 78.89 % in PCA and combination of PCA and LDA.

Receiver operating characteristics curve shown in Fig. 3 gives the verification rate of PCA and combination of PCA and LDA is 75.56 %.

Table 1 shows that recognition rate in PCA and combination of PCA and LDA is 78.89 %. ERR is better in LDA with 5.59 %. But in PCA it is 7.87 %. The combination of both gives 5.59 %. KPCA does not perform well in all the cases.

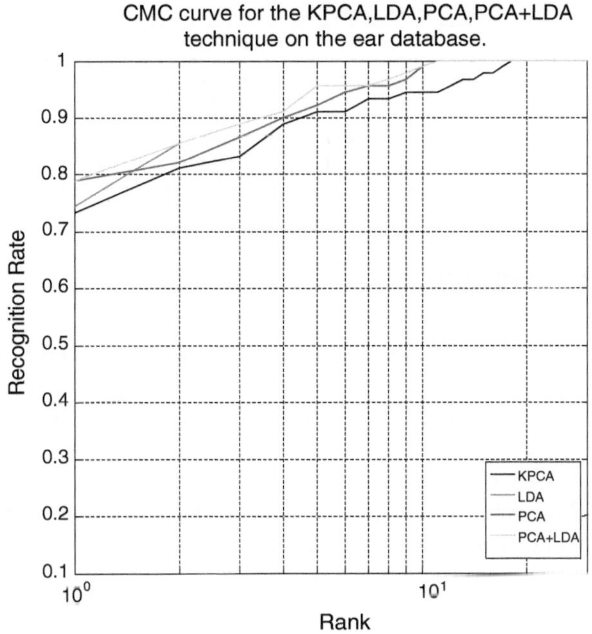

Fig. 2 CMC of PCA, LDA, KPCA, and PCA + LDA

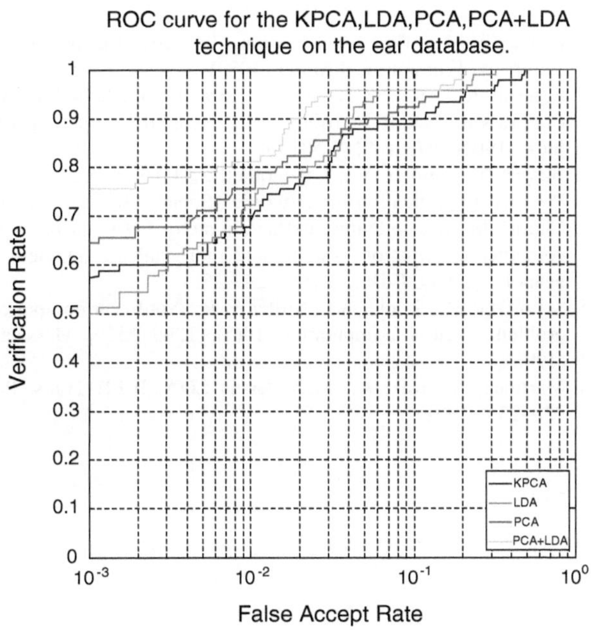

Fig. 3 ROC curves of PCA, LDA, KPCA, and PCA + LDA

Table 1 Performance metrics

Metrics	PCA %	LDA %	KPCA %	PCA + LDA %
Recognition rate	78.89	74.44	73.33	78.89
EER	7.87	5.59	10.04	5.59
Verification rate (1 % FAR)	75.56	72.22	67.78	75.56
Verification rate (0.1 % FAR)	64.44	50	57.78	64.44

Conclusion

We have considered linear algorithms of PCA and LDA, and also have taken very less training samples. Analysis with nonlinear algorithm of KPCA is also performed. PCA, LDA, and combination of PCA and LDA outperform KPCA. This can be improved by considering large datasets.

The results show that combination of PCA and LDA performs well in recognition rate, ERR, and verification rate. Thus, fusion of multialgorithms gives better result than individual algorithm.

References

1. Iannarelli, A.: Ear Identification. Forensic Identification Series. Paramount Publishing Company, Fremont, California (1989)
2. Ramesh, K.P., Rao, K.N.: Pattern extraction methods for ear biometrics - a survey. In: World Congress on Nature & Biologically Inspired Computing, (2009). NaBIC 2009, Coimbatore, India. ISBN: 978-1-4244-5053-4
3. Burger, M., Burger, W.: Ear biometrics. In: Jain A., Bolle R., S. (eds.) Pankanti Biometrics-Personal Identification in Networked Society, Kluwer Academic Publishers, 1999
4. Choras, M.: Image feature extraction methods for ear biometrics–a survey. In: Proceedings of computer information systems and industrial management applications, pp. 261–265, Minneapolis (2007). ISBN: 0-7695-2894-5
5. Scholkopf, B., Smola, A.J., Miller, K.R.: Nonlinear component analysis as a kernel eigenvalue problem. Neural computation **10**(5), 1299–1319, Massachusetts Institute of Technology, (1998)
6. Martinez, A., Kak, A.: PCA versus LDA. IEEE Trans. Pattern Anal. Mach. Intell. **23**(2), 228–233 (2001)

Mobile Agent Model with Efficient Communication for Distributed Data Mining

S. Kavitha, K. Senthil Kumar and K. V. Arul Anandam

Abstract This paper presents the managing of large-scale distributed system with the use of mobile agent concept in distributed environment which is becoming an increasing challenging task. The primary goal is to ensure efficient use of resources, services of computer systems, and networks to provide timely services to the end users using mobile agent with decentralized processing and control in distributed resources so as to minimize the network traffic and speed up management tasks in distributed environment, provide more scalability, and to provide more flexibility in the development and maintenance of the applications. The problem with the traditional techniques is, when the number of visited nodes grows, the mobile agent size also increases and making migration harder. One possible solution to this problem is to visit a fixed number of nodes, return to the agent home or send all data to it (reducing the mobile agent size), and start the task again on the remaining nodes. The initial size of a mobile agent also affects agent performance, since the larger the size, the more difficult the migration.

Keywords Mobile agent · Distributed data mining · Mobile agent model · Communication among agents · Agent architecture

---

S. Kavitha (✉) · K. Senthil Kumar
Department of Computer Applications, SRM University, Chennai, Tamil Nadu, India
e-mail: kavitha.s@ktr.srmuniv.ac.in

K. Senthil Kumar
e-mail: senthil.ku@ktr.srmuniv.ac.in

K. V. Arul Anandam
Govt. Thirumagai Mills College, Gudiyattam, Tamil Nadu, India
e-mail: sakthisivamkva@gmail.com

S. Sathiakumar et al. (eds.), *Proceedings of International Conference on Internet Computing and Information Communications*, Advances in Intelligent Systems and Computing 216,
DOI: 10.1007/978-81-322-1299-7_39,

Introduction

Everyday activity generates large quantities of data like the use of bar code on goods, online business transactions with the advent of Internet, and many more. Because of that, new techniques and tools have to be developed to deal with intelligent and automatic transformation of processed data into useful information and knowledge. The research field of data mining for knowledge extraction is one of the solutions [1].

Distributed data mining (DDM) is the process of mining distributed and heterogeneous datasets. It addresses the specific issues associated with the application of data mining in distributed computing environments, which are typically characterized by the distribution of users, data, hardware, and mining software. DDM is widely seen as a means of addressing the scalability issue of mining large datasets.

There are predominantly two architectural models used in the development of DDM systems, namely, *client–server* (CS) and *software agents* (SA). The "agents" category can be further divided on the basis of whether the agents have the ability to migrate in a self-directed manner or not (i.e., *mobility*) [2].

DDM focuses on analyzing data remotely which are distributed in different interconnected locations. It deals with public datasets available on the Internet, corporate databases within an Intranet, environments for mobile computation, collections of distributed sensor data for monitoring, etc. It offers better scalability and better response time when compared with a centralized model.

Mobile Agent

A mobile agent is a running program that can move from host to host in a network which created a new paradigm for data exchange and resource sharing in rapidly growing and continually changing computer network [3].

It is capable of migrating autonomously and intelligently in various target nodes through network to perform computation in response to changing conditions in the network environment [4]. The objectives of the agent dispatcher are to realize and satisfy the user's goal on behalf of user. It is used for searching information, retrieval of information, filtering, and intruder recognition in networks. Mobile agent suspends its execution, transfer itself from one networked host to another, and resume execution on the new host. Mobile agent technology has the advantages of the bandwidth conservation, reduction of completion time, reduction of network traffic, overcoming of network delay, enabling asynchronous (disconnected) execution/communications, load balancing, and enhancement of dynamic adaptability/deployment [5]. Mobile agents enable flexible, dynamic network building, provide efficient discovery of network elements that violate normal behavior and offer remote maintenance features. It reduces bandwidth consumption by moving the data processing elements to the location of the sensed data.

Mobile agent allows complete mobility of cooperating applications in distributed system. It can either follow a pre-defined path on the network or determine its itinerary based on the data collected from the network. It has highly dynamic movement of code and data. It allows optimizing between the requirements of low bandwidth, high latency, and disconnected network connections [6].

Proposed Model of Mobile Agent

The agent-based model is a popular approach to constructing DDM systems and is characterized by a variety of agents coordinating and communicating with each other to perform the various tasks of the data mining process. Agent technology is seen as being able to address the specific concern of increasing scalability and enhancing performance by moving code instead of data and thereby reducing the communication overhead incurred in the client–server (CS) model. Mobile agents do not waste the bandwidth, because the agent migrates to the server as shown in Fig. 1. The agent performs the necessary sequence of operations locally, and returns just the final result to the client [7, 8].

Drawbacks of Client Server Model

The major drawback in the CS-based DDM model is that huge amount of datasets migrate from the data sources locations to the DM sever to accomplish the required DM process. This results into a considerable waste in the network bandwidth and consequently a big increase in latency. A typical mobile agent-based DDM process begins with a client request for a DM process. The client determines the required data severs for the DM process and multicasts a set of mobile agents data miners (MADMs). The MADMs migrate to the data servers and perform the data mining operations locally and return the final results to the

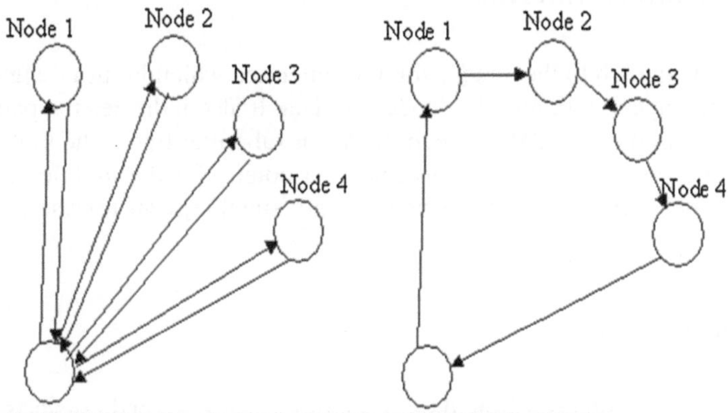

Fig. 1 Client server and mobile agent models

client. Finally, the client uses a knowledge integration (KI) program to integrate the DM results from the different MADMs [7].

Mobile Agent Model

Use of mobile agent technology proves to be an efficient solution [9]. The challenging task is, when the number of visited nodes grows, the mobile agent size also increases and making migration harder. One possible solution of this problem is to visit a fixed number of nodes, return to the agent home or send all data to it (reducing the mobile agent size), and start the task again on the remaining nodes. The initial size of a mobile agent also affects agent performance, since the larger the size, the more difficult the migration. So, instead of fixing the number of nodes, a new model is introduced for avoiding and taking of too much time for traveling across the network. We propose a model, in which the mobile agent is introduced with the use of agent submitter interface and facilitator agent in DDM. After executing the tasks with mobile agent, tasks are stored in local data warehouse. This reduces the work load of mobile [10] agent container's and avoid migrating to home. Facilitator agent is used to mine the tasks from all the warehouses and convert into global mining which are stored in the user information database [11] as shown in Fig. 2.

Graphical User Panel

Data mining tools were added for visualization. It plays an important role in making the discovered knowledge understandable and interpretable by humans.

Agent Submitter Interface

User submits the job to the agent submitter interface which in turn assigns the jobs to the mobile agent with mobility code and data. It also indicates the path to travel on network based on the data collection. Agent submitter tracks the mobile agent's progress and position. It is a communication point with the mobile agent which does pre-process the returned information into suitable form for user.

Facilitator Agent

Facilitator agent collects all the data from various sources and converts them into global data. It reduces the load of the mobile agent for easy movement, speed up the process and visit large scale in distributed environment.

Algorithm

Step 1: Mobile agent (MA) launches from agent submitter interface based on the client.
Step 2: MA initialized with the load state in the server, perform the local functions, save the current states and store the task in the local data warehouse.
Step 3: MA have the number of nodes to visit and does step 2 process in every node till it reaches the last node.
Step 4: Likewise, MA moves to all the server, does all tasks store them in the local data warehouse.
Step 5: If it is the last node and last server, it returns to the home.
Step 6: Facilitator agent is used to mine all the data which are stored in the various server data warehouse, and convert them into global data and display the results.

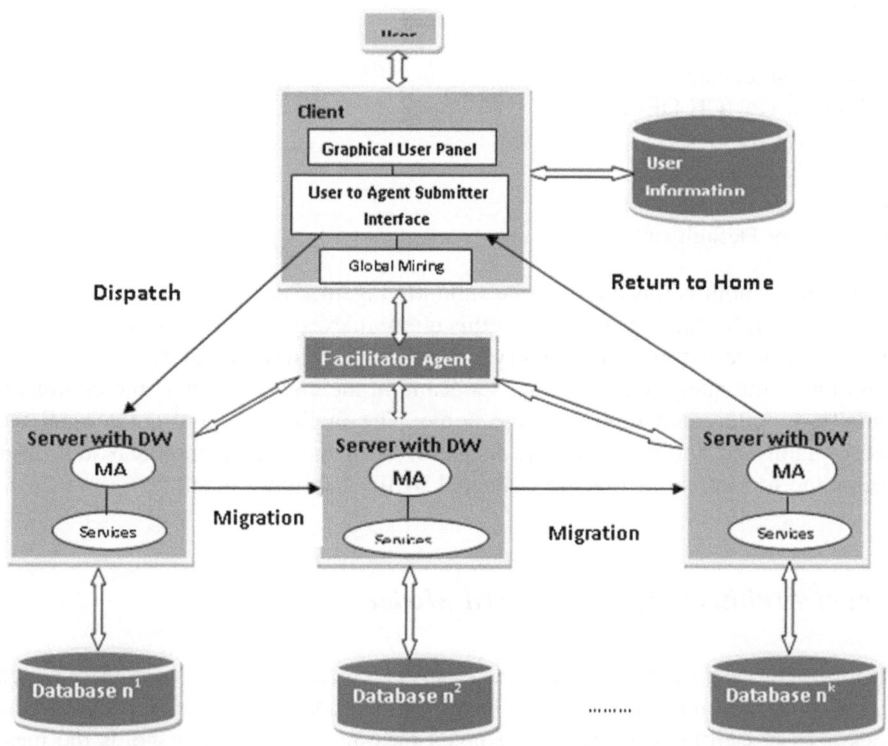

Fig. 2 Proposed model of mobile agent. MA, Mobile agent; DW, Datawarehouse

Communication Among Agents in Proposed Model

Agents are able to interact and communicate with each other which require a common language, an agent communication language or ACL [12]. Adoption of ACL in this model will be able to communicate and coordinate, allow negotiating the follow of information and providing increased transparency.

As an illustration in ACL Language, with Sender as server and Receiver as client,

Agent [13] user1 asks for the price of a share of Dell stock in all share markets in the world.

Ask – one

- Sender user1
- Content (PRICE DELL? price)
- Receiver stock-server
- Reply-with dell-stock
- Language Default-language
- Ontology Default-ontology

The stock server sends a reply to user1

Tell

- Sender stock-server
- Content (PRICE DELL? list)
- Receiver user1
- In-reply-to dell-stock
- Language Default-language
- Ontology Default-ontology

The user submits this query to the submitter agent. This query is assigned to the mobile agent. Mobile agent executes this query in every server, save the processed data (Dell price list in every country's share market) in the local database, moves with this same query and processes the same in the entire server in the continent. Finally, mobile agent returns to the home. The facilitator agent collects all the information and submits to client database. With the use of graphical tool, user views the formations which are collected by the facilitator agent.

Agent Architecture in Proposed Model

Mobile agent is a light-weight agent, because it carries only the instructions which are given by the interface agent on behalf of actor. Message manager handles the messages passed between actors. Delayed message [14] manager holds the messages for mobile agent while they are moving from one site to another site. Actor

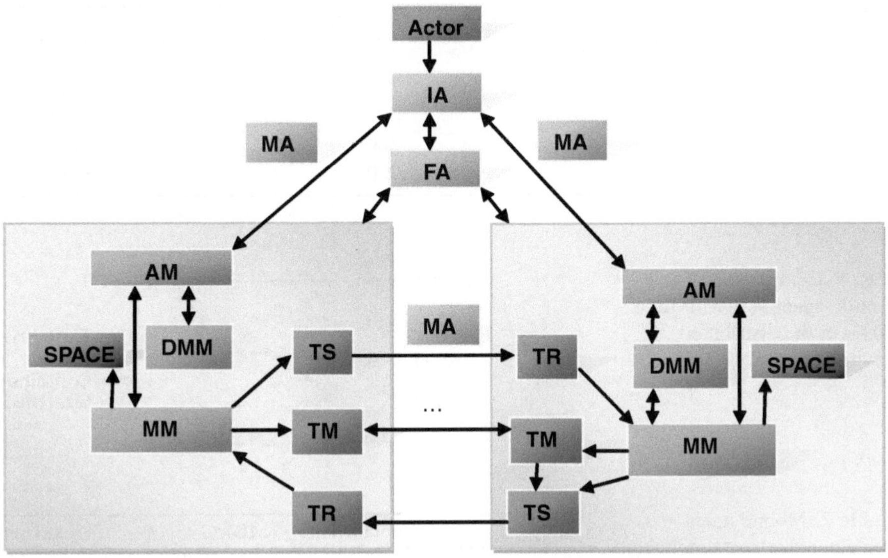

Fig. 3 Agent architecture *MA* Mobile Agent, *IA* Interface Agent, *FA* Facilitator Agent, *AM* Action Manager, *DMM* Delayed Message Manager, *MM* Message Manager, *TS* Transport Sender, *TM* Transport Manager, *TR* Transport Receiver

manager manages the state of actors that are currently executing and also the locations of mobile actors created on the site. Space provides middle agent services such as matchmaking and brokering services. The mail addresses or names of all agents are not globally known in the open agent system and also an agent may not have the other agent's addresses with whom it needs to communicate. For this, middle agent services need to be supported. Transport Manager contains a public port for message passing between different sites. Transport sender on the same platform receives the message from the message manager and delivers it to the transport receiver in the different site. Facilitator agent collects formations from all site's databases and submit to the clients database. (Global mining). The detailed architecture is shown in Fig. 3.

Experimental Analysis and Results

To measure the efficiency of the proposed algorithm, we have set up distributed system with different configurations. Tests were conducted with fixed number of nodes. Table 1 and Fig. 4 show the mobile agent without fixing number of nodes. If the mobile agent container size increases, the time taken to move in various nodes for processing also increases.

Table 1 Mobile agent container size and time taken

S. no	Container size (MB)	Time (ms)
1.	100	1,000
2.	300	2,200
3.	500	3,500
4.	800	4,600
5.	1,000	5,200

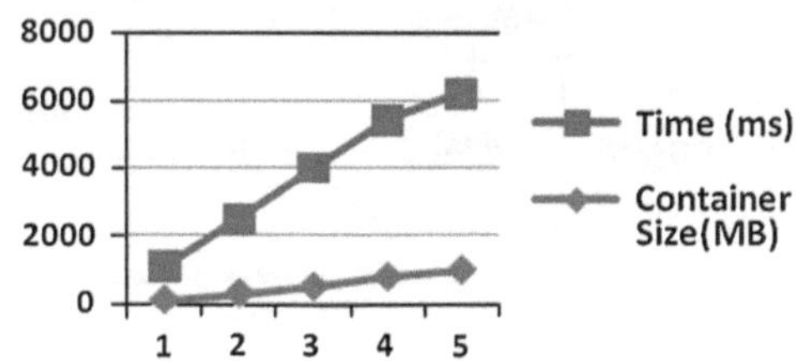

Fig. 4 Comparison of mobile agent size with time taken in the distributed environment using Table 1

Table 2 Mobile agent container size and time taken using proposed model

S. no	Container size (MB)	Time (ms)
1.	100	2,000
2.	100	1,850
3.	100	1,590
4.	100	1,430
5.	100	1,200

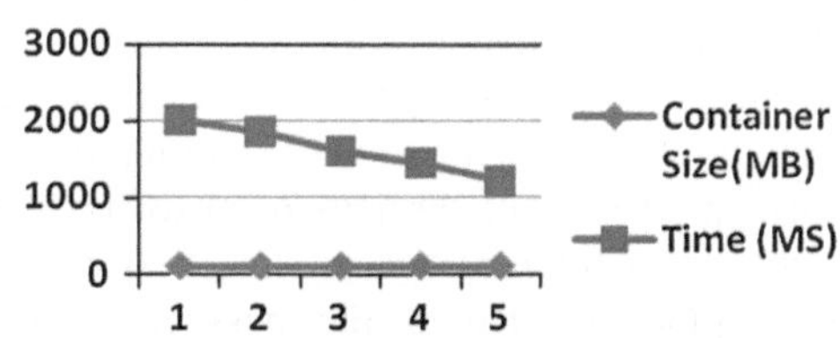

Fig. 5 Comparison of mobile agent size with time taken in the proposed model using Table 2

After that the test was conducted in the proposed model. The mobile agent container size is constant (contains content language), as in Table 2. The speed or time taken to visit the fixed nodes for processing is decreased, because it reduces the load of the mobile agent container with fixed number of nodes visiting for processing as shown in Fig. 5.

Conclusion

This work describes a new mobile agent model, communication among agents, and also a agent architecture for the management of distributed systems. This mobile agent framework differs from other frameworks in the sense that it proposes to reduce the load of the mobile agent container and covers large scales for processing.

References

1. Krishnaswamy, S.: Introduction - Agents and Mobile Agents, Lecturer Notes
2. Krishnaswamy, S., Zaslavsky, A., Loke, S.W. Techniques for estimating the computation and communication costs of distributed data mining. Research work of Department of Industry, Science and Tourism, Commonwealth Govt. of Australia
3. Kulkarni, U.P., Tangod, K.K., Mangalwede, S.R., Yardi, A.R. Exploring the capabilities of mobile agents in distributed data mining. In: Proceeding IDEAS Proceedings of the 10th International Database Engineering and Applications Symposium pp. 277–280 IEEE Computer Society Washington, DC, USA ISBN:0-7695-2577-6 (2006)
4. Mitrovic, D., Budimac, Z., Ivanovic, M., Vidakovic, M.: Improving fault-tolerance of distributed multi-agent systems with mobile network-management agents. In: Proceedings of the International Multiconference on Computer Science and Information Technology, pp. 217–222. Wisła (2006)
5. Ku, H., Luderer, G.W.R., Subbiah, B.: An intelligent mobile agent framework for distributed network management. Network System Laboratory, Telecommunication Research Centre, Global Telecommunications Conference, IEEE, Phoenix (1997)
6. Patel, R.B., Garg, K.: Distributed banking with mobile agents: an approach for e-commerce. WSEAS Trans. Comp. **3**(1), (2004), ISSN:1109-2750
7. Ogunde, A.O., Folorunso, O., Sodiya, A.S., Oguntuase, J.A., Ogunleye, G.O. Improved cost models for agent–based association rule mining in distributed databases. Anale SEria Informatica **IX** fasc 1-, pp. 231–250 (2011)
8. Outtagarts, A.: Mobile agent-based applications: a survey. Int. J. Comp. Sci. Netw. Secur. **9**(11), 331–339 (2009)
9. Dale, J., DeRoure, D.C. A mobile agent architecture for distributed information management. Presented at VIM University of Southampton, UK, March (1997)
10. Zhou, D., Rao, W., Lv, F.: A multi-agent distributed data mining model based on algorithm analysis and task prediction. In: IEEE 2nd International Conference on Information Engineering and Computer Science, Wuham, 25–26 Dec (2010)
11. Kavitha, S., Senthil Kumar, K., Dr., K.S. Anandam, K.V.A.: Mobile agent model for distributed data mining. J. Inform. Commun. Technol**2**(6), (2012)
12. Jang, M.-W., Ahmed, A., Agha, G.: Efficient agent communication in multi-agent systems. In: Proceedings of SELMAS 2004, LNCS 3390, pp. 236–253, Springer-Verlag, Berlin Heidelberg ISSN 2047-3168 (2005)
13. Berna-Koes, M., Nourbakhsh, I., Sycara, K.: Communication efficiency in multi-agent systems. In: International Conference on Robotics and Automation, Vol.**3**, pp. 2129–2134, Apr 26–May 1 (2004)
14. Tarr, B., Nebesh, D., Foster, S.: Introduction to mobile agent systems and applications. In: Tools USA (2000)

Development of Locally Weighted Projection Regression for Concurrency Control in Computer-Aided Design Database

A. Muthukumaravel, S. Purushothaman and R. Rajeswari

Abstract Concurrency control (CC) is the activity of synchronizing operations issued by concurrently executing programs on a shared database. Concurrency control is an important concept for proper transactions on objects to avoid any loss of data or to ensure proper updating of data in the database. This paper presents development of locally weighted projection regression (LWPR) for concurrency control while developing bolted connection using Autodesk inventor 2008. While implementing concurrency control, this work ensures that associated parts cannot be accessed by more than one person due to locking. The LWPR learns the objects and the type of transactions to be done based on which node in the output layer of the network exceeds a threshold value. Learning stops once all the objects are exposed to LWPR. We have attempted to use LWPR for storing lock information when multi users are working on computer-aided design (CAD).

Keywords Concurrency control · Locally weighted projection regression · Transaction locks · Time stamping

A. Muthukumaravel (✉)
Department of MCA, School of Computing Sciences, VELS University,
Chennai 600 117, India
e-mail: muthu14673@gmail.com

S. Purushothaman
PET Engineering College, Tirunelveli District, Tamil Nadu 627 117, India
e-mail: drsppuru@gmail.com

R. Rajeswari
Department of Computer Science, Mother Teresa Womens University,
Kodaikanal, India
e-mail: rajeswaripuru@gmail.com

S. Sathiakumar et al. (eds.), *Proceedings of International Conference on Internet Computing and Information Communications*, Advances in Intelligent Systems and Computing 216,
DOI: 10.1007/978-81-322-1299-7_40,

Introduction

Concurrency control is a set of mechanisms used to maintain consistency in transactions of databases. Proper locking method for controlled transactions. The most common way in which access to items is controlled by “locks-L”. Lock manager is the part of a database management system (DBMS) those records, for each item ‘O’, whether one or more transactions are reading or writing any part of ‘O’. If so, the manager will forbid another transaction from gaining access to ‘O’, provided the type of access (read or write) could cause a conflict. The lock manager can store the current locks in a lock table which consists of records with fields (< object-‘O’ > , < lock type-‘L’ > , < transaction-‘T’ >) the meaning of record (‘O’, ‘L’, ‘T’) is that transaction ‘T’ has a lock of type ‘L’ on object ‘O’ [1-3].

Concurrency problems in data base systems arise from controlling concurrent updates on a shared data base so that (1) no lost updates can occur, (2) data base consistency can be maintained, and (3) noncascade backup is possible.

The advantages of two phase locking (2PL) and time stamping have been taken as the initial starting point and improvements have been achieved by implementing LWPR for locking objects to achieve concurrency control in long time transactions of CAD/CAM database system.

The concurrency control requires proper locking method for controlled transactions. The most common way in which access to items is controlled by “locks”. Locks are used for accessing objects [4-8].

In this research work, LWPR has been implemented. The performances of the algorithm have been compared based on the following criteria.

1. Locking time for each object
2. Releasing time for each object
3. Total Locking time for each transaction group
4. Total Releasing time for each transaction group.

Locally Weighted Projection Regression

LWPR is an algorithm that achieves nonlinear function approximation in high dimensional spaces even in the presence of redundant and irrelevant input dimensions [3, 14, 15]. At its core, it uses locally linear models, spanned by a small number of univariate regressions in selected directions in input space. This nonparametric local learning system

1. learns rapidly with second order learning methods based on incremental training.
2. Uses statistically sound stochastic cross validation to learn.

3. Adjusts its weighting kernels based on local information only.
4. Has a computational complexity that is linear in the number of inputs.
5. Can deal with a large number of—possibly redundant and irrelevant—inputs.

The structure of the event loop is shown in Fig. 1. The algorithm is at one of the four action states at any given point of time. The INITIALIZE phase is used to initialize the LWPR and read in the script variables from the script file and fill in default values for those variables not specified in the script file.

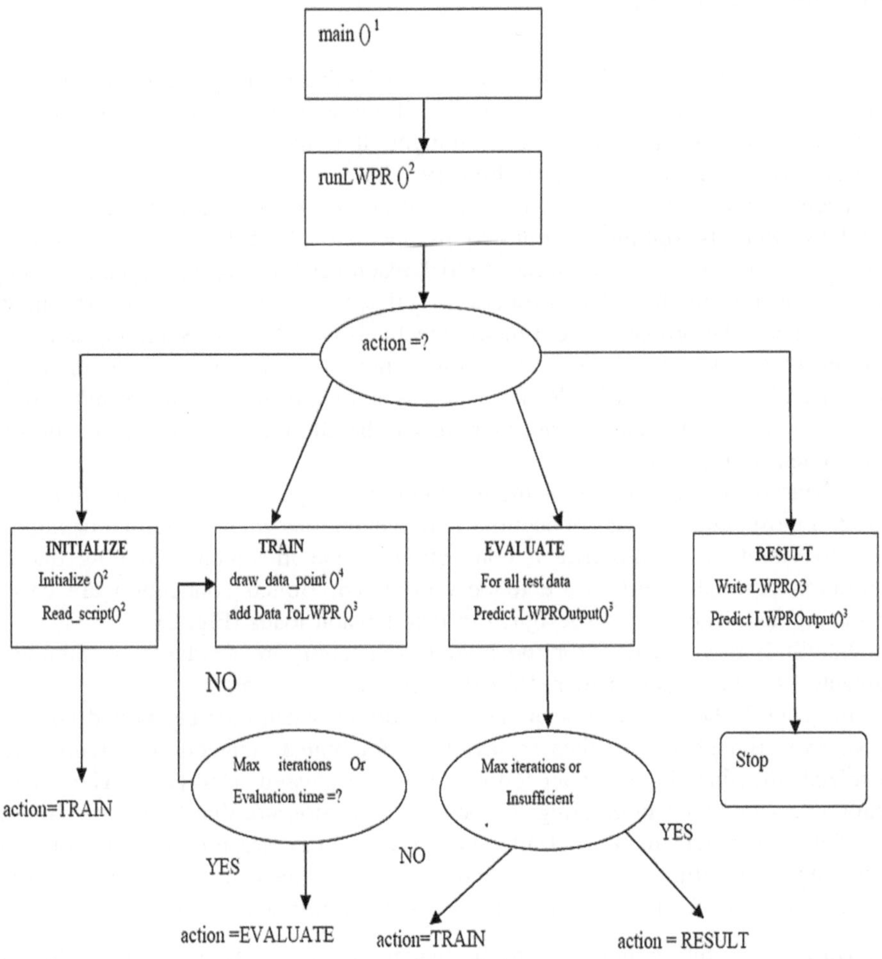

Fig. 1 LWPR training modules Filenames: 1.lwpr_main(), 2.lwpr_test(), 3.lwpr(), 4.utilities()

The TRAIN phase of the algorithm draws data from the training data set file and trains the local model on it. After every 'evaluation', the program goes into the EVALUATE phase where the learned model is tested against the novel (test) data set.

When the number of iterations has exceeded the 'max_iterations' count or the change of normalized mean squared error (nMSE) between the last EVALUATE phase and the current, a checking is done to find if the values is below a THRESHOLD. In such case, the program goes into the RESULT phase during which, it saves the learned LWPR.

Problem Definition

There is inability to provide consistency in the database when long transactions are involved. It will not be able to identify if there is any violation of database consistency during the time of commitment. It is not possible to know, if the transaction is with undefined time limit [9, 10].

There is no serializability when many users work on shared objects. During long transactions, optimistic transactions and two phase locking will result in deadlock. Two phase locking forces to lock resources for long time even after they have finished using them. Other transactions that need to access the same resources are blocked. The problem in optimistic mechanism with Time Stamping is that it causes repeated rollback of transactions when the rate of conflicts increases significantly [11]-[13]. LWPR has been used to manage the locks allotted to objects and locks are claimed appropriately to be allotted for other objects during subsequent transactions.

A bearing-type connection is the most common type of bolted connection. It is used in most simple-shear connections and in situations when loosening or fatigue due to vibration. A slip-critical connection is one in which loosening due to vibration or load reversals are to be considered. Bolted connection are easily disassembled. They can be designed to take tension loads (Fig. 2).

Inbuilt library drawing for the bolted connection are available in Autodesk inventor 9. Drawings considered for this research work are:

In general, the following sequences are formed when creating bolted connection. Even though library files are available for bolted connection drawing, customized drawing bolted connection file is discussed [16]-[19]. The major parameters involved in creating the bolted connection are diameter (outer diameter/inner diameter), depth (hole)/length (extension or extrusion or projection) and chamfering (slanting). The various constraints that have to be imposed during modifications of features by many transactions on this bolted connection are.

1. During chamfering, both length and diameter have to be write locked at the first transaction T1.
2. During diameter modification, chamfering has to be locked.
3. During length or depth modification, diameter has to be locked.

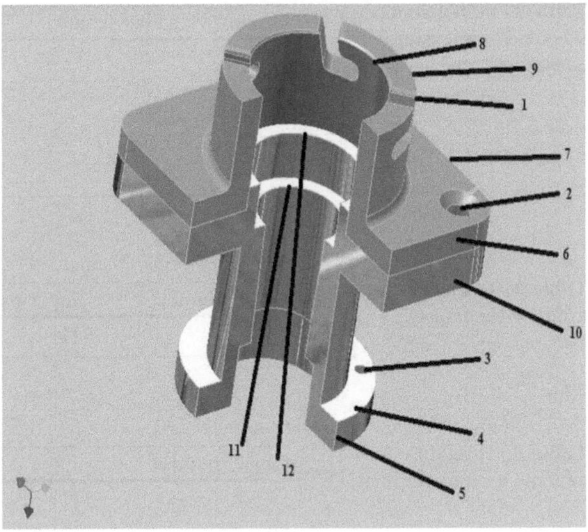

Fig. 2 Bolted connection *1* Locking slot – 4 numbers, *2* Bolt Fixing – 1, *3* Bolt Fixing – 2, *4* Lower base support width of annular, *5* Lower base support height, *6* Height of bolt fixing, 7 Square upper late, *8* Inner diameter, *9* Outer diameter, *10* Square lower plate, *11* Concentric 1, *12* Concentric 2

4. This bolted connection has two major entities.
1. Features 1-2-10-11-12 (set 1)
2. Features 3-4 (set 2)
3. Features 5-6-7-8-9 (set 3)

Set 1, set 2 and set 3 can be made into individual drawing part files (part file 1 part file 2 and part file 3) and combined into one assembly file (containing the part files 1, 2, and 3 are intact). When the transactions are accessing individual part files, then transactions in part file 1 about the type of transactions in part files 2, 3 and vice versa. However, when the part files 1, 2 and 3 are combined into a single assembly file, then inconsistency in the shape and dimension of the set 1, set 2 and set 3, during matching should not occur. Hence, provisions can be made in controlling the dimensions and Shapes with upper and lower limits confirming to standards. At any part of time when a subsequent transaction is trying to access locked features, one can modify the features on the system and store as an additional modified copy of the features with time stamping and version names (allotted by the transaction/allotted by the system).

Let us assume that there are two transactions editing the bolted connection. Transaction 1 edits O_1 and hence O_8 and O_9 will be written locked sequentially (Table 1). Immediately transaction 2 wants to edit O_2 and hence O_6, O_7 will be locked in sequence. Any other transaction can access O_{10} or O_4 or O_3.

Initially, transaction 1 and transaction 2 have opened the same bolted connection file from the common database. The following steps shows sequence of execution.

Step 1: For the first time, a pattern is generated as soon as a transaction accesses the object. T_1 access O_1 with write mode. Table 2 shows pattern formed for the OL training.

Table 1 Bolted connection shape and dimension consistency management

Group	First feature	Remaining feature to be locked
G1	1	8,9
G2	2	6,7
G3	10	12
G4	4	5,11
G5	3	4

Table 2 T_1 First pattern used for training OL

Object number	Input pattern	Target output pattern
O_1	[1]	[0 1 0]

Table 3 T_1 Additional patterns used for training OL

Object number	Input pattern	Target output pattern
O_1	[1]	[0 1 0]
O_8	[8]	[0 1 0]
O_9	[9]	[0 1 0]

Table 4 T_2 Additional patterns used for training OL

Object number	Input pattern	Target output pattern
O_1	[1]	[0 1 0]
O_8	[8]	[0 1 0]
O_9	[9]	[0 1 0]
O_2	[2]	[0 1 0]

Step 2: The transaction manager locks objects mentioned in the third column of Table 1 Now, repeat step 1 with the patterns given in Table 3.

Step 3: A new transaction T_2 accesses O_2. A pattern is formed to verify if lock has been assigned to O_2 and its associated objects O_6, O_7. Only when the locks are not assigned to O_6 and O_7 then T_2 is allowed.

The patterns of additional transactions are presented to the OL testing module to find if the output [0 0 0] is obtained in the output layer. During OL testing, the final weights obtained during OL training will be used. Otherwise, it means that lock has been assigned to either O_2 or O_6 or both. In such case, transaction is denied for T_2. Else the following patterns in Table 4 are presented in step 1.

Step 4: To know the type of lock value assigned to an object and for a transaction, OL testing is used. OL testing uses the final weights created by OL training.

Results

Figures 3, 4, 5, and 6 shows the preformance of CC during locking and releasing of bolted connection object. For bolted connection object the performances have been evaluated using LWPR under controlled environment.

The performance of the LWPR algorithm has been presented on the following criteria.

1. Releasing time for each object in bolted connection (Fig. 3).
2. Locking time for each object in bolted connection (Fig. 4).
3. Total Locking time for each transaction group in bolted connection (Fig. 5).
4. Total Releasing time for each transaction group in bolted connection (Fig. 6).

Bolted connection drawing

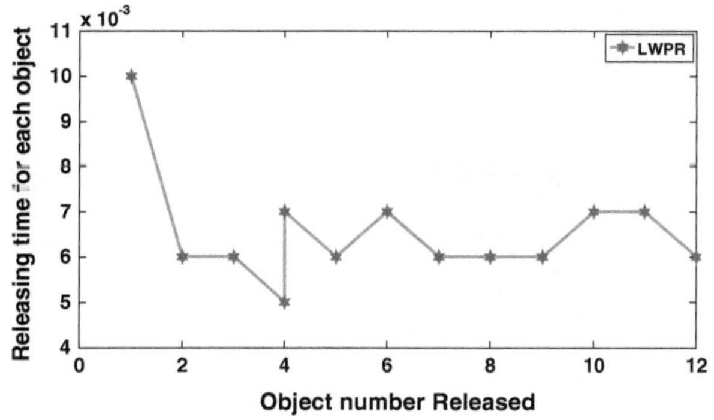

Fig. 3 Releasing time for each object in bolted connection

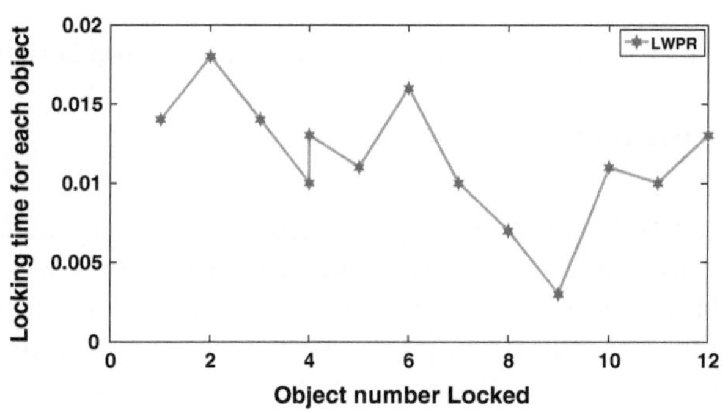

Fig. 4 Locking time for each object in bolted connection

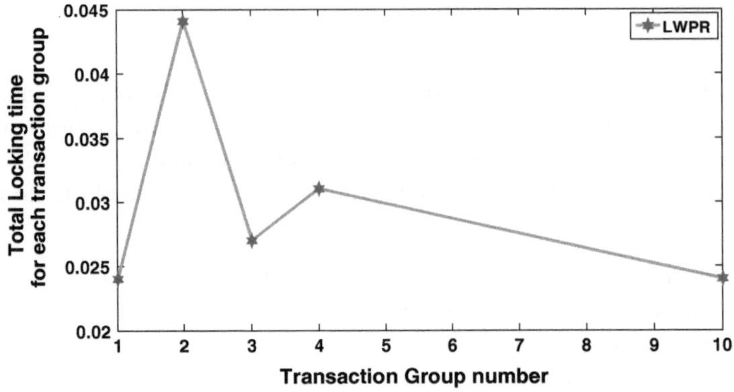

Fig. 5 Total locking time for each transaction group in bolted connection

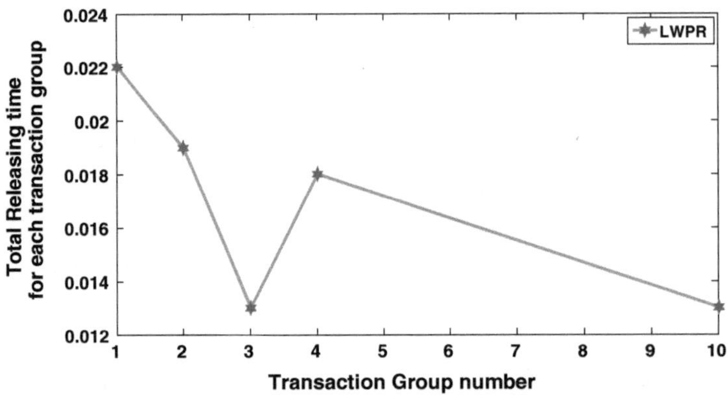

Fig. 6 Total releasing time for each transaction group in bolted connection

Conclusion

A LWPR has been implemented for providing concurrency control to maintain consistency in the CAD database. The performance of the LWPR algorithm is based on the Locking time and releasing time for each object, Total Locking time for each transaction group, Total Releasing time for each transaction group and Computation complexity. The LWPR algorithm takes less number of iterations to reach a stable state.

References

1. Buhr, P.A., Harji, A.S., Lim, P.E., Chen, J.: Object-oriented real-time concurrency. In: Proceedings of the 15th ACM SIGPLAN Conference on Object-Oriented Programming Systems, Languages and Applications, pp. 29–46, ACM New York, NY, USA (2000)
2. Raviram, P., Wahidabanu, R.S.D., Purushothaman, S.: Concurrency control in CAD with KBMS using counter propagation neural network. IEEE International Advance Computing Conference, pp. 1521–1525, 6–7 March, Patiala (2009)
3. Purushothaman, S., Elango, M.K., Nirmal Kumar, S.: Application of Hilbert Huang transform with locally weighted projection regression method for power quality problems. Int. Rev. Elect. Eng. **5**(5) 2405–2412 (2010)
4. Nizamuddin, M.K., Sattar, S.A.: Data count driven concurrency control scheme with performance gain in mobile environments. J. Emerg. Trend. Comput. Inform. Sci. 2(2), 106–112 (2010)
5. Moiz, S.A., Rajamani, L.: An algorithmic approach for achieving concurrency in mobile environment. In: INDIACom, pp. 209–211, (2007)
6. Lingam, K.M.P.: Analysis of real-time multi version concurrency control algorithms using serialisability graphs. Int. J. Comp. Appl. (0975 - 8887) **1**(21), 57–62 (2010)
7. Han, Q., Pan, H.: A concurrency control algorithm access to temporal data in real-time database systems. In: International multi symposiums on computer and computational sciences, pp. 168–171, IEEE Computer Society Washington, DC, USA (2008)
8. Choe, T.-Y.: Optimistic concurrency control based on cache coherency in distributed database systems. Int. J. Comp. Sci. Netw. Secur. 8(11), 148–154 (2008)
9. Nizamuddin, M.K., Sattar, S.A.: Adaptive valid period based concurrency control without locking in mobile environments. In: Recent Trends in Networks and Communications, vol. 90, part 2, pp. 349–358. Springer CCIS, Springer, Heidelberg (2010)
10. Yadav, A.V., Agarwal, A.: An approach for concurrency control in distributed database system. Int. J. Comp. Sci. Commun. 1(1), 137–141 (2010)
11. Arumugam, G., Thangara, M.: An efficient locking model for concurrency control in OODBS. Data Sci. J. **4**(31), 59–66 (2005)
12. Singh, P., Yadav, P., Shukla, A., Lohia, S.: An extended three phase commit protocol for concurrency control in distributed systems. Int. J. Comp. Appl. **21**(10), 0975–8887 (2011)
13. Guo, J.: An exploratory environment for concurrency control algorithms. Int. J. Comp. Sci. **1**(3), 203–211 (2006)
14. Vijayakumar, S., Schaal, S.: Locally weighted projection regression: an O(n) algorithm for incremental real time learning in high dimensional spaces. In: Proceedings ICML, vol. 1, pp. 288–293, Los Angeles, USA (2000).
15. Vijayakumar, S., Schaal, S.: Locally weighted projection regression: an *O(n)* algorithm for incremental real time learning in high dimensional space. In: Proceedings of 17th International Conference on Machine Learning, pp. 1079–1086, Los Angeles, USA (2000)
16. Ghosh, S.K., Islam, M.S., Lee, S.-Y., Liou, R.-L.: A multi-granularity locking model for concurrency control in object – oriented database systems. IEEE Trans. Knowl. Data Eng. 8(1), 144–155, (1996)
17. Kuo, T.-W., Wu, J., Hsih, H.-C.: Real-time concurrency control in a multiprocessor environment. IEEE Trans. Parallel Distrib. Syst. **13**(6), 659–671 (2002)
18. Park, S.-K.: Seismic performance test of bolted connections between precast-concrete column and H-beam. In: 8th Russian-Korean International Symposium on Science and Technology 2004 Proceedings, vol. 2, pp. 335–339, Ulsan University, South Korea
19. Rahman, M.A.: On analytical performance measurement of concurrency control protocols in DBMS. Int. J. Comp. Elect. Eng. **1**(3), 284–287 (2009)

Performance Study of Combined Artificial Neural Network Algorithms for Image Steganalysis

P. Sujatha, S. Purushothaman and R. Rajeswari

Abstract Steganalaysis is a technique for detecting the presence of hidden information. Artificial neural network (ANN) is a widespread method for steganalysis. Back propagation algorithm (BPA), radial basis function (RBF), and functional update back propagation algorithm (FUBPA) are some of the popular ANN algorithms for detecting hidden information. Training and testing performance is improved when two algorithms are combined instead of using them separately. This paper analyzes the performance of combined algorithms of BPARBF and FUBPARBF. Among the two combinations FUBPARBF provides promising results than BPARBF since FUBPA uses less number of iterations for the network to converge. But still organizing the retrieved information is a challenging task.

Keywords Back propagation algorithm · Functional update back propagation algorithm · Radial basis function · Carrier image · Information image

Introduction

The technology of covert communication based on steganography develops increasing attention in academics and government. An important subdiscipline of information hiding is steganography. Steganalysis mainly discusses the quality of

P. Sujatha (✉)
Department of Computer Science, VELS University, Chennai 600117, India
e-mail: sujinagi@gmail.com

S. Purushothaman
PET Engineering College, Tirunelveli, Tamil Nadu 627117, India
e-mail: drsppuru@gmail.com

R. Rajeswari
Mother Teresa Women's University, Kodaikanal, India
e-mail: rajeswaripuru@gmail.com

S. Sathiakumar et al. (eds.), *Proceedings of International Conference on Internet Computing and Information Communications*, Advances in Intelligent Systems and Computing 216,
DOI: 10.1007/978-81-322-1299-7_41,

breaking the steganography methods. Image steganography is defined as the covert embedding of data into digital pictures. The scope of using various types of carrier file formats is high, but digital images are the most popular as carriers because of their frequency on the Internet. Two early steganographic methods are: replacing the least significant bit (LSB) plane of an image with a message; and adding a message bearing signal to image [1]. Image steganography can be grouped into three categories: (1) spatial domain, (2) transformation domain, and (3) quantization index modulation [2]. The LSB insertion/modification is considered as a difficult one to detect. Popular transformations include the two-dimensional discrete cosine transformation (DCT) [3], discrete Fourier transformation (DFT) [4], and discrete wavelet transformation (DWT) [5], that are commonly used in image steganography.

Literature Survey

Supervised learning methods construct a classifier to differentiate between stego and nonstego images using training examples. Supervised learning methods using neural networks as classifiers, gained much importance in recent studies on steganalysis [6, 7]. Describing the supervised learning steganalysis method in a general scenario, some image features are first extracted then given as training input to a learning machine. These inputs include both stego and nonstego messages. The learning classifier iteratively updates its classification rule based on its prediction and the ground truth. Upon convergence, the final stego classifier is obtained. Some of the major advantages using supervised learning-based steganalysis are: (1) construction of universal steganalysis detectors using learning techniques and (2) usage of several freely available software packages.

An early approach using neural networks was proposed [8], to detect arbitrary hiding schemes. They design a feature set based on image quality metrics, metrics designed to mimic the human visual system. A variation of passive steganalysis is active steganalysis that deals in estimating the length of the secret message and the extraction of actual contents of the message [9]. Most of the present literature on steganalysis follows either a blind model [10, 11] or a parametric model. Significant work has been done in detecting steganography using image statistical observations. LSB insertion in raw pixels results in specific changes in the image gray-scale histogram, which can be used as the basis for its detection [12]. Using genetic algorithm (GA) and higher order statistics, feature generation and classification approach is proposed for universal steganalysis [13]. Binary similarity method is used for steganalysis by capturing 7th and 8th bit planes of the nonzero DCients from JPEG images. Also 14 different features were computed to construct a support vector machine classifier [14].

Using SVM classifier, the characteristic vector input is classified into SVM derived by character substitution in texts [15].

Methodologies

In this research work, a systematic approach has been developed to train different artificial neural network (ANN) algorithms with different architectures for steganalysis. Algorithms proposed are:

1. Back propagation algorithm (BPA) neural network
2. Functional update back propagation algorithm (FUBPA)
3. Radial basis function (RBF) neural network

ANN requires training and testing process. In the training process, patterns are generated from the covert image. During the testing process, the same or different covert images are presented to the ANN topology to estimate the presence of any useful hidden information (Fig. 1).

Back Propagation Algorithm

The BPA uses steepest-descent method to reach a global minimum. The number of layers and number of nodes in the hidden layers are decided. The connections between nodes are initialized with random weights. A pattern from the training set is presented in the input layer of the network and the error at the output layer is calculated. The error is propagated backwards. Toward the input layer and the weights are updated. This procedure is repeated for all the training patterns. This forms one iteration. At the end of all iterations, test patterns are presented to ANN, and the prediction performance of ANN is evaluated. Further training of ANN is continued till the desired prediction performance is reached.

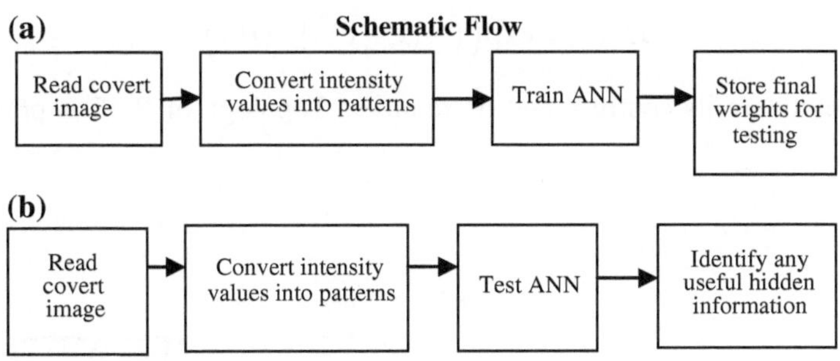

Fig. 1 Schematic flow (**a**) Training phase (**b**) Testing phase

Functional Update Method Algorithm

When the network is trained with analog data, the number of iterations is large for the objective function (J) to reach the desired mean squared error (MSE) while using conventional (BPA). The objective function does not reach the desired MSE due to some local minima, whose domains of attractions are as large as that for the global minimum. The network converges to one of those local minima, or the network diverges. The updating of the weights will not stop, unless every input is outside the significant update region (0.1–0.9), and the outputs of the network will be approaching either 0 or 1. This requires much iteration for the network to converge. To overcome these difficulties, a functional criterion, which results in faster convergence of the network, is used.

The main idea of this method is that the weights of this network are updated only when any one of the nodes in the output layer of the network is misclassified. Even if one of the nodes in the output layer is not misclassified, no updating of weights is done. A node in the output layer is misclassified, if the difference between the desired output and the network is greater than 0.5.

Radial Basis Function

A RBF is a real-valued function whose value depends only on the distance from the origin. If a function ‘h’ satisfies the property h(**x**) = h(||**x**||), then it is a radial function. Their characteristic feature is that their response decreases (or increases) monotonically with distance from a central point. The centre, the distance scale, and the precise shape of the radial function are parameters of the model, all fixed if it is linear.

A typical radial function is the Gaussian which, in the case of a scalar input, is

$$H(\mathrm{x}) = \exp\left(\left(-(x-c)^2\right)/\left(r^2\right)\right)$$

Its parameters are its centre c and its radius r. Topology of RBF is presented in Fig. 2.

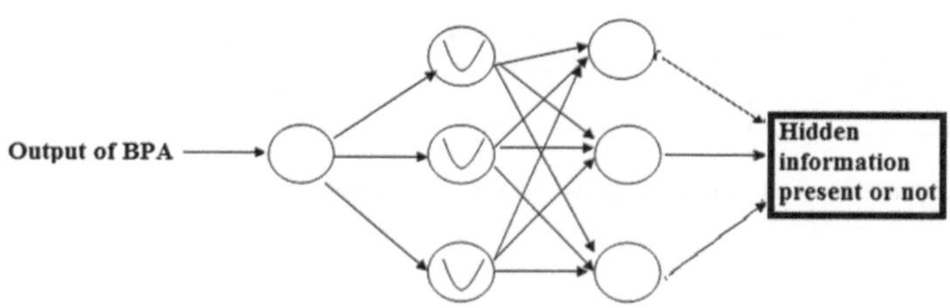

Fig. 2 Radial basis function network

Experimental Results

The steganalysis process has been implemented using MATLAB 2011a. Sample sets of images considered are gray and true color images. The different sets of cover images considered in the simulation are given in Fig. 3. The information image is given in Fig. 4. The different ways the secret information scattered in the cover images are given in Fig. 5.

An approach to combine BPA with RBF for steganalysis of covert information: The performance of BPARBF is appreciable compared to the performance of BPA and performance of RBF. The direction of information lay into cover image is presented. Information present is identified; however it is difficult to make it useful information if the information is randomly hidden.

In this simulation, the information (Fig. 5) is separated into red plane, green plane, and blue plane. Figure 5a shows red plane, Fig. 5b shows green plane, and Fig. 5c shows blue plane. The image with Bit-1 is chosen either from red plane or green plane or blue plane. Region of interest is chosen randomly from any of the

Fig. 3 Cover images

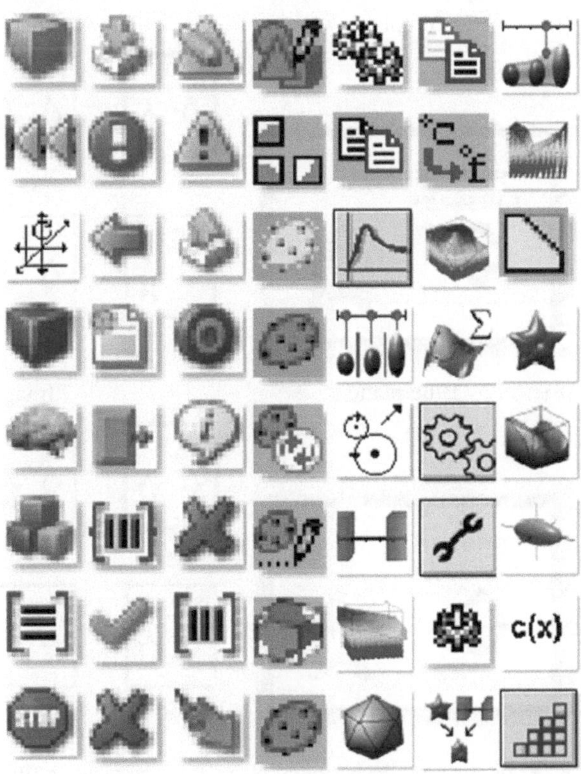

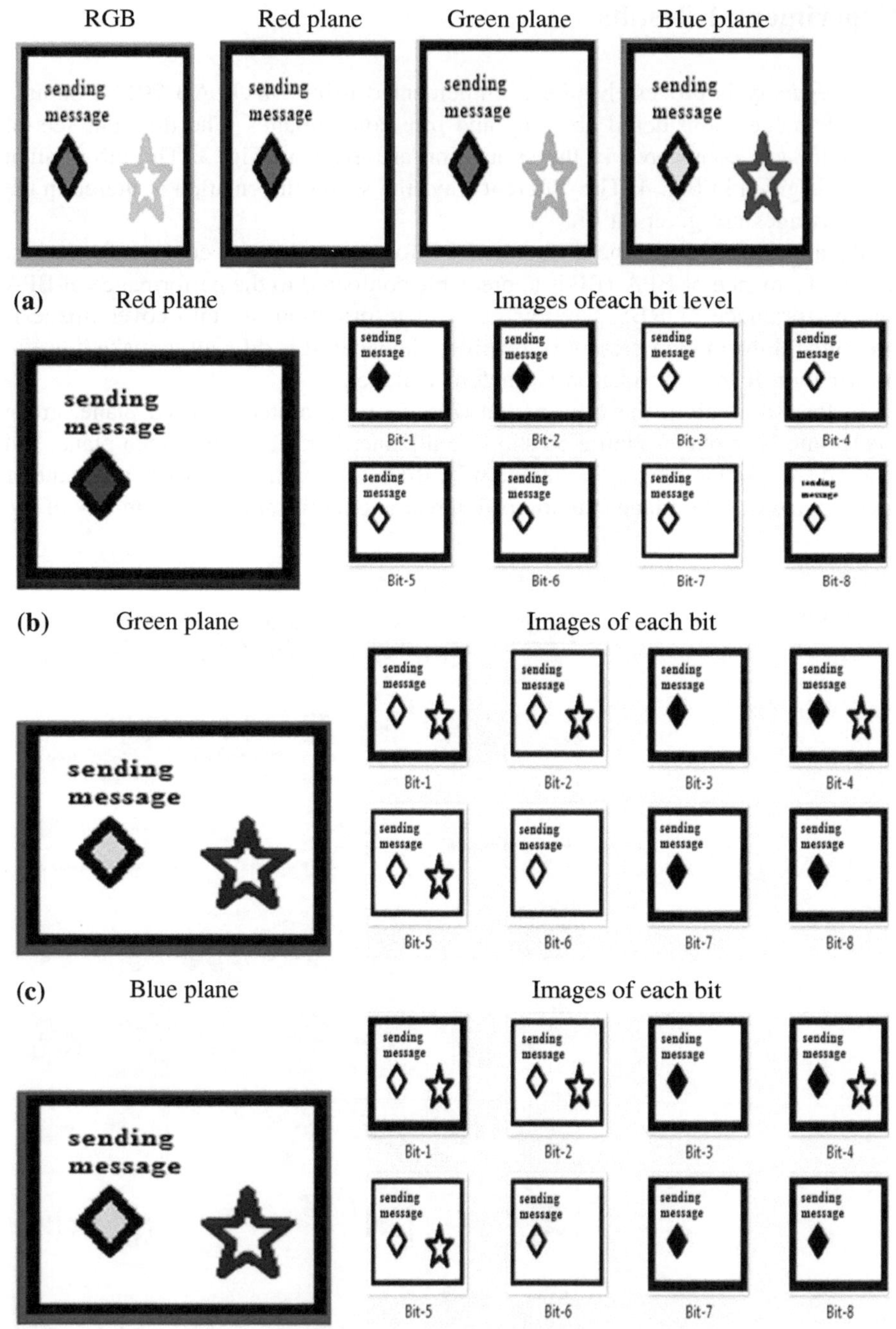

Fig. 4 Information image **a** Red plane, **b** Green plane, **c** Blue plane

Fig. 5 Distribution of information image in cover image

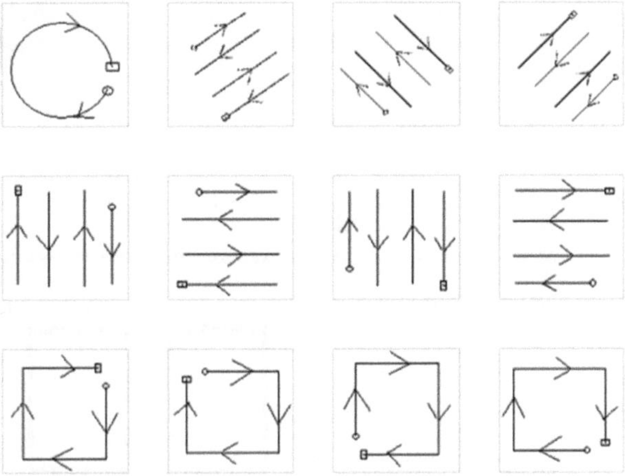

cover image given in Fig. 4 (MATLAB 2011a). In case of bit embedding, image of Bit-1 is considered for embedding in cover image. In case of DCT, the red plane or green plane or blue plane image is used to get DCT coefficients, which is further processed to get bits for embedding. Figure 6 presents, one of the directions in which information is hidden into cover image. In each subfigure of Fig. 6, the beginning and ending of the information hiding is given.

An approach to combine FUBPA with RBF for steganalysis of covert information: The performance of FUBPARBF is appreciable compared to the performance of FUBPA and performance of RBF. The direction of information lay into cover image is presented. Information present is identified; however it is difficult to make it useful information if the information is randomly hidden.

In this simulation, the information (Fig. 6) is separated into red plane, green plane, and blue plane. Figure 6a shows red plane, Fig. 6b shows green plane and Fig. 6c shows blue plane. The image with Bit-1 is chosen either from red plane or green plane or blue plane. Region of interest is chosen randomly from any of the cover image given in Fig. 3 (MATLAB 2011a). In case of bit embedding, image of Bit-1 is considered for embedding in cover image. In case of DCT, the red plane or green plane or blue plane image is used to get DCT coefficients, which is further processed to get bits for embedding. Figure 7 presents the information identification by only RBF and Fig. 8 presents the information identification by FU-BPARBF (Fig. 9).

Conclusion

This paper starts with identifying group of cover images and an information image that analysis and modifies the existing algorithms and proposed their implementation for steganalysis. ANN with supervised BPA has been considered initially

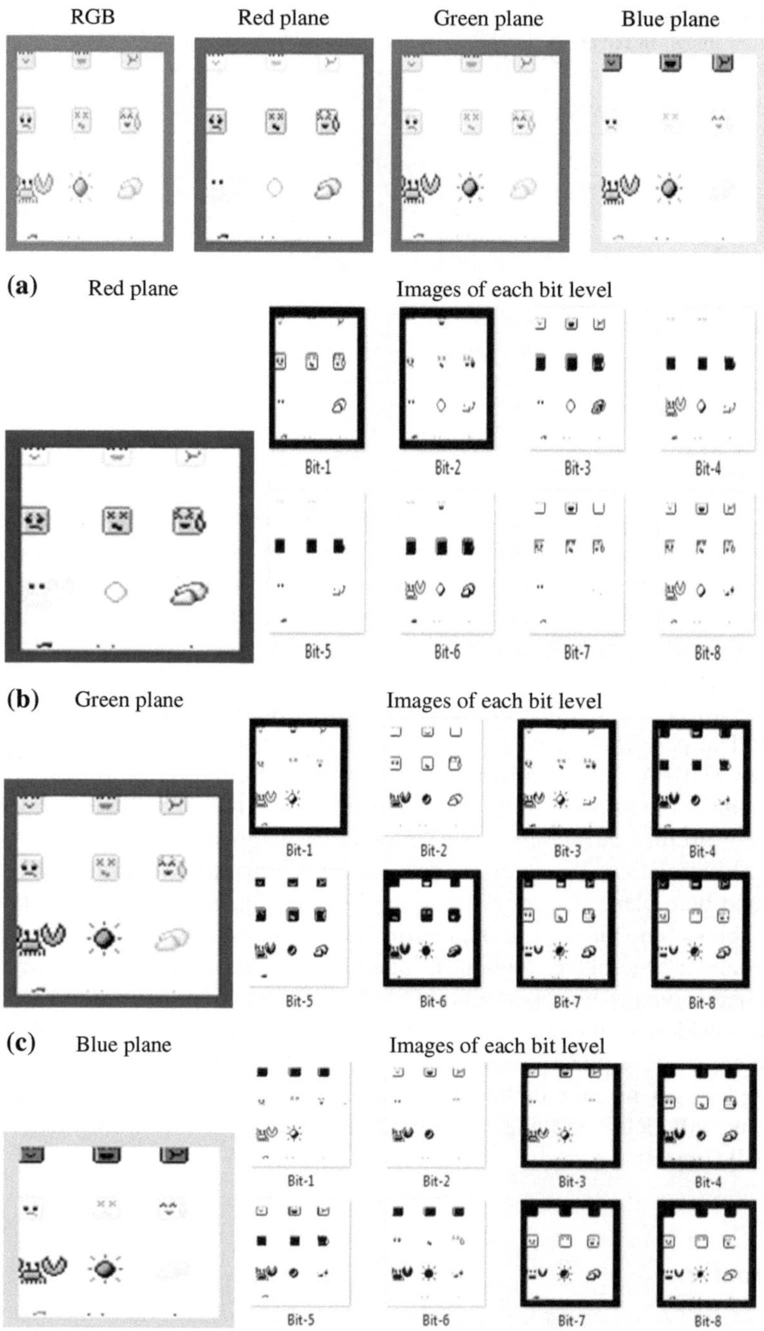

Fig. 6 Information image **a** Red plane, **b** Green plane, **c** Green plane

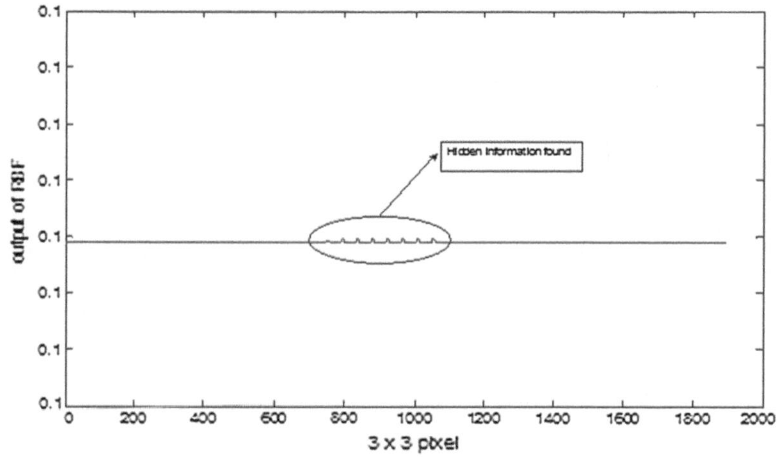

Fig. 7 Presence of hidden information using RBF

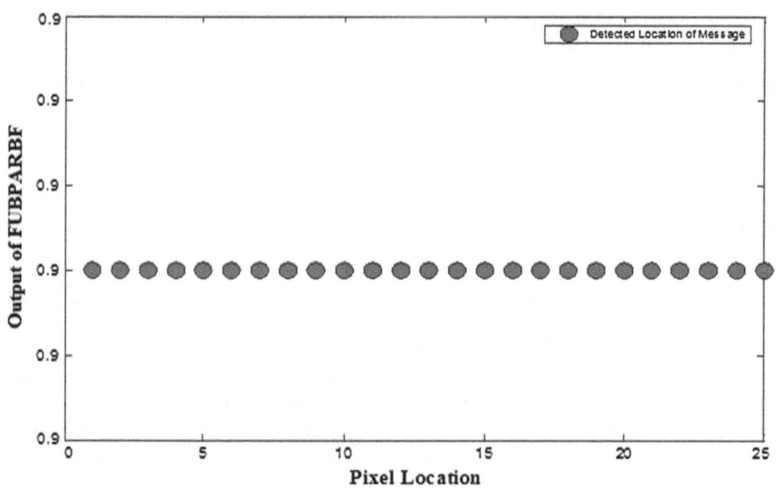

Fig. 8 Information identification by FUBPARBF

and learning of the generated data has been done. Training of the network with different learning factors has been tried and finally a value for learning factor with value 1 has been selected. This indicates a standard convergence. Identifying the correct way of normalizing the data itself was a challenging task. Another supervised RBF has been used as it is very much suitable for nonlinear data. The steganographed image can be considered as nonlinear data and hence RBF has been chosen. Finally, the performance of combined algorithms of BPARBF and

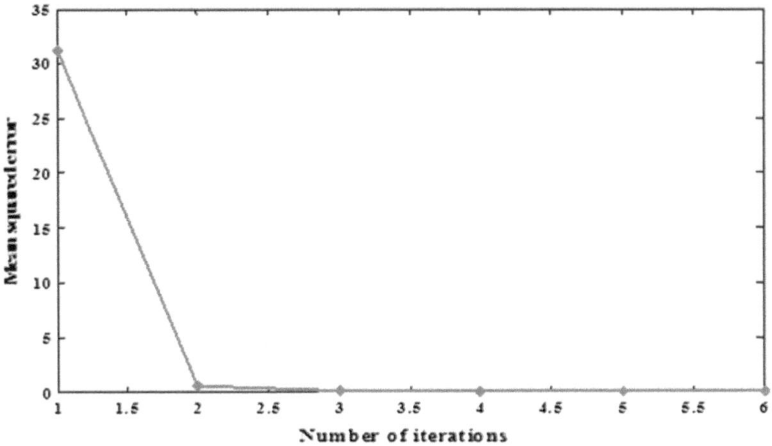

Fig. 9 FUBPA convergence curve

FUBPARBF was analyzed. It is observed that training and testing performance is improved when two algorithms are combined instead of using them separately that leads to promising results. FUBPA uses less number of iterations for the network to converge, and hence it provides better performance than using conventional BPA. Certainly, the combined ANN algorithms of FUBPA and RBF provide better performance than combining BPA and RBF. But still organizing the retrieved information is a challenging task.

References

1. Martin, A., Sapiro, G., Seroussi, G.: Is image steganography natural? IEEE Trans. Image Process. **14**(12), 2040–2050 (2005)
2. Sullivan, K., Bi, Z., Madhow, U., Chandrasekaran, S., Manjunath, B.S.: Steganalysis of quantization index modulation data hiding. In: Proceedings of IEEE International Conference on Image Processing (ICIP), pp. 1165–1168 Singapore (2004)
3. Fu, D., Shi Y.Q., Zou, D., Xuan, G.: JPEG steganalysis using empirical transition matrix in block DCT domain. In: Proceedings of 8th IEEE International workshop on Multimedia Signal Processing, pp. 310–313 Victoria, BC (2006)
4. Shi, Y.Q., Xuan, G., Zou, D., Gao, J., Yang, C., Zhang, Z., Chai, P., Chen, W., Chen, C.: Image steganalysis based on moments of characteristic functions using wavelet decomposition, prediction-error image, and neural network. IEEE International Conference on Multimedia and Expo, ICME, Newark pp. 1-4, 6–8 July 2005
5. Mehrabi, M.A., Faez, K., Bayesteh, A.R.: Image steganalysis based on statistical moments of wavelet sub band histograms in different frequencies and support vector machine. In: 3rd International Conference on Natural Computation,Haikou, vol. 1, pp. 587–590 August 2007
6. Wang, Y., Moulin, P.: Optimized feature extraction for learning-based image steganalysis. IEEE Trans. Inform. Forens. Secur. **2**(1), 31–45 (2007)
7. Gul, G., Kurugollu, F.: SVD based universal spatial domain image steganalysis. IEEE Trans. Inform. Forens. Secur. **5**(2), 349–353 (2010)

8. Avcibas, I., Memon, N., Sankur, B.: Steganalysis using image quality metrics. IEEE Trans. Image Process. **12**(2), 221–229 (2003)
9. Zhi, L., Fen, S.A. Detection of random LSB image steganography. In: 60th IEEE International Conference on Vehicular Technology, China vol. 3, pp. 2113-2117 September 2004
10. Chandramouli, R.: A mathematical framework for active steganalysis. ACM Multimedia Syst. **9**(3), 303–311 (2003)
11. Farid H.: Detecting hidden messages using higher-order statistical models. In: Proceedings of IEEE International Conference on Image Processing, pp. 905–908 New York (2002)
12. Wu, X., Dumitrescu, S., Wang, Z.: Detection of LSB steganography via sample pair analysis. In: 5th International Workshop on Information Hiding, London pp. 355–372 (2002)
13. Yu, X.Y., Wang, A.: Steganalysis based on regression model and Bayesion network. In: International Conference on Multimedia Information Networking and Security, MINES '09.Hubei, vol. 1, pp.41–44 (2009)
14. Lin, J.-Q.,Zhong, S.-P.: JPEG image steganalysis method based on binary similarity measures. International Conference on Machine Learning and Cybernetics. vol. 4, 2238–2243 (2009)
15. Zhao, X., Huang, L., Li, L., Yang, W., Chen, Z., Yu, Z.: Steganalysis on character substitution using support vector machine knowledge discovery and data mining, Second International Workshop, Moscow pp. 84–88 (2009)

Implementation of Radial Basis Function Network in Content-Based Video Retrieval

S. Prasanna, S. Purushothaman and R. Rajeswari

Abstract This paper presents retrieving a video from a given database using radial basis function (RBF) network method. The features of the videos are used by RBF for training and testing RBF in the algorithm developed. The features of frames of a video are extracted from the contents in the form of text, audio, and image. In this analysis, RBF is programmed to retrieve the words spoken by four different speakers in video presentation.

Keywords Content-based retrieval · Radial basis function · Video retrieval

Introduction

Video will be one of the key issues in the upcoming Internet evolution in infotainment and education. Converting raw video streams into highly and thoroughly structured and indexed, web ready, database-driven information entities are a must. Information databases have evolved from simple text to multimedia with video, audio, and text. The query mechanism for such a video database is similar in concept compared to a textual database. Object searching is analogous to word searching; scene browsing is similar to paragraph searching; video indexing is comparable to text indexing or bookmarking. Implementation for video

S. Prasanna (✉)
Department of MCA, VELS University, Chennai 600117, India
e-mail: prasannabalalak05@gmail.com

S. Purushothaman
PET Engineering College, Tirunelveli, Tamil Nadu 627117, India
e-mail: drsppuru@gmail.com

R. Rajeswari
Research Scholar Mother Teresa Women's University Kodaikanal, Kodaikanal, India
e-mail: rajeswaripuru@gmail.com

S. Sathiakumar et al. (eds.), *Proceedings of International Conference on Internet Computing and Information Communications*, Advances in Intelligent Systems and Computing 216,
DOI: 10.1007/978-81-322-1299-7_42,

content-based searching is very different and much more difficult than the query mechanism for a textual database. Scene change detection technique can often be applied for scene browsing and automatic and intelligent video indexing of video sequences for video databases. Once individual video scenes are identified, we use content-based indexing mechanisms (such as indexing by object texture, shape, color, motion, etc.) to index and query image contents in each video scene.

Related Works

There are many information carriers in a video stream, as is the visual content, the narrative or speech part, text captions. Visual content remains the most important. We prefer representative image from a long scene with little -or no- change. This process is the key frame extraction. Two kinds of key frame extraction strategies have been developed and used by various researchers. The simplest way is to select one or several frames from each segmented shot. Some use the first frame of each shot as the representative frame, i.e., the key frame of the shot. Others may use a random one, the last one, or the middle one as the prototypical frame.

Lots of research works have been done by earlier researchers on retrieval of videos. Web browsers and e-mail service providers have developed many indexing methods and retrieval algorithms for retrieving videos. In addition to the existing development, we are focusing on implementing intelligent method like radial basis function (RBF).

Erol and Kossentini [1] use object-based video representation, such as the one suggested by the MPEG-4 standard, offers a framework that is better suited for object-based video indexing and retrieval. In such a framework, the concept of a "key frame" is replaced by that of a "key video object plane." Approaches mainly differ in the set of acoustic features used to represent the audio signal and the classification technique applied [2]. In speech research, features such as Mel frequency cepstral coefficients (MFCCs) or linear prediction coefficients (LPCs) have been demonstrated to provide good representations of a speech signal allowing for better discrimination than temporal- or frequency-based features alone. Holmes and Holmes [3] and Liu and Wan [4] conclude from their study that cepstral features, such as MFCCs perform better than temporal- or frequency-based features and advocate their use for general audio tasks particularly when the number of audio classes is large.

Semantic filtering and retrieval of multimedia content is crucial for efficient use of the multimedia data repositories. Naphade and Huang [5] and Slaney [6] proposed a state-of-the-art system which incorporates a mapping between audio and semantic spaces. Methods are developed to describe general audio with words (and also predict sounds given a text query) using a labeled sound set. Adams et al. [7] have focused on semantic classification through the identification of meaningful intermediate-level semantic components using both audio and video features.

Fang et al. [8] proposed indexing and retrieval system of the visual contents based on feature extracted from the compressed domain. Direct possessing of the compressed domain spares the decoding time, which is extremely important when indexing large number of multimedia archives. Benmokhtar and Huet [9] proposed an improved version of RBF network based on Evidence Theory (NNET) using one input layer and two hidden layers and one output layer, to improve classifier combination and recognition reliability in particular for automatic semantic-based video content indexing and retrieval. Many combination schemes have been proposed in the literature according to the type of information provided by each classifier as well as their training and adaptation abilities.

Snoek et al. [10] identified three strategies to select a relevant detector from thesaurus, namely: text matching, ontology querying, and semantic visual querying for a given query. They evaluate the methods against the automatic search task of the TRECVID 2005 video retrieval benchmark, using news video archive of 85 h in combination with a thesaurus of 363 machine learned concept detectors. They assessed the influence of thesaurus size on video search performance, evaluated and compared the multimodal selection strategies for concept detectors, and finally discuss their combined potential using oracle fusion.

Lu et al. [11] surveyed some of the existing techniques and systems for content based indexing and retrieval of two main types of multimedia data images and videos. In content-based image retrieval, they have proposed multiscale color histograms by incorporating color and spatial information.

Gao et al. [12] discussed video sequences as temporal trajectories via scaling and lower dimensional representation of the video frame luminance field, and a video trajectory indexing and matching scheme was developed to support video clip search. Simulation results demonstrated that the proposed approach achieved excellent performance in both response speed and precision-recall accuracy.

Slaney [6] proposed a state-of-the-art system which incorporates a mapping between audio and semantic spaces. Methods are developed to describe general audio with words (and also predict sounds given a text query) using a labeled sound set. Adams et al. have focused on semantic classification through the identification of meaningful intermediate-level semantic components using both audio and video features.

Methods and Materials

Material

Table 1 shows the number of speeches recorded at two different instances

In this work, the audio track of videos is extracted from each shot and stored as patterns. When an user gives audio as input query, it will match with the stored patterns using RBF.

Table 1 Recordings of speech at different instances

Person name	Speech1	Speech2
Prasanna	√	√
Purushothaman	√	
Rajeswari	√	
Shwetha	√	

Fig. 1 Training and testing of radial basis function (RBF)

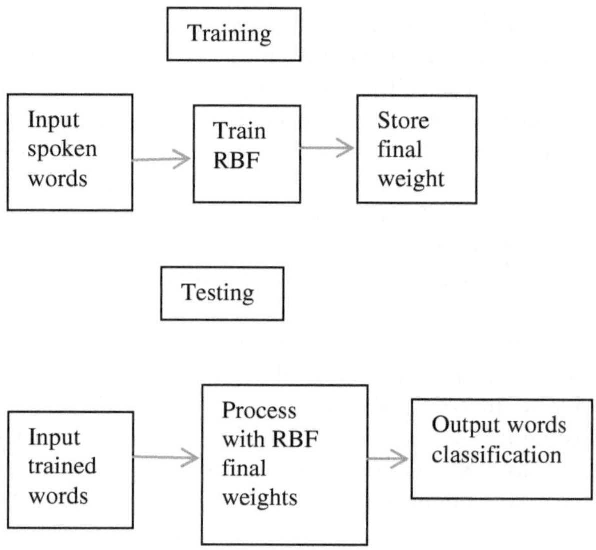

Schematic Diagram

Figure 1

Radial Basis Function Network

Figure 1 represents the block diagram of radial basis function (RBF) algorithm. A RBF is a real-valued function whose value depends only on the distance from the origin. If a function 'h' satisfies the property $h(x) = h(||x||)$, then it is a radial function. Figure 2 depicts the training method used through neural networks algorithm using a single hidden layer method. Tables 2 and 3 provide the training and testing procedure employed in doing so. Their characteristic feature is that their response decreases (or increases) monotonically with distance from a central point. The center, the distance scale, and the precise shape of the radial function are parameters of the model, all fixed if it is linear. Here the distance is found between a pattern and each center. The center is also one of the patterns

Fig. 2 Training with RBF

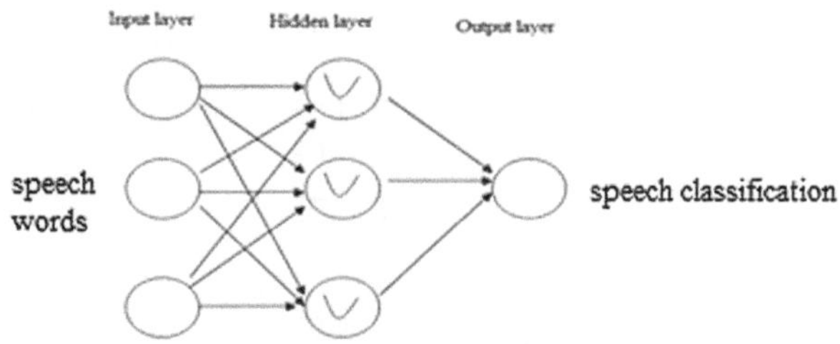

Table 2 Training RBF

Step 1: radial basis function (RBF) is applied
No. of input = 4
No. of patterns = 10
No. of center = 10
Calculated RBF as
RBF = exp (-X)
Calculated Matrix as G = RBF
$A = G^T * G$
Calculated $B = A^{-1}$
Calculated $E = B * G^T$
Step 2: Calculated the Final Weight F = E * D
Step 3: Stored the Final Weight in a file.

Table 3 Testing RBF

Step 1: Read the input
Step 2: Read the final weights
Calculated Numerals = F * E
Step 3: Input to RBF
Step 4: Classification of Speech by RBF

predefined. The square of the distance is a node in the hidden layer. An exponential function is used as an activation function which will be the output of the particular node. The number of nodes in the hidden layer is based on the number of centers decided in an implementation.

Results and Discussion

Figure 3 represents the speech patterns versus the prediction of words by the various speakers.

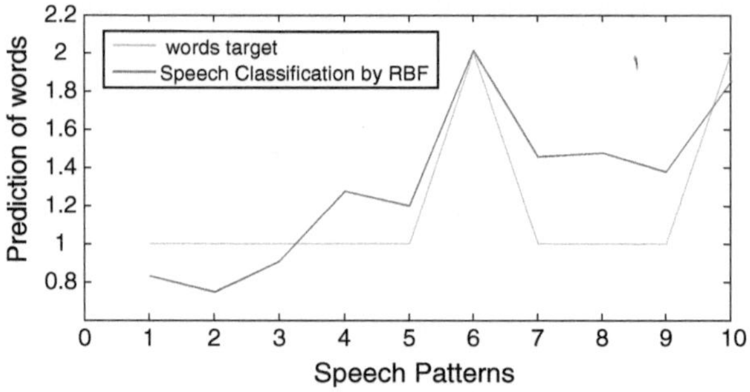

Fig. 3 Speech patterns vs prediction of words

Conclusion

In this paper, we discuss the implementation of RBF and the distance between pattern and centers are found. Final weights are calculated in Tables 2 and 3. The RBF network learns the patterns through iteration.

References

1. Erol, B., Kossentini, F.: Automatic key video object plane selection using the shape information in the Mpeg-4 compressed domain. IEEE Trans. Multimedia **2**(2), 129–138 (2000)
2. Tsekeridou, S., Pitas, I.: Content-based video parsing and indexing based on audio–visual interaction. IEEE Trans. Circuits Syst. Video Technol. **11**(4), 522–535 (2001)
3. Holmes, J.N., Holmes, W.J.: Speech Synthesis and Recognition, 2nd ed. Taylor & Francis, London (2001)
4. Liu M., Wan C.: A study on content-based classification and retrieval of audio database. In: Proceedings of International Database Engineering and Applications Symposium, Grenoble, France. IEEE Comput. Soc., Washington DC pp. 339–345 (2001)
5. Naphade, M.R., Huang, T.S.: A probabilistic framework for semantic video indexing, filtering, and retrieval. IEEE Trans. Multimedia **3**(1), 141–151 (2001)
6. Slaney M.: Mixtures of probability experts for audio retrieval and indexing. In: Proceedings of IEEE International Conference on Multimedia and Expo, Goteborg, Sweden, vol. 1, pp. 345–348 (2002)
7. Adams, W.H., Iyengar, G., Lin, C.Y., Naphade M.R., Neti, C., Nock, H.J., Smith J.R.: Semantic indexing of multimedia using audio, text and visual cues, EURASIP J. Appl. Signal Process. **2**, 170–185 (2003)
8. Fang, H., Qahwaji, R., Jiang, J.: Video indexing and retrieval in compressed domain using fuzzy-categorization, ISCV 2006, LNCS 4292, Springer-Verlag Berlin Heidelberg, pp. 1143–150 (2006)
9. Benmokhtar, R., Huet, B.: Neural network combining classifier based on Dempster-Shafer theory. In: Proceedings of the International Multiconference on Computer Science and Information Technology, Singapore, pp. 3–10 (2006)
10. Snoek, C.G.M., Huurnink, B., Hollink, L., De Rijke, M., Schreiber, G., Worring, M.: Adding semantics to detectors for video retrieval. IEEE Trans. Multimedia **9**(5), 975–986 (2007)

11. Lu, H., Xue, X., Tan, Y.-P.: Content-Based Image and Video Indexing and Retrieval, pp. 118–129. Springer Verlag, Berlin Heidelberg (2007)
12. Gao, L., Li, Z., Katsaggelos, A.: An efficient video indexing and retrieval algorithm using the luminance field trajectory modeling. IEEE Trans. Circuits Syst. Video Technol., **19**(10), 1566–1570 (2009)

Implementation of Textile Image Segmentation Using Contextual Clustering and Fuzzy Logic

R. Shobarani and S. Purushothaman

Abstract This paper presents the segmentation analysis on textile images. These images have innumerable textures. The content of the images are regularly arranged or repeated or random in a tessellated fashion. It is not necessary that the entire image has to be compulsorily segmented. However, at least one full object has to be segmented correctly in an image. In this work, a systematic approach has been developed to extract textures from the given texture images. The features of the textile images are extracted and used for segmenting those images using contextual clustering and fuzzy logic. The proposed methods combine to improve the segmentation accuracies and to analyze the effects of parameters of the proposed algorithms in segmentation of textures.

Keywords Contextual clustering · Segmentation · Textile textures · Fuzzy logic · K-means algorithm

Introduction

In textile manufacturing companies, database of images are already available. If the company people are interested in adding new images, they have to make sure that the new image is not present already in the database. They can also retrieve the existing images for any modifications. It is not really possible to compare the new image with the image in the database visually. For example, we cannot count

R. Shobarani (✉)
Research Scholar, Mother Teresa Women's University, Kodaikanal, India
e-mail: rshobarani@gmail.com

S. Purushothaman
PET Engineering College, Tirunelveli District, Tamil Nadu 627 117, India
e-mail: drsppuru@gmail.com

S. Sathiakumar et al. (eds.), *Proceedings of International Conference on Internet Computing and Information Communications*, Advances in Intelligent Systems and Computing 216,
DOI: 10.1007/978-81-322-1299-7_43, © Springer India 2014

comfortably the presence of number of '1' in a text '1111111111111111111111 111111' with easiness. If this is the case, it is still more difficult to compare patterns of designs like that of textile images.

Image texture has a number of perceived qualities that play an important role in describing the texture. Davis [1] MITES algorithm represents an alternative to the traditional pixel classification approach to texture image segmentation, because it makes explicit use of the spatial coherence of uniformly textured regions. Conners et al. [2] utilized texture analysis methods to detect defects in lumber wood automatically. Spann and Wilson [3] A Quad Tree Approach combining statistical and spatial information to the problem of image segmentation.

Panjwani and Healey [4] proposed a method for unsupervised segmentation of color textures using Markov random fields. These models characterize a texture in terms of spatial interaction within each color plane and interaction between different color planes. The models are used by a segmentation algorithm based on agglomerative hierarchical clustering. The heart of agglomerative clustering is a stepwise optimal merging process that at each iteration maximizes a global performance functional based on the conditional pseudo likelihood of the image.

Structural texture analysis consists of two major steps: (a) Extraction of the texture elements and (b) Inference of the placement rule. Texture elements consist of regions in the image with uniform gray levels.

Zucker [5] proposed a method in which he examines the structure underlying textural patterns. Textures are developed on the basis of both perceptual processes of the observer and structural properties of the patterns. Local constructs called primitives serve as generators for globally perfect, but unobservable, ideal textures. The graph representation of these ideal textures reveals the basic regularity underlying visual textures. These perfect patterns are then transformed into observable textures by transformations, which must not violate psychophysically based restrictions. Cohen et al. [6] described textile fabric inspection using the visual textural properties of the fabric. The problem is to detect and locate the various kinds of defects that might be present in a given fabric sample based on an image of the fabric. Stochastic models are used to model the visual fabric texture. The authors use the Gaussian Markov random field to model the texture image of non defective fabric. Chen and Jain [7] used a structural approach to detect defects in textured images. They extracted a skeletal structure from images, and by detecting anomalies in certain statistical features in these skeletons, defects in the texture are identified.

Hanmandlu et al. [8] describe the texture segmentation techniques are diversified by the existence of several approaches. For this purpose, a membership function is constructed to represent the effect of the neighboring pixels on the current pixel in a window. Using these membership function values, we find a feature by weighted average method for the current pixel. Kaloyan Krastev and Georgieva [9] deal with an application of fuzzy logic (FL) for leather surface defects identification is investigated. The defects create a discontinuity at the surface, which leads to regions with different intensities than the background.

Binary images are obtained after the segmentation. They contain non defective (regular texture) and defective structural elements. The objective is to classify the elements and to detect the defective ones. Arasteh and Hung [10] proposed Uniform Local Binary Pattern (ULBP) method to extract texture features. Image segmentation is carried out using the K-means algorithm on feature vectors, including color and texture features. The distance measure is defined as a function of the color and texture feature vector distances from the K-means defined centers. The weighting parameter is used to adjust the relative contribution of the color and texture features. Mohand Saïd Allilli and Djemel Ziou [11] proposed an automatic segmentation of color–texture images with arbitrary number of regions. The approach combines region and boundary information and uses active contours to build a partition of the image. Houhou et al. [12] present an efficient approach for unsupervised segmentation of natural and textural images based on the extraction of image features and a fast active contour segmentation model. Sujaritha and Annadurai [13] present a simple and efficient modified Gaussian mixture model-based clustering algorithm for color–texture segmentation. Law et al. [14] develop a robust and effective algorithm for texture segmentation and feature selection. Shaabany and Jamshidi [15] present texture segmentation concept using supervised method in FL.

Algorithms Proposed for Image Texture Segmentation

Contextual Clustering

Contextual clustering algorithm (Salli et al. [16]) segments a data into two categories: category 1 (ω_0) and category 2 (ω_1). They are assumed to be drawn from standard normal distribution. The following steps are adopted for implementing contextual clustering:

1. Define decision parameter T_{cc} (positive) and weight of neighborhood information β (positive). Let N_n be the total number of data in the neighborhood. Let z_i be the data itself, *'I'*.
2. Classify data with $z_i > T_\alpha$ to ω_1 and data to ω_0. Store the classification to C_0 and C_1.
3. For each data *'I'*, count the number of data u_i, belonging to class ω_1 in the neighborhood of data *'I'*. Assume that the data outside the range belong to ω_0.
4. Classify data with $z_i + \frac{\beta}{T_{cc}}(u_i - \frac{N_n}{2}) > T_\alpha$ to ω_1 and other data to ω_0. Store the classification to variable C_2.
5. If $C_2 \neq C_1$ and $C_2 \neq C_0$, copy C_1 to C_0, C_2 to C_1, and return to step 3, otherwise stop and return to C_2.

The implementation of the *CC* is given as follows:

Step 1. Read a pattern (texture image feature).
Step 2. Sort the values of the pattern.
Step 3. Find the median of the pattern C_m.
Step 4. Find the number of values greater than the median values, U_m.
Step 5. Calculate *CC* using $C_m + (\beta/T_{cc}) * (U_m - (bs/2))$.
Step 6. Assign *CC* as the segmented values.

Fuzzy Logic

FL is an extension of multivalve logic. However, in a wider sense FL is almost synonymous with the theory of fuzzy sets, a theory which relates to classes of objects with unship boundaries in which membership is a matter of degree.

The training and testing FL is to map the input pattern with target output data. During testing, the membership function is used to test the pattern.

Training Fuzzy Logic

Step 1. Read the pattern (texture image feature) and its target value.
Step 2. Create fuzzy membership function.
Step 3. Create clustering using K-means algorithm.
Step 4. Process with target values.
Step 5. Obtain final weights.

Testing Fuzzy Logic for Texture Segmentation

Step 1. Input a pattern (texture image feature).
Step 2. Process with fuzzy membership function.
Step 3. Find the cluster to which the pattern belongs.
Step 4. Obtain estimated target values.
Step 5. Classify the texture.

Results and Discussions

This paper presents the results of segmentation for 100 textile images. These images fall under the category of artificial regular, natural, and stochastic textures. Table 1 presents the methods and the parameters used for segmenting the textures.

Table 2 presents 20 sample textile images. Column 2 shows the original true color image. Column 3 presents the contextual clustering segmentation output. The

Table 1 Algorithms and parameters used

Serial no.	Proposed methods	Parameters used
1.	Contextual clustering	Block of pixels, beta, and threshold
2.	Fuzzy logic	Radii of the clustering

image contains '0' (black) and '255' (white). Similarly, the fuzzy segmented outputs (Column 4) are presented.

Figure 1 presents image segmented by contextual clustering. The x-axis shows the textile image 100 numbers. The y-axis shows the percentage of the respective image segmented.

Figure 2 presents image segmented by FL. The x-axis shows the textile 100 image numbers. The y-axis shows the percentage of the respective image segmented.

Table 2 SAMPLE IMAGES

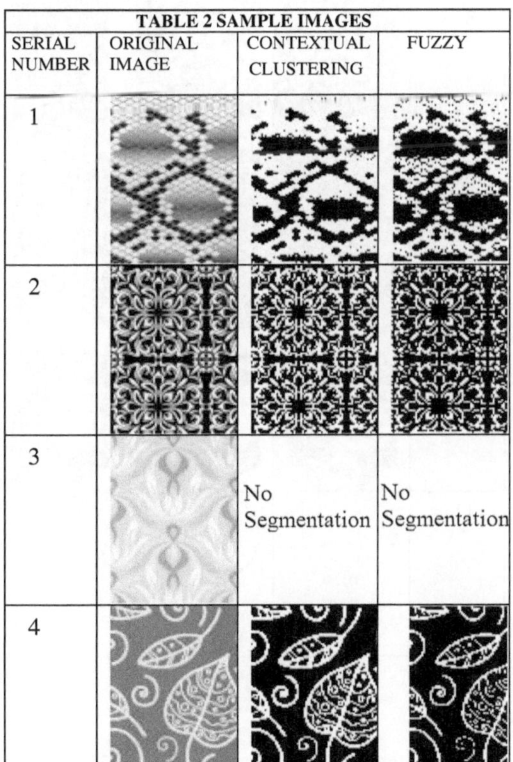

TABLE 2 SAMPLE IMAGES			
SERIAL NUMBER	ORIGINAL IMAGE	CONTEXTUAL CLUSTERING	FUZZY
1			
2			
3		No Segmentation	No Segmentation
4			

(continued)

5			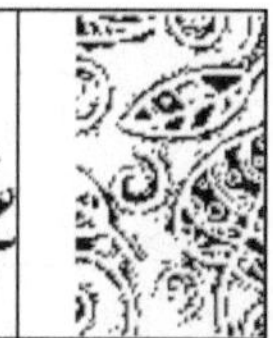
6			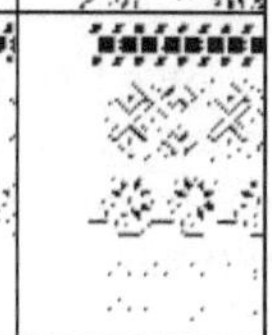
7	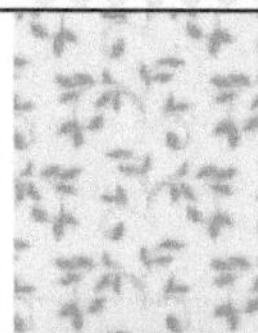	No Segmentation	No Segmentation
8			
9			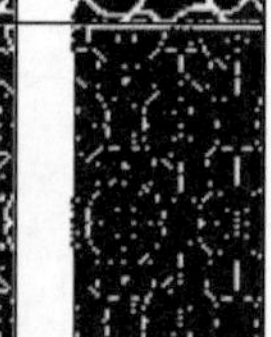
10			
11	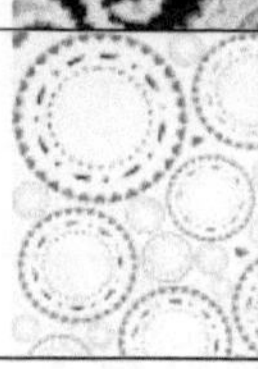	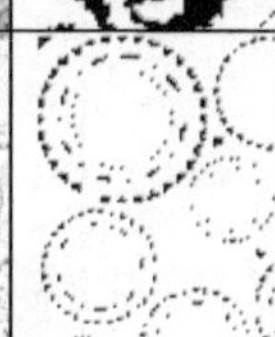	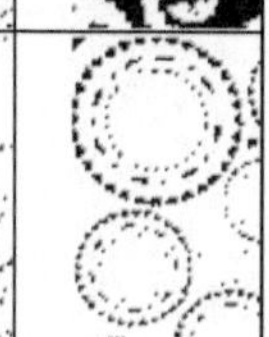
12			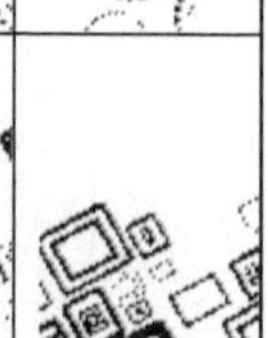

(continued)

13		No Segmentation	
14	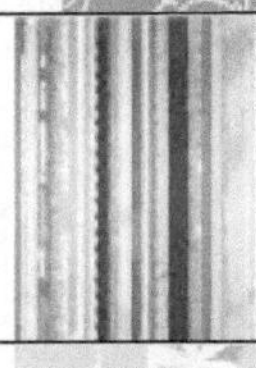	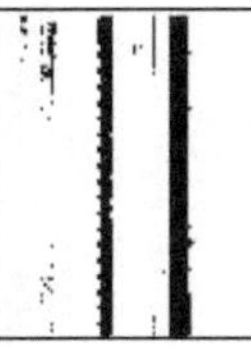	
15	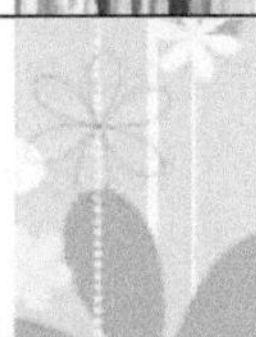	No Segmentation	No Segmentation
16			
17			
18			
19			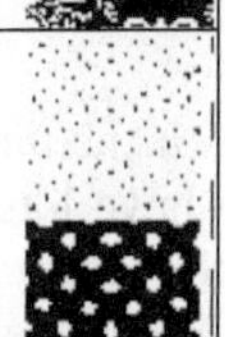
20			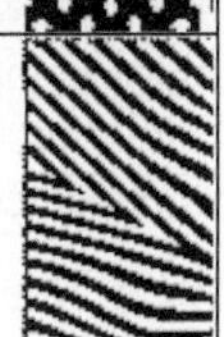

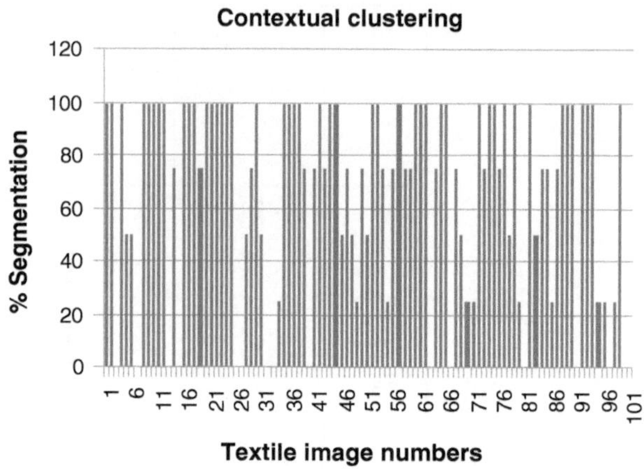

Fig. 1 Segmentation by contextual clustering

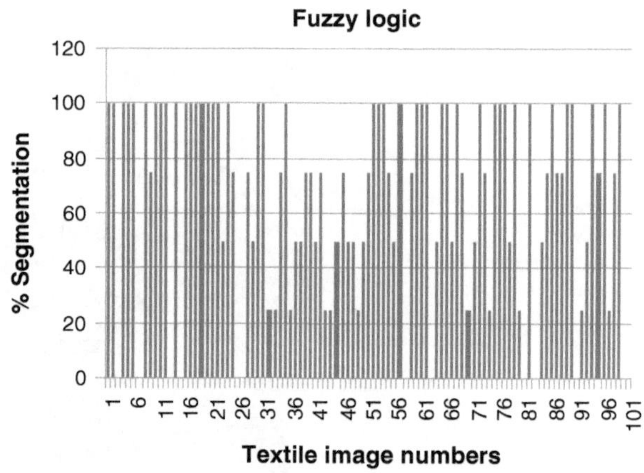

Fig. 2 Segmentation by Fuzzy Logic

Conclusions

This paper focuses in texture segmentation of textile images. These images are considered as they have lots of objects in the images. Textile images have been downloaded from the Internet resources. Randomly, 100 textile images have been considered for the research work. Contextual clustering and FL have been used for segmenting the images. Contextual clustering is used for segmenting the features generated by the contextual clustering algorithm are used for training and testing the FL method. The paper concludes the following achievements:

1. Image enhancement like intensity adjustment or histogram equalization or brightening is a must for dull appearance of the image.
2. Higher threshold is required for very bright images as the bell of the histogram lies on right side.
3. Medium threshold of 140 is sufficient for segmenting majority of the images.

4. Whenever objects are clearly distinct in the images, the segmentation is also good. If the objects are messy in the images, then segmentation is possible only at higher thresholds.
5. Majority of textile images and mosaic images have complex objects and hence thresholds cannot remain same for all images.
6. Contextual clustering performs good segmentation.
7. Fuzzy Logic performs equally good segmentation as that of contextual clustering.

References

1. Larry, S.Davis: MITES (mit-æs): a model-driven, iterative texture segmentation algorithm. Comput. Graphics Image Process. **19**(2), 95–110 (1982)
2. Conners, R.W., McMillin, C.W., Lin, K., Vasquez-Espinosa, R.E.: Identifying and locating surface defects in wood: part of an automated lumber processing system. IEEE Trans. Pattern Anal. Mach. Intell. *PAMI-5*(6), 573–583 (1983)
3. Spann, M., Wilson, R.: A quad-tree approach to image segmentation which combines statistical and spatial information. Pattern Recogn. **18**(3/4), 257–269 (1985)
4. Panjwani, D.K., Healey, G.: Markov random-field models for unsupervised segmentation of textured color images. IEEE Trans on PAMI **17**(10), 939–954 (1995)
5. Zucker, S.W.: On the structure of texture. **5**(4), 419–436 (1976)
6. Cohen, F.S., Fan, Z., Attali, S.: Automated inspection of textile fabrics using textural models. IEEE Trans. Pattern Anal. Mach. Intell. **13**(8), 803–808 (1991)
7. Chen, J., Jain, A. K.: A structural approach to identify defects in textured images. In: Proceedings of IEEE International Conference on Systems, Man, and Cybernetics, Beijing, Las vegas, Nevada, pp. 29–32 (1988)
8. Hanmandlu, M., Madasu, V.K., Vasikarla, S.: A fuzzy approach to texture segmentation. In: Proceedings of the International Conference on Information Technology: Coding and Computing (ITCC'04), **1**, 636–642 (2004)
9. Krastev, K., Georgieva, L.: Identification of leather surface defects using fuzzy logic. International Conference on Computer Systems and Technologies – Comp Sys Tech', Varna, Bulgaria, pp. IIIA. 12-1-12-6 (2005)
10. Arasteh, Sara, Hung, Chih-Cheng: Color and texture image segmentation using uniform local binary patterns. Mach. Graph. Vis. Int. J. **15**(3), 265–274 (2006)
11. Allilli, M.S., Ziou, D.: Globally adaptive region information for automatic color-texture image segmentation. Pattern Recogn. Lett. **28**(15), 1946–1956 (2007)
12. Houhou, Nawal, Thiran, Jean-Philippe, Bresson, Xavier: Fast texture segmentation based on semi-local region descriptor and active contour. Numer. Math. Theor. Meth. Appl. **2**(4), 445–468 (2009)
13. Sujaritha, M., Annadurai, S.: A new modified gaussian mixture model for color-texture segmentation. J. Comp. Sci. **7**(2), 279–283 (2011)
14. Law, Yan Nei, Lee, Hwee Kuan, Yip, Andy M.: Subspace learning for Mumford–Shah-model-based texture segmentation through texture patches. Appl. Opt. **50**(21), 3947–3957 (2011)
15. Shaabany, A., Jamshidi, F.: Texture segmentation concept using fuzzy logic. Int. J. Multidiscipl. Sci. Eng. **3**(1), 26–29 (2012)
16. Salli, Eero, Aronen, Hannu J., Savolainen, Sauli, Korvenoja, Antti, Visa, Ari: Contextual Clustering for Analysis of Functional MRI Data. IEEE Trans. Med. Imaging **20**(5), 403–414 (2001)

Detections of Intima-Media Thickness in B-Mode Carotid Artery Images Using Segmentation Methods

V. Savithri and S. Purushothaman

Abstract This study presents the investigations carried out on carotid artery to identify the intima-media thickness of carotid artery that affected with plaques. B-mode ultrasound image video of the artery has been used as the data for processing. The frames of the video are processed to know the plaque properties of the artery. In order to achieve this, two segmentation processing techniques have been used on each frame. The features extracted from the frames are consolidated to know the conditions of the artery. Information of a frame are converted into features. The values of the features are estimated by artificial neural network (ANN) algorithm. ANN has not been used extensively by the past. ANN is used in estimating the plaque thickness in the carotid artery.

Keywords B-mode · Back propagation network · Pattern recognition · Artificial neural network · Clustering

Introduction

Intima-media thickness (IMT) is a well-known indicator of the physical properties of carotid artery in vascular disease examinations. The brief study of an existing methods applied to detect IMT boundary of carotid artery is discussed. Santhiyakumari and Madheswaran [1] presented intelligent medical decision system for identifying ultrasound carotid artery images with vascular disease. The study reveals

V. Savithri (✉)
Department of Computer Science, Mother Teresa Women's University, Kodaikanal, India
e-mail: savi3_8@yahoo.co.in

S. Purushothaman
PET Engineering College, Tirunelveli, TN, India
e-mail: dr.s.purushothaman@gmail.com

S. Sathiakumar et al. (eds.), *Proceedings of International Conference on Internet Computing and Information Communications*, Advances in Intelligent Systems and Computing 216,
DOI: 10.1007/978-81-322-1299-7_44,

that the MBPN-based medical decision system developed offers higher classification and actual efficiency.

Cai et al. [2] incorporated local spatial and gray information together, a novel fast and robust fuzzy c-means framework for image segmentation, fast generalized fuzzy c-means clustering algorithms is proposed. Lei et al. [3] introduced an adaptive based on statistical learning theory as proposed to enhance fuzzy electrocardiogram (ECG) classifier with orientation to smart ECG interpreters. Markos et al. [4] presented fuzzy rule-based decision support system for the diagnosis of coronary artery disease based on easily and noninvasively acquired features, and is able to provide interpretation for the decision made. Li et al. [5] proposed a new fuzzy level set algorithm and confirm its effectiveness for medical image segmentation. Bing et al. [6] examined an integrating spatial fuzzy clustering with level set methods for automated medical segmentation. The approaches proposed in this paper are the Mumford-shah level set methods for image segmentation.

Amartur et al. [7] examined the segmentation of magnetic resonance images by optimizing neural networks. This study has demonstrated the applicability of Hopfield net for the tissue classification in MRI. Levinski et al. [8] describe the approach for correcting the segmentation errors in 3D modeling space, implementation, and principles of the proposed 3D modeling space tool and illustrates its application. Paragios [9] introduces a knowledge-based constraints, able to change the topology, capture local deformations, surface to follow global shape consistency while preserving the ability to capture using implicit function. Suri [10] attempts to explore geometric methods, their implementation and integration of regularizers to improve the robustness of independent propagating curves/surfaces. Ozbay and Ceylan [11] revealed the 100 % classification accuracy of carotid artery Doppler signals using complex-valued artificial neural network. Wendelhag et al. [12, 13] results show variations secondary to subjective parameters when manual measurement methods are employed. A thorough computerized system is necessary to detect the IMT of carotid artery to detect plaque area. Our proposed method acts as a tool to detect the IMT of b-mode ultrasound images effectively and efficiently with less time and less memory allocation. Our proposed method can also be treated as secondary observer to detect IMT.

Materials and Methods

In this study, two different segmentation techniques, fuzzy clustering and dynamic programming algorithms were examined to extract the boundary of IMT. The results of these two methods were compared using back propagation algorithm. The features like size, age, sex, circularity, internal diameter, wall thickness, height, area, and length of segmented regions traced by fuzzy clustering algorithm were compared with the same parameters traced by the dynamic programming algorithm. This methodology provides a reliable tool to detect the IMT in ultrasound images and can be used efficiently as secondary observer in clinical decision making. A thorough evaluation of this method in the clinical environment shows that inter observer variability is evidently decreased and so is the overall analysis time. The results demonstrate that it has the potential to perform qualitatively better than applying existing methods in IMT detection on b-mode images.

Clustering refers to the classification of objects into groups according to certain properties of these objects. Clustering is an unsupervised classification technique that identifies some inherent structure present in a set of objects based on a similarity measure. Clustering is also widely used for vector quantization, image segmentation, function approximation, and data mining. In clustering technique, an attempt is made to extract a feature vector from local areas in the image. Clustering is used to select a distance measure which will determine how the similarities of the element are calculated. It also refers to the classification of objects into groups according to certain properties of these objects in the clustering techniques; an attempt is made to extract a feature vector from local areas in the image. A standard procedure for clustering is to assign each pixel to the class of the nearest cluster mean. Clustering is an unsupervised classification technique that identifies some inherent structure present in a set of objects based on a similarity measure. Partition-based clustering uses an iterative optimization procedure that aims at minimizing an objective function, which measures the goodness of clustering. One among the partitional-based clustering method is c-means clustering. In proposed method, three different clustering methods are implemented and compared and as a result fuzzy c-means clustering is concluded as the best method and the same applied for pattern recognition and training and testing in back propagation algorithm.

Fuzzy Clustering

Fuzzy clustering is an important class of clustering algorithms. The one among these algorithms is fuzzy *c*-means clustering. Fuzzy clustering methods can be considered to be superior to those of their hard counterparts since they can represent the relationship between the input pattern data and clusters more naturally. Fuzzy clustering seeks to minimize a heuristic global cost function by exploiting the fact that each pattern has some graded membership in each cluster. The clustering criterion allows each pattern for multiple assignments of clusters. The fuzzy algorithm iteratively updates the cluster centroid and estimates the class membership function using the gradient descent approach.

Fuzzy C-Means Clustering

The discreteness of each cluster endows the c-means algorithm with analytical and algorithmic intractabilities. Partitioning the data set in a fuzzy manner helps to circumvent such difficulties. FCM cluster consider each cluster as a fuzzy set and each feature vector may be assigned to multiple cluster with some degree of certainty measured by the membership function taking values in the internal [0,1] iteratively updates the cluster centroid and estimates the class membership function by using the gradient descent approach. In this study, the two approaches applied are (i) estimated the distribution and risk factors for IMT by threshold. (ii) testing and validation are done. The integration of fuzzy clustering and neural networks has emerged as a promising field of research in recent years. This has led to the development of a new branch called neuro-fuzzy computing. Neuro-fuzzy system combines the advantages of both the uncertainty handling capability of fuzzy systems and the learning ability of neural networks.

Dissimilarity Function

An essential quantity to measure the distinction between data patterns is the dissimilarity function. The dissimilarity function must be carefully chosen due to the diversity and scale of feature inhabited in patterns. Different choices of clustering dissimilarities lead to distinct clustering results. A common method to formulate the dissimilarity of two data vectors is the Minkowski metric which is given by

$$d_{p(xi,xj)} = \frac{[\sum_{x=1}^{\dim} ||xi, s - xj, s||p]1}{p} \tag{1}$$

where d_p (x_i, x_j) is the pth root of summarization norm of (x_is, x_js) and (x_is, x_js) is applied to calculate the distance between x- and y-axis. If $p = 1$, it implies the L1 distance or Manhattan distance. If $p = 2$, it implies the Euclidean distance or L2 distance.

A metric should satisfy the following requirements of a distance function:

1. $d(x,y) \geq 0$
2. $d(x,y) = 0 => x = y$
3. $d(x,y) = d(y,x)$
4. $d(x,y) \leq d(x, z) + d(z, y)$

The definition of metrics is also restricted by a rigorous condition, the triangular inequality, which is given by

$$d(x,z) <= d(x,y) + d(y,z) \text{ and } d(x,y) = 0 = x = y \tag{2}$$

In case with $1 = \infty$ in Minkowski metric comes to the maximum metric the distance is measured by

$$d(x_i, x_j) = \max |(x_is - x_js)| \tag{3}$$

where max is $\leq s \leq$ dim. These measures are invariant to translation but variant to scaling.

Dynamic Programming

Dynamic programming is a technique that deals with situations where decisions are made in stages, with the outcome of each decision being predictable to some extent before the next decision is made. A key aspect of such situations is that decisions cannot be made in isolation. Rather, the desire for a low cost at the present must be balanced against the undesirability of high costs in the future. This is a credit assignment problem because credit or blame must be assigned to each one of a set of interacting decisions. The Image is scanned and all the possible boundary lines are considered polylines with N vertices represented as a vector p

$$P = (p_1, p_2, \ldots, p_{t-1}, p_t, \ldots p_N) \tag{4}$$

where $p_{t\text{-}1}$ and p_t are horizontal neighbors. N is the horizontal length of a contour line.

The criteria for judging a polyline to be a valid boundary line are a cost function which is a sum of the local energies along this line. The optimum line is the one that minimizes the cost function:

$$E_{\text{sum}} = \sum\nolimits_{t=1}^{N} E(p_t) \tag{5}$$

The local cost $E(p_t)$ is a weighted sum of cost components

$$E(p_t) = c_1 E_{\text{discount}}(p_{t-1}, p_t) - c_2 E_{\text{int}}(p_t) - c_3 E_{\text{grad}}(p_t) \tag{6}$$

where C_1, C_2, and C_3 are weighting factors C_3.

E_{discount} is the value of polylines discontinuity (measurement of smoothness).

E_{int} is the average brightness below the tested pixel.

E_{grad} is the value of intensity gradient (the rate of change of the intensity).

In proposed method one can add other components to the cost function which may improve the result, such as parallelism. The added terms must only exploit features that generally characterize a boundary.

An image grid is denoted as $I(x,y)$, of size M X N, where M and N are the number of rows and columns, respectively. Let $g(x,y)$, be the normalized image feature of

$$I(x,y), \text{with } 0 <= \text{g } (x,y),) <= 1. \tag{7}$$

Suppose the dual curves run from left to right on I I(x, y), with two y-coordinates $y1$ and $y2$ in each column, 1 <= $y1$, $y2$ <= M. The node pair ($g(x, y1)$, $g(x, y2)$) is considered here unlike the single node in the traditional dynamic programming (TDP). We assume that the dual curves have the minimal and maximal distances to each other, defined by $d_{\min}$ and $d_{\max}$ respectively, where $0 <= d_{\min} < d_{\max} \ll M$ with the definition

$$d_{\min} <= y1 - y2 <= d_{\max}. \tag{8}$$

With this design, the complexity of QP is reduced and two geometric properties are embedded simultaneously. Since $d_{max} \ll M$, the number of pair combinations is tremendously decreased, resulting in a reduced computation complexity. Notably, a special case is given when $d_{min} = d_{max} = 0$, then the algorithm corresponds to TDP. The embedded geometric properties are (1) the dual curves are limited to some specified range. The first property guarantees that the intima and adventitia form quasi-parallel boundaries while the second one helps the system to detect the correct boundaries.

Single Curve Continuity and Smoothness

A parameter for continuity is defined. Let d_r denote the range that the nodes in column $x-1$ are allowed to jump onto the next column x, i.e., the nodes in column $x-1$ can move at most d_r positions in either the up or down direction. For instance, if $d_r = 1$, then node $g(x–1,y)$ has three candidate successors in column x. Since node pairs are considered, there are $(2\ x\ d_r + 1)$^2 candidates in the next column. The smoothness of a single curve can be qualified by

$$|yk(x) - \mathrm{yk}(x - 1)| \tag{9}$$

where $k = 1,2$ denotes the curve number and $yk(x)$ represents the y-coordinate of the kth curve in the xth column. Notably, the number of candidates decreases if $y1$ and $y2$ are adjoined based on the design $y1 >= y2$. This parameter plays a role similar to the curve smoothness and continuity. If dr is set larger, the curve's roughness and computation time are both increased.

Dual Curves Smoothness

Another consideration is the width changes between the dual curves. The dual curves have totally nine possible strikes for $d_r = 1$. The changes of strikes are calculated as a factor of smoothness. Let $wx = (y1-y2)x$ denote the width between the dual curves in column x. Then the width changes can be modeled as:

$$|wx - wx - 1| \tag{10}$$

All the above-mentioned considerations are embedded into the cost function listed next.

Cost Function

The cost function of the node pair in column x denoted as $C(x,y1,y2)$ is defined as $C(x,y1,y2) = \min\ \{C(x - 1,\ y1 + j1,\ y2 + j2)\ j1,\ j2€\ \{-\ d_r\ ....\ d_r\} + g(x,y1) + g(x,\ y2) + \lambda 1\ |wx - wx\ \text{-}1| + \lambda 2(|j1| + |j2|)$ subject to

$$d_{\min} <= y1 + j1 - y2 - j2 <= d_{\max} \text{ and } 2 <= x <= N \tag{11}$$

$\lambda 1$ and c $\lambda 2$ are weighting factors of the curve smoothness. All tuples ($y1$,$y2$) are tested if they fit the constraint given in the optimal index ($j1$*, $j2$*) could be determined by the following equation:

$$(j1*, j2*) = \text{argC}(x, y1, y2) \tag{12}$$

Therefore, the index can be stored in the coordinate matrix $x1$ and $x3$ as follows:

$$x1(x, y1, y2) = y1 + j1* \tag{13}$$

$$x2(x, y1, y2) = y2 + j2* \tag{14}$$

In this construction, small values indicate higher likely edge locations. Therefore, the position with the minimum value in the cost map $C(N,*,*)$ is searched, which specifies the last points of the dual contours. With a backward search from N to 1 in $x1$ and $x2$, the complete coordinates of the dual contours can be obtained, which the optimal solutions are corresponding to the global minimum of the optimization function.

Back Propagation Network

A back propagation neural network is a multi-layer feed-forward neural network consisting of an input layer, a hidden layer, and an output layer. The neurons present in the hidden and output layers have biases, which are connections from the units whose activation is always one. The training of the back propagation network is done in three stages.

(1) Input training pattern.
(2) Error calculation.
(3) Updation of weights.

Back Propagation Algorithm

The back propagation learning procedure has become the single most popular method to train networks. It has been used to train networks in problem domains including prediction of elasticity detection of carotid artery recognition. The algorithm is developed considering supervised method for measuring minimum error value and steepest-descent method to examine a global minimum.

BPA helps to classify a plaque based on nine characteristics of sample ultrasound segmented images. The dataset consists of 200 samples. Carotid dataset input is an 9 x 200 matrix, whose characteristics are age, size, height, sex, length, internal diameter, wall thickness, and circularity.

A total of 200 samples of carotid dataset are randomly used for validation and testing. In that 70 % for training, (140 samples), 15 % for validation (30 samples), and 15 % for testing (30 samples) are considered in BPN.

Results and Discussion

A brief study of two different segmentation methods is discussed. Features are extracted from segmented images using pattern recognition of neural network. Back propagation algorithm is applied to train the network and to classify the inputs according to the outputs. To validate the output confusion matrix is implemented for training, testing, and validation done to refer true positive rate versus false positive rate.

Table 1 gives the output of segmented carotid artery images.

Figures 1 and 2 show the high accuracy in training confusion matrix with dynamic programming using back propagation algorithm.

Figures 3 and 4 show the high accuracy in validation with dynamic programming using back propagation algorithm.

Table 2 gives the output of segmented carotid artery image using fuzzy *c*-means clustering. To validate the desired output training, testing, and validation

Table 1 Carotid image segmented based on dynamic programming

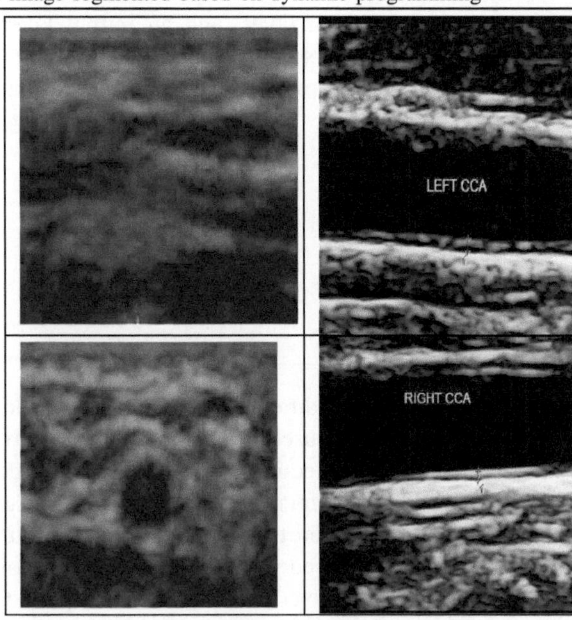

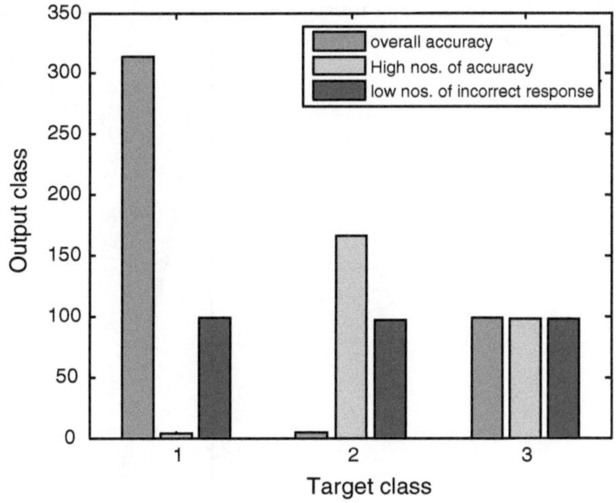

Fig. 1 High accuracy in training confusion matrix

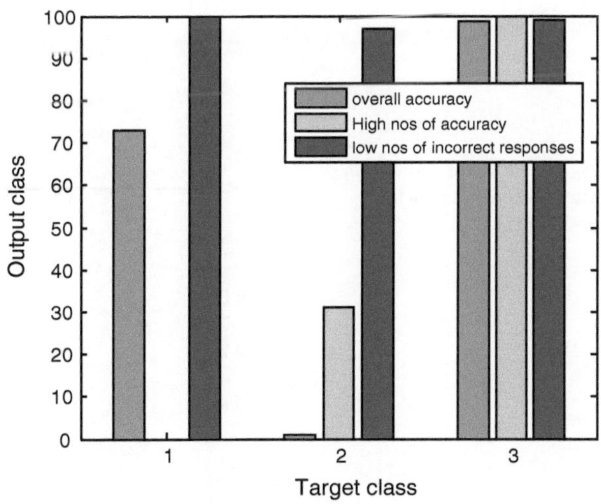

Fig. 2 Shows high accuracy in validation confusion matrix

are done by using confusion matrix network. Confusion matrix is applied for training, validation, and testing ROC (Region of characters) using true positive rate versus false positive rate. Figures. 3 and 4, show the 99.51 % accuracy and no number of incorrect responses to predict the elasticity of normal and abnormal carotid artery is validated.

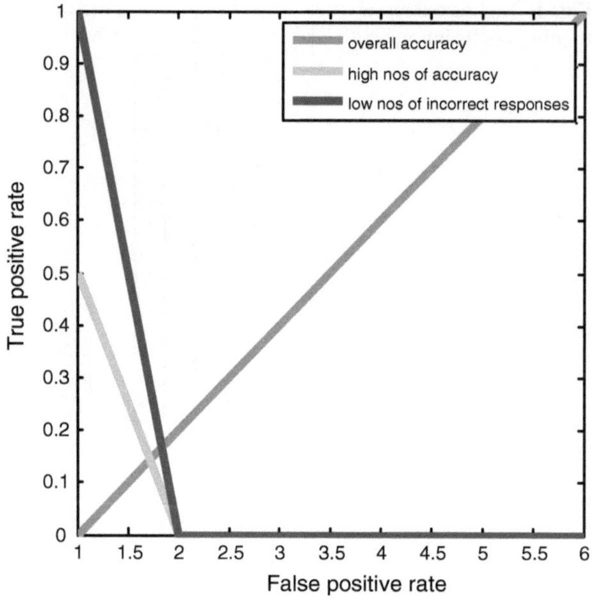

Fig. 3 ROC for training true positive rate vs false positive rate

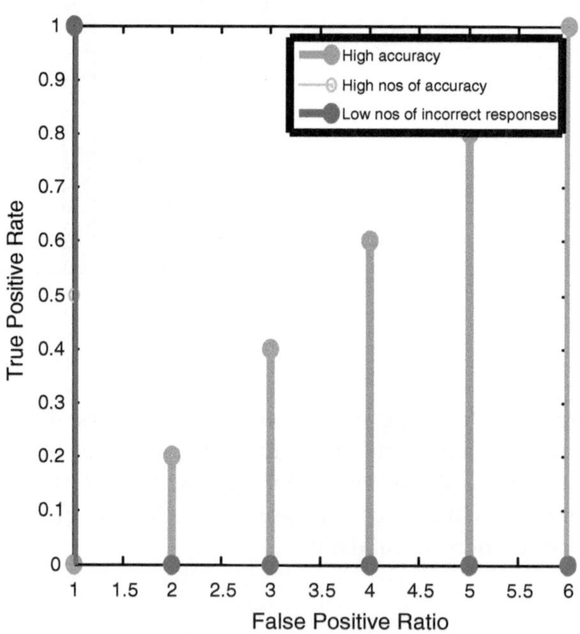

Fig. 4 ROC for validation true positive rate vs false positive rate

Table 2 Segmentation based on fuzzy c-means clustering

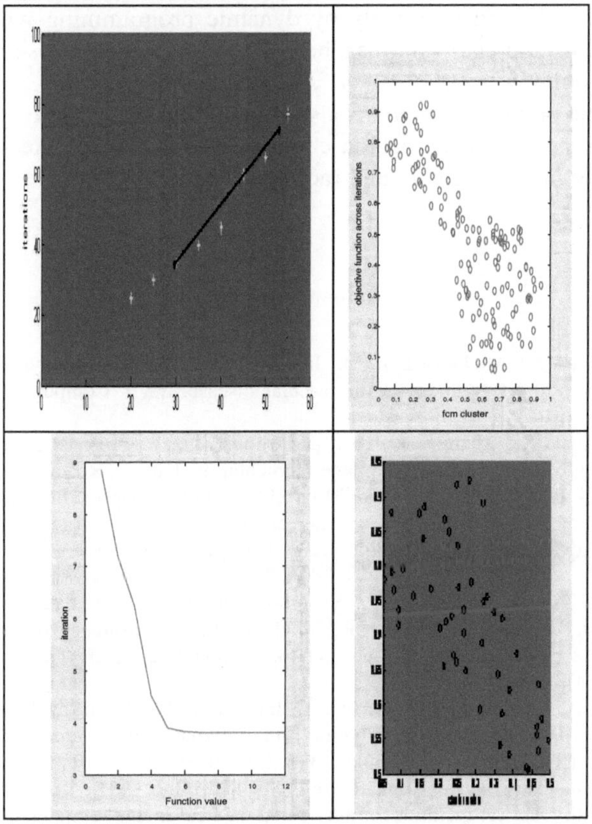

Conclusion

A real-time measurement of the elasticity of the arterial wall was detected by dynamic programming through extracting the internal diameter of normal and diseased carotid artery image and classified by back propagation network which produces more accurate results than fuzzy c-means clustering. It also helps the practitioner to detect and conclude the extent of the normal and abnormal plaque region to diagnose for further observation and treatment.

It is also believed the proposed methods may assist the physicians to predict the future possibility of normal subject becoming abnormal based on IMT value. This will provide a faster solution and effective way for classification of diseased carotid artery, where it reduces the burden of the conventional way of manual observation through ultrasonic images.

Thus concluded that prediction of elasticity of carotid artery plaque can be detected effectively and accurately by dynamic programming with back propagation algorithm and the same can be used as the second observer apart from practitioner's opinion.

In future, the proposed method is also suitable for other medical applications to detect closer contours. This method can also be suitable to detecting vessel and spine boundary in angiography and radiography.

References

1. Santhiyakumari, N., Madheswaran, M.: Intelligent medical decision system for identifying ultrasound carotid artery images with vascular disease. Int. J. Comput. Appl. **1**(13), 32–39 (2010)
2. Cai, W., Chen, S., Zhang, D.: Fast and robust fuzzy *c*-means clustering algorithms incorporating local information for image segmentation. Pattern. Recogn. **40**, 825–838 (2007)
3. Lei, W.K., Li, B.N., Dong, M.C., Vai, M.I.: AFC-ECG: an adaptive fuzzy ECG classifier. In: Proceedings of the 11th World Congress on Soft Computing in Industrial Applications (WSC11). Advances in Soft Computing,Springer Berlin,Heidelberg vol. 39, pp. 189–199 (2007)
4. Markos, G., Tsipouras, Themis, P., Exarchos, Dimitrios, I., Fotiadis, Anna, P., Kotsia, Konstantinos, V., Vakalis, Naka, K.K., Michalis, L.K.: Automated diagnosis of coronary artery disease based on data mining and fuzzy modeling. IEEE Trans. Inform. Technol. Biomed. **12**(4), 447–456 (2008)
5. Li, B.N., Chui, C.K., Ong, S.H., Chang, S.: Integrating FCM and level sets for liver tumor segmentation. In: Proceedings of the 13th International Conference on Biomedical Engineering, (ICBME 2008) Singapore, 3–6, December 2008
6. Li, Bing Nan, Chuti, Chee Kong, Chang, Stephen, Ong, S.H.: Integrating spatial fuzzy clustering with level set methods for automated medical image segmentation. Comput. Biol. Med. **41**(1), 1–10 (2011)
7. Amartur, S.C., Piraino, D., Takefuji, Y.: Optimization neural networks for the segmentation of magnetic resonance images. IEEE Trans. Med. Imaging **2**(2), 215–220 (1992)
8. Levinski, K., Sourin, A., Zagorodnov, V.: Interactive surface-guided segmentation of brain MRI data. Comput. Biol. Med. **39**(12), 1153–1160 (2009)
9. Paragios, N.: A level set approach for shape-driven segmentation and tracking of left ventricle. IEEE Trans. Med. Imaging **22**, 773–776 (2003)
10. Suri, J.S.: Two-dimensional fast magnetic resonance brain segmentation. IEEE Eng. Med. Biol **20**, 84–95 (2001)
11. Ozbay, Y., Ceylan, M.: Effects of window types on classification of carotid artery Doppler signals in the early phase of atherosclerosis using complex-valued artificial neural network. Ultrasound Med. Biol. **37**(3), 287–295 (2006)
12. Wendelhag, I., Gustavsson, T., Suurkula, M., Berglund, G., Wikstrand, J.: Ultrasound measurement of wall thickness in the carotid artery: Fundamental principles and description of a computerized analysing system. Clin. Physiol. **11**, 565–577 (1991)
13. Wendelhag, I., Liang, Q., Gustavsson, T., Wikstrand, J.: A new automated computerized analyzing system simplifies readings and reduces the variability in ultrasound measurement of intima-media thickness. Stroke **28**, 2195–2200 (1997)

An Effective Segmentation Approach for Lung CT Images Using Histogram Thresholding with EMD Refinement

Khan Z. Faizal and V. Kavitha

Abstract Image segmentation is an important step in extracting information from medical images. Segmentation of pulmonary chest computed tomography (CT) images is a precursor to most pulmonary image analysis. The purpose of lung segmentation is to separate the voxels corresponding to lung tissue from the surrounding anatomy. This paper presents an automated CT lung image segmentation. The approach utilizes histogram-based thresholding with Earth Mover's Distance (HTEMD)-based refinement methods. The final segmented output is further refined by morphological operators. The performance of HTEMD is compared with Otsu's, *K*-Means, and histogram thresholding using fuzzy measures.

Keywords Computed tomography · Lung nodule · Earth mover's distance · Histogram thresholding

Introduction

Image segmentation is important for isolating and extracting details of an image. Many image segmentation methods have been proposed by early researchers. This process partitions the image into different meaningful regions with homogeneous characteristics [1]. Segmentation has been used in the past for quantification of

K. Z. Faizal (✉)
Research Scholar, Anna University of Technology, Tirunelveli, India
e-mail: faizalkhan_111@yahoo.co.in

V. Kavitha
Department of Computer Science and Engineering, Anna University of Technology, Tirunelveli, India
e-mail: kavinayav@gmail.com

S. Sathiakumar et al. (eds.), *Proceedings of International Conference on Internet Computing and Information Communications*, Advances in Intelligent Systems and Computing 216,
DOI: 10.1007/978-81-322-1299-7_45,

tissue volumes, computer integrated surgery, reconstruction of volume of medical image slices. Chest CT image helps in evaluating the stages of the unwanted material growth in lung. This process has been used for the diagnosis of various pulmonary diseases (lung cancer, tuberculosis, and pulmonary embolism (PE)). This paper presents four stages for achieving effective segmentation of CT lung image.

Stage 1: The left and right lung of the CT image is segmented using fuzzy logic. *Stage 2*: Apply EMD for refinement of segmented image. *Stage 3*: Apply morphological operation in order to obtain the correctly segmented right and left lung image. *Stage 4*: Performance comparison is presented.

The remaining sections of this paper are organized as follows. Section "Related Works" presents an overview of existing lung segmentation methods. Section "Materials and Methods" explains the proposed methodology in detail. Section "Segmentation Based on Fuzzy Logic" depicts the experimental results, and Sect. "Refinement Process" provides the conclusions on this work and suggests some possible future enhancements.

Related Works

Armato and Sensakovic [2] proposed an accurate lung region segmentation procedure called a multilevel thresholding for identifying the regions of interest (ROI) lung nodule. 5–17 % of the lung nodules are undetected by their approach.

Segmentation of lung high resolution CT (HRCT) images using a pixel-based approach was suggested by Garnavi et al. [3]. These authors have used global-threshold segmentation, mathematical morphology, edge detection, noise reduction, and geometrical computations to achieve the defined ROIs.

Ye et al. [4] presented a CAD method for detecting both solid nodules and ground-glass opacity nodules in the lung image. A fuzzy-based thresholding approach is used to segment the lung region. Their method results in fast computation.

Zheng et al. [5] proposed a method for segmenting the lung region by an adaptive threshold based on value distributions. Component analysis and gray-level thresholding are used to segment the lung volume.

Sang et al. [6] proposed a CAD-based lung segmentation algorithm to segment a 3-D lung. Their algorithm searches for suspicious PE regions.

From the literature mentioned, there is still a requirement for improving segmentation of lung in a CT image in terms of fast computation and improved segmentation accuracy.

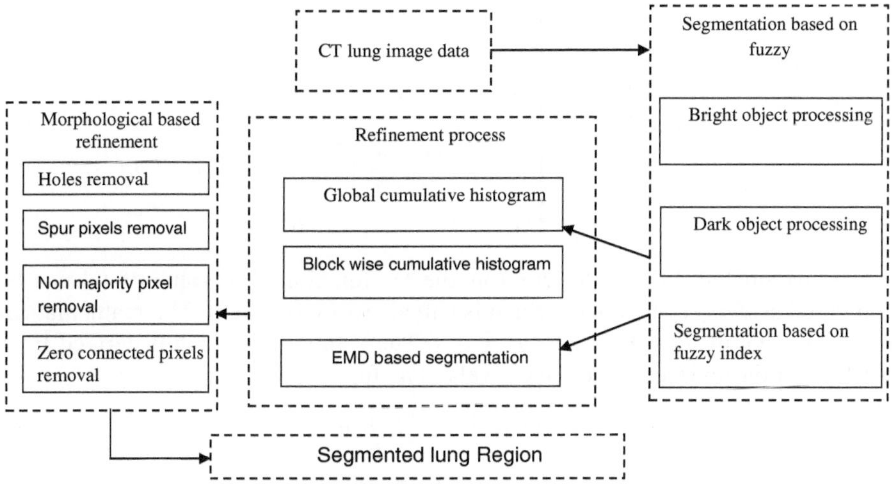

Fig. 1 Block diagram of the proposed system

Materials and Methods

Figure 1 is the prototype for the proposed system of pulmonary Lung region segmentation on CT Image which consists of four stages: (1) segmentation based on fuzzy (2) refinement based on EMD (3) morphological-based refinement, and (4) comparison of segmentation performance.

Segmentation Based on Fuzzy Logic

A fuzzy set is any set that allows its members to have different degree of membership, called membership function within the interval [0, 1]. Which can be mathematically expressed as $\mu A(x)$ which was assigned to each element on the set with in the same closed unit intervals [7]. Let X is the universal space and x is an element of the space X. A fuzzy set an in space X is defined as

$$A = \{\{x, A(X)\}|x \in X\}. \tag{3}$$

Where $A(x)$ is the value of the function 'A' for element 'x'. In this work, the 'S' function which denotes the set of brighter objects is used for modeling the membership function [7]. The S function can be defined as

$$\mu As(x) = S(x; a, b, c) \tag{4}$$

$$S(x;a,b,c) = \begin{cases} 0 & x \le a \\ 2\left\{\frac{(x-a)}{(c-a)}\right\}^2 & a \le x \le b \\ 1 - 2\left\{\frac{(x-a)}{(c-a)}\right\}^2 & b \le x \le c \\ 1 & b \ge c \end{cases} \tag{5}$$

The parameters 'a' and 'c' controls the 'S' function. The value of 'b' can be computed as $b = (1/2)(a + c)$ which is called cross-over point. The mathematical function value of 'b' is 0.5 ($\mu As(b) = 0.5$).The Z-function, which is derived from the S-function represents the dark pixels is as follows:

$$\mu Az(x) = Z(x;a,b,c) \tag{6}$$

$$= \begin{cases} 1 & x \le a \\ 2\left\{\frac{(x-a)}{(b-a)}\right\}^2 & a \le x \le \frac{a+b}{2} \\ 1 - 2\left\{\frac{(x-b)}{(b-a)}\right\}^2 & \frac{a+b}{2} \le x \le c \\ 0 & x \ge b \end{cases} \tag{7}$$

Refinement Process

Improving the quality of segmented output based on fuzzy methodology is carried in this stage. Histogram-based analysis was done in order to improve the segmented result. The segmented output has been further refined by EMD [8]-based refinement to reduce the issues in the existing methods. For dark and white objects, the refinement process is carried using the following steps:

Step 1: Compute the inside and outside histogram 'α' and 'β'.

Step 2: Compute the inside and outside cumulative sum 'I_c', 'O_c'.

$$I_c(x_i) = \sum_{k=0}^{i} \alpha_k \tag{8}$$

$$i \to 0 \text{ to } 255$$

$$O_c(x_i) = \sum_{k=0}^{i} \beta_k \tag{9}$$

Step 3: For all black and white pixels x in the fuzzy region, compute the block histogram 'γ'.

Step 4: For all black and white pixels, compute the foreground distance δ_f

$$\delta_f = \sum_{i=0}^{255} \text{abs}\left\{ (\sum_{k=0}^{i} \gamma_k) - (\sum_{k=0}^{i} \alpha_k) \right\} \tag{10}$$

where 'γ_k' and 'α_k' are, respectively, the block histogram and inside histogram of white and dark objects.

The background distance 'δ_b' is as follows:

$$\delta_b = \sum_{i=0}^{255} \text{abs}\left\{ (\sum_{k=0}^{i} \gamma_k) - (\sum_{k=0}^{i} \beta_k) \right\} \tag{11}$$

Step 5: The membership value presents the state of final marked image as '1' or '0' based on the below condition.

$$\text{Membership value} = \begin{cases} 1 & \text{if } \delta_f(x_i) < \delta_b(x_i) \\ 0 & \text{else} \end{cases} \tag{12}$$

Morphological-Based Refinement

We now use the morphological-based refinement for further improving the quality of initial refined output, which contains the holes, spur, non-majority pixels, and zero connective pixels. Predefined MATLAB morphological operators such as Fill, Majority, clean were used.

Results and Discussion

In order to represent the performance of our proposed method, we have chosen 15 CT images from LIDC Lung database. A ground truth image has been generated manually for each image. The ground truth image was combined with its corresponding original image, to segment the CT lung image. The performance of our proposed segmentation method has been compared with other conventional segmentation algorithms like 'Otsus', '*K*-means clustering', 'Histogram thresholding using fuzzy measures (HDFM)', 'Log', 'Zerocross', and 'our earlier work using fuzzy logic' [9]. The performance of individual methods over Lung image was carried out and the results of the techniques are depicted in Tables 1 and 2.

The segmentation accuracy A_s is computed using the following equation to evaluate the quality of the segmentation results:

Table 1 Original results for four algorithms: Otsu's technique, K-Means, histogram thresholding, fuzzy logic alone, and proposed method

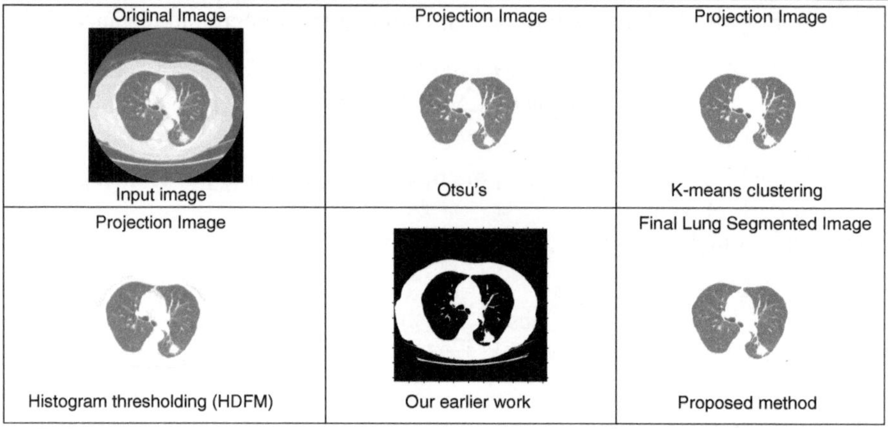

Table 2 Comparison accuracy of Segmentation techniques

Algorithms	Number of objects	'Area'	'Solidity'	'Perimeter'	Accuracy of segmentation
Otsus	40	2,300	34.3	1903	60.0909
K-means	42	2,300	38.6	2103	62.8727
HDFM	54	2,166	40.1	2440	65.5964
Our earlier work using fuzzy logic	49	2,280	44.9	3036	84.7055
Fuzzy with HTEMD	49	2,490	48	3078	87.9091

$$A_s = \frac{S_I}{O_I} * 100 \tag{4}$$

where S_I denotes the ratio between O_I and the number of objects presented in the segmented region. O_I denotes the total of solidity, Area and Perimeter of the pixels present in the unsegmented region.

Conclusions

In this paper, we have proposed an automated Lung segmentation scheme based on fuzzy logic along with EMD-based refinement. The main advantage of this work is that it reduces false segmentation. From the experimentation results, it is proved that the proposed method ensures accuracy level of segmentation. Our method is useful in assisting for further diagnosis of Lung nodules in LDCT by reducing the mortality rate. Further works in this direction can be the application of a suitable classification algorithm for identification of Lung Nodules.

References

1. Cao, W., Yan, Y., Li, S.: Unsupervised color-texture image segmentation based on a new clustering method. J. Next Gener. Inform. Technol. **1**(2), 784–787 (2010)
2. Armato, S.G., Sensakovic, W.F. Automated lung segmentation for thoracic CT. Acad. Radiol. **11**(9), 1011–1021 (2004)
3. Garnavi, R., Baraani-Dastjerdi, A., Abrishami Moghaddam H, 2005, A new segmentation method for lung HRCT images. In: Proceedings of the Digital Imaging Computing: Techniques and Applications, p. 8. IEEE CS Press, Cairns Convention Centre, Brisbane, Australia. doi:10.1109/DICTA.2005.5
4. Ye, X., Lin, X., Dehmeshki, J., Slabaugh, G., Beddoe, G.: Shape-based computer-aided detection of lung nodules in thoracic CT Images. IEEE Trans. Biomed. Eng. **56**(7) 1810–1820 (2009)
5. Faizal Khan, Z., Kavitha, V.: Pulmonary lung segmentation in computer tomography using fuzzy logic. Eur. J. Sci. Res. **81**(3), 329–337 (2012)
6. Park S.C., Kim W.P., Zheng B., Leader J.K., Pu J., and Gur D.: Pulmonary airway tree segmentation from CT examinations using adaptive volume of interest. Proc. SPIE Med. Imag. **7259**, 72593U-1–72593U-9 (2009)
7. Lopes, N.V., Mogadouro do Couto, P.A., and Bustince, H.: Automatic histogram threshold using fuzzy measures. IEEE Trans. Image Process. **19**(1), 199–204 (2010)
8. Adam, A., Kimmel, R., Rivlin, E.: On scene segmentation and histograms-based curve evolution. IEEE Trans. Pattern Anal. Mach. Intell. **31**(9), 1708–1714 (2009)
9. Zheng, B., Leader, J.K., Maitz, G.S., Chapman, B.E., Fuhrman, C.R., Rogers, R.M., Sciurba, F.C., Perez, A., Thompson, P., Good, W.F., Gur, D.: A simple method for automated lung segmentation in X-ray CT images. Proc. SPIE **5032**, 1455–1463 (2003)

Analysis of Mobile Agents Applications in Distributed System

K. Senthil Kumar and S. Kavitha

Abstract With the increasing demand to extend data mining technology to datasets inherently distributed among a large number of autonomous and heterogeneous sources over a network, there comes a new innovative technology that leads the system based on the mobility code especially mobile agents. Primary focus in distributed systems is that mobile agents have a number of key features that allow them to reduce the network load and overcome network latency. Mobile agents offer a more uniform approach to handling code and data in a distributed system. They can encapsulate protocols, and they can work remotely, even asynchronously and disconnected from a network. It improves the latency and bandwidth of client–server applications and reducing vulnerability to network disconnection. This paper presents an analysis of the various applications of mobile agents and some of the benefits and challenges of this new technology.

Keywords Mobile agent · Mobile commerce · Mobile computing · E-commerce · Network management · Cloud computing

Introduction

The rapid development of various research institutions, corporations, and governments work with vast amount of data that originate from several different sources. This growth leads to the revolutionary advancement in the information world. Mining information and knowledge from huge data sources such as weather

K. Senthil Kumar (✉) · S. Kavitha
Department of MCA, SRM University, Chennai, India
e-mail: senthil.ku@ktr.srmuniv.ac.in

S. Kavitha
e-mail: kavitha.s@ktr.srmuniv.ac.in

S. Sathiakumar et al. (eds.), *Proceedings of International Conference on Internet Computing and Information Communications*, Advances in Intelligent Systems and Computing 216,
DOI: 10.1007/978-81-322-1299-7_46, © Springer India 2014

databases, financial data portals, or emerging disease information systems has been recognized by industrial companies as an opportunity for major revenues from applications such as warehousing, process control, and customer services. Mobile agents with mobile code are very important in the digital world today, because it has the flexibility of accessing the information from different sources.

Mobile Agent

Mobile agent is an autonomous software entity that can suspend its execution, transfer itself from one networked host to another and resumes execution on the new host. It is programmed to perform certain tasks. The agent may traverse numerous hosts to perform the task. The agent dispatcher is considered as the Home platform [1].

All data and code travel through the network in order to ensure a transparent execution of the task at the destination without introducing any software requirements (i.e., it is not necessary to install or copy required libraries or source codes). Agent mobility capabilities are used to implement this functionality.

A mobile agent executes on a sequence of machines, where a place pi $(0 < i < n)$ provides the logical execution environment for the agent. Figure 1 shows mobile agent's execution with the three stages [2].

A mobile agent must have an initializing method to activate itself on a new agent host. Once activated, the mobile agent collects the necessary information through interaction with local resources. After completion of all the specified tasks for that particular managed entity, the results are saved into the container of the mobile agent (Mobile agent does not have to return to its launcher (if at all) until it finishes visiting the last node. Figure 2 shows the main loop of the mobile agent [3] (Table 1).

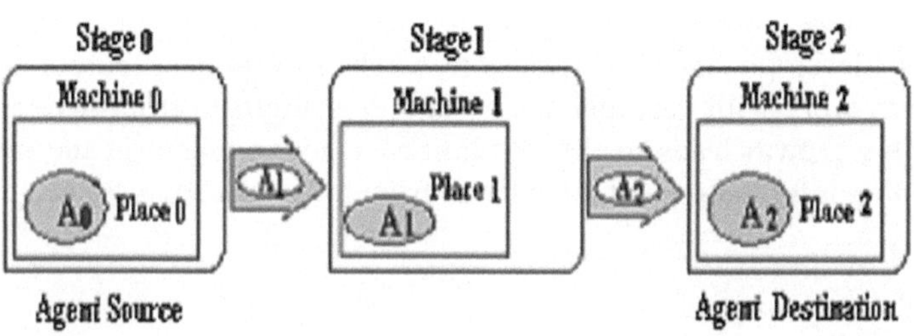

Fig. 1 Model of mobile agent with execution

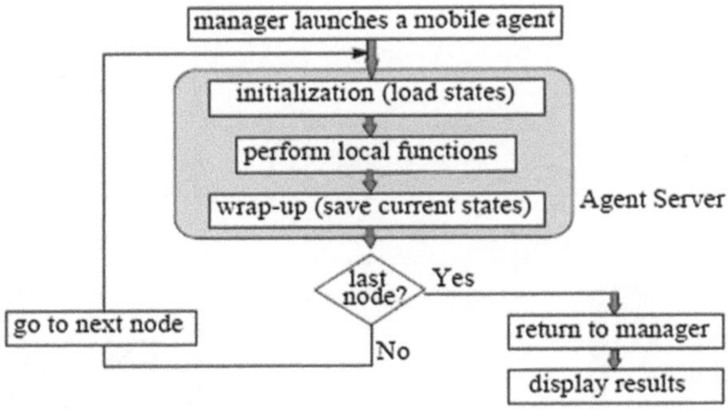

Fig. 2 Main loop of a mobile agent

Table 1 (Franklin and Graesser) shows the several properties

Property	Other names	Meaning
Reactive	Sensing and acting	Responds in a timely fashion to changes in the environment autonomous exercises control over its own actions
Goal oriented	Pro-active purposeful	Proactive purposeful: does not simply act in response to the environment
Temporally continuos		Is a continuously running process
Communicative	Socially able	Communicates with other agents, perhaps including people
Learning	Adaptive	Changes its behaviour based on its previous experience
Mobile		Able to transport itself from one machine to another
Flexible		Actions are not scripted
Character		Believable "personality" and emotional state

Applications of Mobile Agent

Mobile Commerce

The transaction of vending machine is managed by mobile agents enabled phones, Personal Digital Assistant PDAs, or other wireless devices who communicate securely and directly with a remote Internet server through the Internet connection of the handheld device, completing the entire transaction and delivering the goods. Figure 3 shows the wireless mobile vending solutions using eCapsule technology [4].

Internet

Web contains unstructured information resources. The information can be located anywhere in the Web, dynamically moved, rearranged, get distributed, and represented in many different ways. A mobile agent is a new paradigm for retrieving

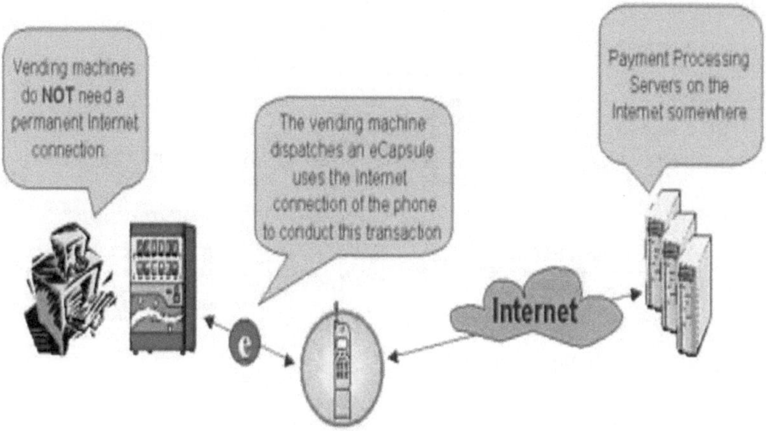

Source: Warp9 Inc. [http://ecapsule.warp9inc.com/]

Fig. 3 Vending machine and mobile commerce

information from Web and avoids redundant information. Figure 4 shows the leading trends of mobile agents [5].

1. Mobile agents can significantly save bandwidth by moving locally to the resources they need.
2. They can carry the code to manage remote resources.
3. They do not require continuous network connections [6].

Electronic Commerce

All commercial transaction in real time, access to remote resources such as stocks quotes and also agent-to-agent negotiation. Mobile agents might embody the intentions of their creators and negotiate on their behalf [4]. A mobile agent can be programmed to bid in an online auction on behalf of the user. Mobile agent in e-commerce search and filter information the interest from electronic markets. Figure 5 describes the sequence of processes carried out during the agent's lifetime [7].

Grid Computing and Grid Service

In order to get common goals, Grid computing allows sharing resources which is distributed geographically. The standard interface used for delivering of services. For migration of agents in the JADE platform Agent Communication Language (ACL) message is used. The agent transmitted as a SOAP message to enter the grid

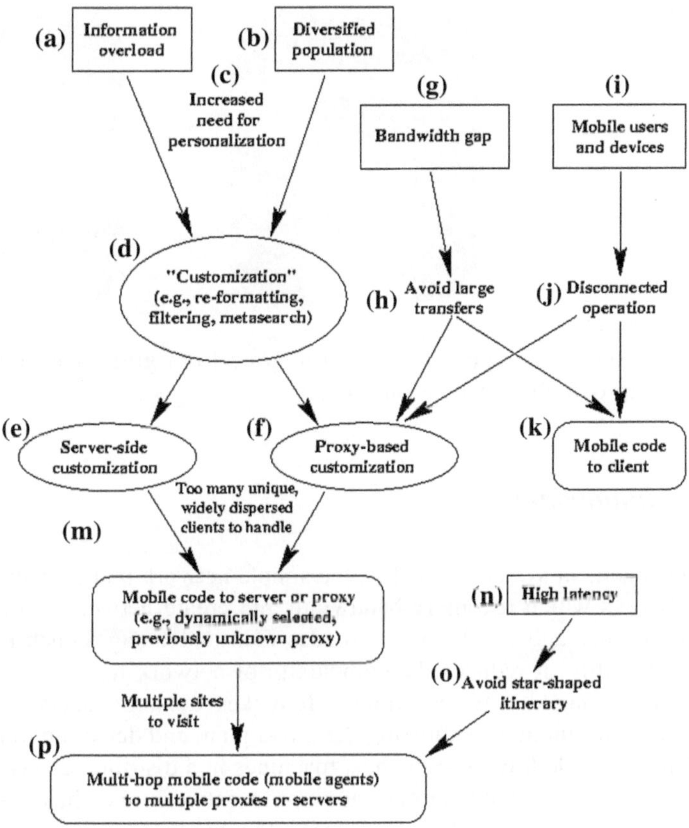

Fig. 4 The trends leading to mobile agents [5]

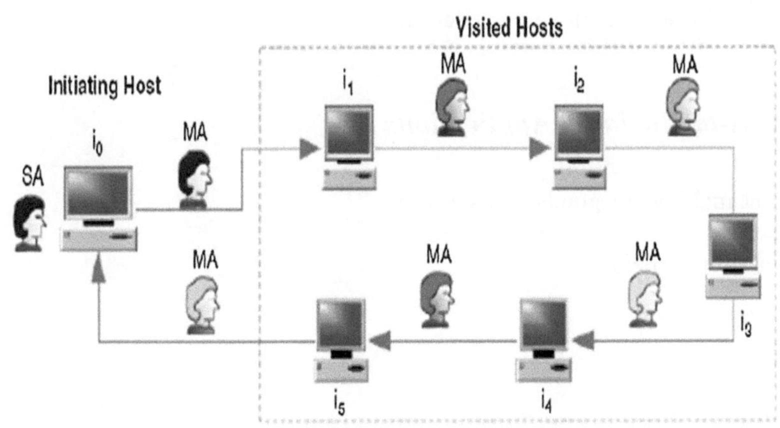

Fig. 5 System architecture

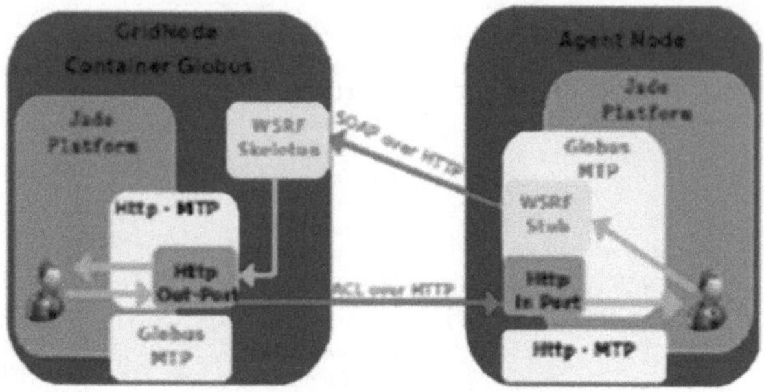

Fig. 6 Agent migration to a grid node [7]

node. Figure 6 shows the agent which can migrate from grid node to an outside node by the http default transmission protocol.

Network Management

Network management systems based on the simple network management protocol for data networks which use more bandwidth and create network traffic. Mobile agent with mobility and intelligence provides many advantages such as extensibility and portability. It reduces the complexity of network management responsibility to the managed network entities. It makes dynamic decisions such as finding the next destination, optimizing the travel plan, and detecting link failures. As the agent with code travels to the host machines in a distributed network, there is no need to bring intermediate data across the network and thus a significant amount of network bandwidth use, reduces the number of necessary human interactions and communication delay can be avoided. It manages the tasks easily with preprogrammed structure. In many ways, mobile agents are more efficient than the traditional client server [Ber96].

Business-to-Business Applications

- Market making for goods and services.
- Brokering.
- Team management.

Personal Agents

- E-mail and news filters.
- Personal schedule management.
- Personal automatic security.

Parallel Computing

As mobile agents may create a cascade of clones in the network, this technology is to administer parallel processing tasks and reducing the time required to solve the problem.

Telecommunication Networks Services

Common management information protocol is used for telecommunication network management. It is designed for centralized model with lack of distribution, a low degree of flexibility, reconfigurability, efficiency, scalability, and fault tolerance. The primary goal of using mobile agents in management of telecommunication network is to reduce network traffic by using load balancing and building scalable and reliable distributed network management system [8]. Mobile agent support and management of telecommunications services are designated by dynamic network reconfiguration and customization.

Mobile Computing

Mobile agents are launched in the field of notebook computers, PDAs, laptops etc. Mobile agent leaves the mobile platform which roam through the wired network to accomplish the user's task or representing the users while they are disconnected. Figure 7 illustrates the physically moved agents in the different hosts of the network [9].

Data Processing

Data processing is a graphical application. In the client–server approach, shown in Fig. 8a, client can request image server, client retrieved, and processed locally.

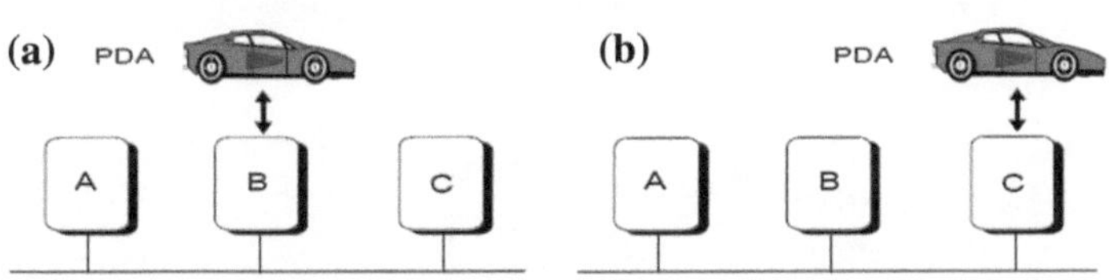

Fig. 7 Mobile computing **a** The PDA is connected to the host B. **b** The PDA changes its connection to the host C

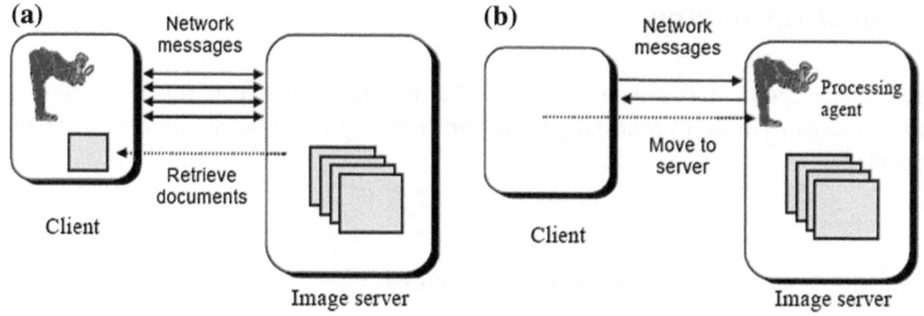

Fig. 8 Image processing **a** Traditional client-server approach. **b** Mobile agent approach [9]

This strategy needs large bandwidth. In the mobile agent approach, shown in Fig. 8b, a processing agent send the user request to server image and gets back the results only thereby it uses significant amount of bandwidth.

Distributed Information Retrieval

Instead of moving large amounts of data to the search engine so that it may create search indexes, we dispatch agents to remote information sources, where they locally create search indexes that can be shipped later to the origin. In this system shown in Fig. 9, mobile agent is tasked with searching for information and consulting with the planning module first, then the planning module asks a yellow page for possible locations, where the mobile agent finds the desired information [10]

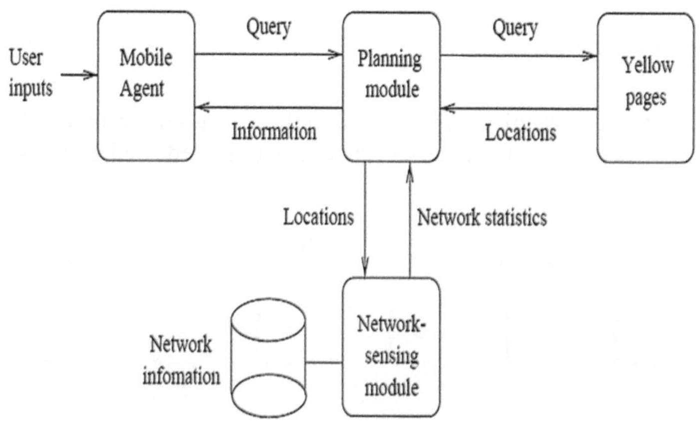

Fig. 9 Architecture of the planning system for information retrieval from distributed system [10]

Energy Efficiency and Metering

Mobile agent technology has been applied in for controlling power of distributed generations in an isolated micro grid. In their solutions, the mobile agent has three roles:

- Acquire operation data and equipment parameters from all the distributed power,
- Determine the output power order for each power source, and
- Distribute the output power order to all the power sources [11].

Cloud Computing

Load balancing is a legacy application of process migration and mobile agent technologies. Computers tend to be numerous and their computational loads are different in cloud computing system. Computers may also be dynamically added to or removed from the system. Mobile agents can migrate to other computers, tasks that are implemented as mobile agents can be relocated at suitable computers whose processors can execute the tasks.

Distance Education

Mobile agent approach has been used for designing, implementing, and deploying a system for distance evaluation of students. For example, the entire examination process like

1. Paper Setting, where the examiners spread over the Internet collaborate to produce a question paper,
2. Examination conduction, where the question papers are distributed and the answer papers are collected, and
3. Answer paper evaluation, result compilation, and publishing [6].

Military

Mobile agents are used in the military tactical operations like information push, information pull, and sentinel information monitoring. The objective is to increase the network reliability and bandwidth, domain-dependent information processing, and complex autonomous information processing that involves large heterogeneous data resources.

Table 2 Mobile agent platforms characteristics

Mobile agent systems	Programming language	Operating system
Voyager	Java.NET, C++	Unix, Windows
TACOMA	C, Tcl/Tk, Perl, Python, and Scheme	Unix
PIAX	Java	All (with JRE)
Trinity	.NET	Windows
JADE	Java	All (with JRE)
SMARD	Java	All (with JRE)
Float	Borland Delphi	Windows
SensorWare	TCL/TK	Unix
MAF	Python	Unix
TAgent	Java	All (with JRE)
Aglet	Java	All (with JRE)

Mobile Agent Platforms

Several platforms are implemented for mobile agents. Table 2 shows the various mobile agent platforms with characteristics. Various research laboratories and even commercial products use these platforms.

Agent-Based Publications

Agent-based publications like Journal Articles and Chapters, Conference Proceedings, Thesis are improving year by year. The following chart shows the publications of the above from 2008 to 2011 (Fig. 10).

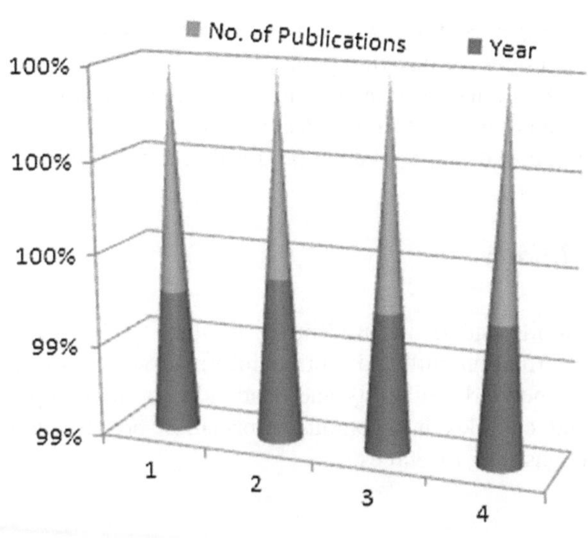

Fig. 10 Agent-based publications from 2008 to 2011

Future Work

As mobile agent is known to be used in various fields and domains, our future work would be to address the security issues in various levels of applications in the distributed system environment.

Conclusion

This paper describes the applications of mobile agent in different domains. The mobile agent major functions are: gathering, filtering, sharing, monitoring, recommending, comparing information, guiding Web surfers, e-mail filtering, auto responders, and negotiating for applying in different areas in distributed system. Mobile agent technology is applied in all the fields which are geographically distributed in nature.

References

1. Kapse, P.: Security in mobile agents. Secure system research group, Oct 31 (2003)
2. Hami, H.: Modeling and evaluation fault-tolerant and reliable mobile agent execution. The Second International Conference on Innovations in Information Technolony (IIT'05), Dubai, UAE. Sep 26–28 (2005)
3. Ku, H., Luderer, G.W.R., Subbiah, B.: An intelligent mobile agent framework for distributed network management. CiteSeerX (1997)
4. Braz, C.: Mobile Agents for Wireless E- Commerce Applications (MAWA). Artificial Intelligence IFT 6261 (Winter 2003)
5. Kotz, D., Gray, R.S.: Mobile agents and the future of the Internet. Workshop Mobile Agents in the Context of Competition and Cooperation (MAC3) at Autonomus Agents, Seattle, Washington, May 1 (1999)
6. Rajguru, P.V., Deshmukh, S.B.: Current trends and analysis of mobile agent application. In: Proceedings of National Conference on Emerging Trends in Computer Technology (NCETCT-2012)
7. Outtagarts, A.: Mobile agent-based applications: survey. Int. J. Comp. Sci. Netw. Secur. **9**(11), Nov (2009)
8. Makki, S., Wunnava, S.V.: Application of mobile agents in managing the traffic in the network and improving the reliability and quality of service. Int. J. Comp. Sci. **32**:4, IJCS_32_4_16
9. Cabri, G., Leonardi, L., Zambonelli, F.: Mobile agent technology: current trends and perspectives. CiteSeerX (1998)
10. Brewington, B., Gray, R., Moizumi, K.: Mobile agents in distributed information retrieval. In: Klusch, M. (ed.) Intelligent Information Agents, Chapter 15, pp. 355–395. Springer Verlag, Germany (1999)
11. Intellient Agent Laboratory. Department Systems, The Melbourne School Of Engineering, Melbourne

Author Index

S. Sathiakumar et al. (eds.), *Proceedings of International Conference on Internet Computing and Information Communications*, Advances in Intelligent Systems and Computing 216,
DOI: 10.1007/978-81-322-1299-7, © Springer India 2014